W0111350

NATO ASI Series

Advanced Science Institutes Series

A series presenting the results of activities sponsored by the NATO Science Committee, which aims at the dissemination of advanced scientific and technological knowledge, with a view to strengthening links between scientific communities.

The Series is published by an international board of publishers in conjunction with the NATO Scientific Affairs Division

A Life Sciences	Plenum Publishing Corporation
B Physics	London and New York
C Mathematical and Physical Sciences	Kluwer Academic Publishers Dordrecht, Boston and London
D Behavioural and Social Sciences	
E Applied Sciences	
F Computer and Systems Sciences	Springer-Verlag Berlin Heidelberg New York
G Ecological Sciences	London Paris Tokyo
H Cell Biology	

The ASI Series Books Published as a Result of
Activities of the Special Programme on
CELL TO CELL SIGNALS IN PLANTS AND ANIMALS

This book contains the proceedings of a NATO Advanced Research Workshop held within the activities of the NATO Special Programme on Cell to Cell Signals in Plants and Animals, running from 1984 to 1989 under the auspices of the NATO Science Committee.

The books published as a result of the activities of the Special Programme are:

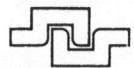

Series H: Cell Biology Vol. 24

Bacteria, Complement and the Phagocytic Cell

Edited by

Felipe C. Cabello

Department of Microbiology and Immunology
New York Medical College
Valhalla, NY 10595, USA

Carla Pruzzo

Institute of Microbiology, University of Genova
Viale Benedetto XV 10, 16132 Genova, Italy

Springer-Verlag
Berlin Heidelberg New York London Paris Tokyo
Published in cooperation with NATO Scientific Affairs Division

Proceedings of the NATO Advanced Research Workshop on Bacteria, Complement and the Phagocytic Cell held in Acquafreda di Maratea, Italy, April 6–9, 1987

ISBN 978-3-642-85720-1   ISBN 978-3-642-85718-8 (eBook)
DOI 10.1007/978-3-642-85718-8

© Springer-Verlag Berlin Heidelberg 1988
Softcover reprint of the hardcover 1st edition 1988

2131/3140-543210 – Printed on acid-free paper

PREFACE

This volume contains the manuscripts of the presentations at the NATO Workshop "Bacteria, Complement and the Phagocytic Cell" held in Maratea, Italy from April 5 to 9, 1987. The goal of the meeting was to bring together investigators working in the fields of genetics of bacterial virulence, complement and phagocytosis.

The main lectures covered the topics of bacterial structure, genetics of virulence, mechanisms of complement lysis and phagocytosis, and analysis of their relevance in bacteria-host interactions. As the Table of Contents indicates, the meeting did not cover any one topic exhaustively because it was the intention of the organizers to allow plenty of time for discussion and for interaction between the lecturers and students attending the workshops.

The lectures, poster presentations and discussions put in evidence the vast array of structures and mechanisms evolved by bacteria to evade host defences, and the danger of making indiscriminate generalizations and extrapolating knowledge acquired with one experimental system to another. Moreover, some of the presentations indicated that the coordinate use of bacterial genetics, recombinant DNA technology, <u>in</u> <u>vitro</u> and animal experiments are the most fruitful approach to discern the different mechanisms used by bacteria to invade the host. The discussions also pointed to the inability of results from experimental work to explain, sometimes, epidemiological and clinical findings and the need to strengthen the collaboration between people working at the experimental level with those working at the clinical level.

Thanks are due to the Consiglio Nazionale delle Ricerche (C.N.R.), Italy for providing funds that allowed us to expand the number of lecturers and to Hoechst-Roussel and Ciba-Geigy for providing additional funds. We wish to thank M.J. Mroczenski-Wildey, M.E. Fernandez-Beros and L. Delgado for their assistance in the preparation of the volume. We also wish to thank H.V. Harrison for her excellent secretarial assistance.

Felipe C. Cabello
Carla Pruzzo

CONTENTS

M.E. Agüero, M. Binns, G. de la Fuente,
E. Vivaldi and F.C. Cabello

THE MUREIN SACCULUS, THE BACTERIAL EXOSKELETON-STRUCTURE AND FUNCTION IN THE BACTERIUM AND POSSIBLE ROLE IN THE HOST ORGANISM

U. Schwarz

Department of Biochemistry
Max-Planck-Institut fur Entwicklungsbiologie
Spemannstrasse 35
D-7400 Tubingen
Federal Republic of Germany

It is the cell wall with which bacteria interact with their outside world. It also the wall with which bacteria protect themselves against their outside world. Both functions, contradictory to each other to some extent, are fulfilled by different elements of the complex cell wall. The surface layer, with its components exposed on it, contacts the cellular environment. The aspects of this interaction shall be discussed in the following chapters. Another vital function of the cell wall is the mechanical protection against the environment. This is effected by a defined structural element in the envelope.

A giant macromolecule encloses the cell completely, it maintains cellular shape and withstands the differences in osmotic pressure betwen the cytoplasm and the cellular environment. This exoskeleton, the sacculus (review: 1), encloses the cytoplasmic membrane completely as an integrated part of the complex cell envelope. Most bacteria, with only a few exceptions, such as the halobacteria, are endowed with a sacculus. The principle according to which sacculi are constructed is the same throughout, differences in detail, however, do exist (review: 2). In the following the sacculus of E. coli and its metabolism shall be discussed as an exemplary case.

Sacculi are tailored from a polymer, peptidoglycan or murein. For long, E. coli murein was assumed to be tailored from a single-layered net in which glycan strands with short peptide chains are interlinked by peptide bridges. The only variation known was in the length of the peptide side chains, varying from two to five amino acids (3,4,5). The enlargement of

NATO ASI Series, Vol. H24
Bacteria, Complement and the Phagocytic Cell
Edited by F. C. Cabello und C. Pruzzo
© Springer-Verlag Berlin Heidelberg 1988

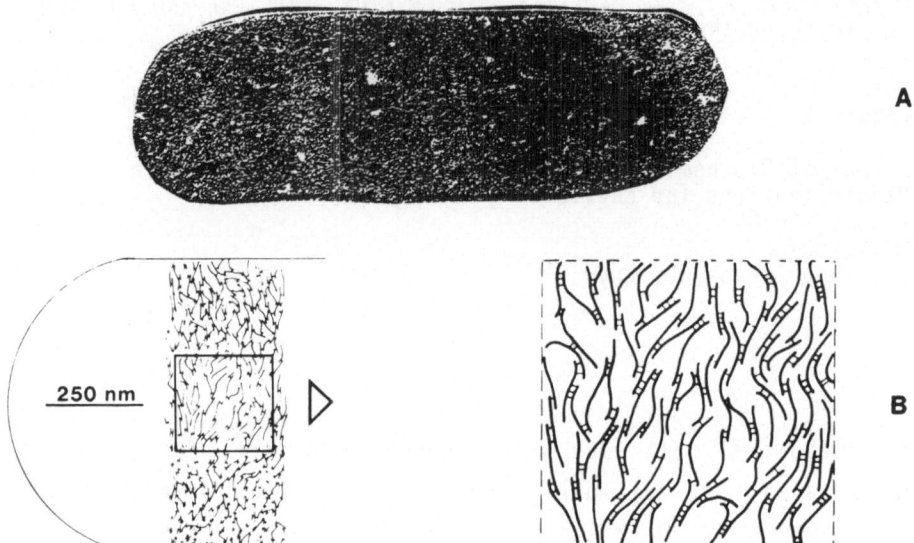

Fig. 1. The murein sacculus of E. coli. An isolated sacculus is shown (A), reflecting the shape of the cell from which it was isolated. The arrangement of glycan chains in the sacculus (B) is assumed to be irregular, however, with a long-range orientation perpendicular to the long axis of the cell. Only the peptide side chains involved in the crosslinkage of glycan chains are shown - about 50% of all side chains are bridged with each other. The individual glycan chains with an average of 30 disaccharide units have an average length of 30 nm (review: 1). A close-up of a flattened-out region of the sacculus (C) shows the alternating sequence of β-1, 4-linked amino sugars N-acetyl glucosamine and N-acetyl muramic acid in the glycan chains with a peptide side chain on each muramic acid residue. The peptide bridges shown are between the diaminopimelic acid residue (Dpm) of one chain and the alanine of its neighbour. About 7% of the peptide bridges found are between the two Dpm residues of neighboured peptide chains (8, not shown). The three-dimensional structure of murein, the addition of new subunits through apposition to preexisting murein on the inner side of the sacculus and the release of subunits as a result of recycling are schematically represented in D.

such a structure and its modification during cell growth and division were assumed to essentially need only two types of synthetic activities; the elongation of glycan chains by the addition of new subunits would require a transglycosylase activity. The crosslinking between the peptide side chains of adjacent glycan strands would occur through transpeptidation. The chemical energy for the formation of the glycosidic and the peptide bonds would be carried in the low molecular weight precursors presynthesized in the cytoplasm (review:2). This simple picture, however, changed by time.

The enzyme systems engaged in murein metabolism turned out to be unexpectedly complex. There was not only one synthetic transglycoslylase and one transpeptidase engaged in murein biosynthesis, but rather a collection of different proteins, the so called penicillin-binding proteins (6), targets of the antibiotic action of penicillins. Furthermore, a large set of murein hydrolases was revealed by time. For each type of chemical bond found in murein, at least one, in a few cases several specific hydrolases have been identified. In the case of E. coli the list of murein hydrolases contains nine different proteins (review: 1).

For quite some time, the complexity of the enzyme systems participating in the metabolism of murein was in marked contrast to the simplicity of known murein structure in E. coli. A careful reexamination of murein chemistry with high-pressure liquid chromatography has finally solved this paradox. The separation of murein subunits obtained from isolated sacculi by complete enzymatic degradation with muramidase on a reverse phase column yields a pattern of surprising complexity (7). Instead of only the few types of murein subunits known so far, around 80 different compounds were identified. They all were isolated, purified and their chemical structure was determined. Their analysis showed very clearly that they all are indeed native components of murein.

Some of the novel subunits found are of special relevance for our conceptions about murein structure and murein metabolism. One of them is a group of oligomers, in which all peptide side chains are crosslinked with each other. The other class of subunits encloses components with a new sort of peptide bond. In this case, other than in the normal type of crossbridge in which the terminal alanine of one side is linked to the diaminopimelic acid residue of the other side chain, the crosslinking bond is between the two diaminopimelic acid residues of the side chains (review: 8).

The existence of the novel oligomeric subunits forces us to reconsider the assumption of E. coli murein being a monolayered net. In a monolayered murein these subunits could exist only at chain ends if one accepts the data based on X-ray diffraction studies of murein, on infra-red spectroscopy and of ^{15}N-NMR relaxation studies (reviews: 1; 9). The analysis of murein composition, however, shows that about 80% of the trimers exist within the glycan chains. This is possible only when the glycan chains are arranged in different planes. The same conclusion is reached when the mode of insertion of new building blocks into the sacculus during cell growth is studied. With the murein-specific alpha, epsilon-diaminopimelic acid as a label, pulse-chase experiments have shown that peptide crossbridges in the murein have different life times (8). Crosslinks formed between new and preexisting glycan strands are cleaved and reformed during the ageing of murein (10). The crosslinks formed among newly incorporated glycan chains, however, are relatively stable (8). The most straightforward interpretation of this observation is again the notion of E. coli murein as a multilayered net growing according to an inside-to-outside mechanism as shown for gram-positive species (2). Thus, there exists no principal difference in of gram-negative and gram-positive organisms as assumed for quite some time.

Other recent observations are in strong support of the assumption of E. coli murein being a multilayered net. The electronmicroscopy of cell-envelopes using new techniques of fixation revealed murein as a thick gel-like structure with a high water content (11) rather than a very thin layer. Furthermore, when envelopes are viewed after partial degradation of murein, again a multilayered structure of murein is revealed (12).

Is there enough murein available per cell to make a multilayered sacculus? This, in fact, is the case if one accepts recent stereochemical considerations and physico-chemical data on murein structure: the density of murein is lower than originally assumed, murein is not a chitin-like polymer, the peptide chains are helically arranged around the glycan chains and finally, in stretched form the distance between adjacent sugar chains is considerably longer than assumed so far.

In summary, all data available at present are very much in favor of a multilayered murein, also in the case of E. coli. According to our calculations, the murein found in a cell would suffice to make, in the average, 3 layers. From an evolutionary point of view, the structure of the E. coli murein may be regarded as an improvement of the concept found in gram-positive species. The thickness of the murein seems to be optimized to the minimal number of layers necessary to guarantee the structural integrity of the wall.

If the addition of new murein in the growing cell is achieved through apposition to the preexisting sacculus, the participation of murein hydrolases in murein metabolism is strictly required. When new material is added at the inner side of the sacculus - inside to outside growth (16) - the enlargement of the surface area of murein requires the hydrolysis of bonds in the preexisting murein. In addition, part of it necessarily has to be removed from the outside to avoid the situation that after some generations the cell consists only of murein and nothing else.

In the growing cell peptide bridges once formed during the integration of new material in the sacculus indeed are cleaved and reformed (10). More striking, however, is the liberation of murein subunits from the sacculus and their reuse. In this recycling process about 50% of the murein of a cell are cut out and reintegrated in the sacculus in the course of one generation. The cells are quite efficient in reusing the components set free in the cells, nevertheless, a constant leakage of cell wall peptides into the culture medium is observed. About 6 to 8% of the murein per cell is lost this way in a generation (18).

The recycling of murein in the cell and the release of muropeptides in the environment is a constant source of muropeptides set free in the bacterial host. These compounds may act as a special class of signaling molecules with important biological functions for the host not only as a response to bacterial infection but also in normal life. A host of data about striking biological effects of murein degradation products in higher organisms has accumulated in the past 25 years. To name a few examples, muropeptides interact in many ways with macrophages, are involved in immunomodulation and promote slow wave sleep (reviews: 19,20).

The biological effects of murein degradation products in higher organisms are striking, almost as striking, however, is the lack of information on the metabolism of murein in the bacterial host. What is the major source of these compounds, is it the bacterial population in the intestinal tract? What is the contribution of bacterial autolytic enzymes and of murolytic enzymes of the host in the liberation of muropeptides? How do muropeptides enter the blood stream, what are their physiological targets and how do they interact with them, etc. etc.? Here opens a wide field, we have the tools to cultivate it and we should use them.

References

1. Holtje J-V, Schwarz U (1985) Biosynthesis and growth of the murein sacculus, p 77-119. In Nanniga (ed), Molecular Cytology of Escherichia coli. Academic Press Inc London
2. Rogers HJ, Perkins HR, Ward JB (eds) (1980) Microbial walls and membranes, p 190-214. Chapman and Hall London New York
3. Primosigh J, Pelzer H, Maass D, Weidel W (1961) Chemical characterization of muropeptides released from the E. coli B cell wall by enzymic action. Biochim Biophys Acta 46:68-80
4. De Pedro MA, Schwarz U (1981) Heterogeneity of newly inserted and preexisting murein in the sacculus of Escherichia coli. Proc Natl Sci USA 78:5856-5860
5. Gmeiner J, Kroll H-P (1981) N-acetylglucosaminyl-N-acetyl-muramyl-dipeptide, a novel murein building block formed during the cell division of Proteus mirabilis. FEBS Letters 129:142-144
6. Spratt GB (1977) Properties of the penicillin-binding proteins of Escherichia coli K12. Eur J Biochem 72:341-352
7. Glauner B, Schwarz U (1983) The analysis of murein composition with high-pressure-liquid chromatography, p 29-34. In Hakenbeck R, Holtje J-V, Labischinski H (eds), The target of penicillin. Walter de Gruyter Berlin New York
8. Schwarz U, Glauner B (1987) Murein structure data and their relevance for the understanding of murein metabolism in Escherichia coli. Proc ASM Conf on Antibiotic Inhibition of Bacterial Cell Surface. In press
9. Labischinski M, Barnickel G, Naumann D, Keller P (1985) Conformational and topical aspects of the three-dimensional architecture of bacterial peptidoglycan. Ann Microbiol 136A:45-50

10. Goodell EW, Schwarz U (1983) Cleavage and resynthesis of peptide cross bridges in Escherichia coli murein. J Bacteriol 156:136-140
11. Hobot JA, Carlemalm E, Villinger W, Kellenberger E (1984) Periplasmic gel: New concept resulting from the reinvestigation of bacterial cell envelope ultrastructure by new methods. J Bacteriol 160:143-152
12. Leduc M, van Heijenoort J (1985) Correlation between degradation and ultrastructure of peptidoglycan during autolysis of Escherichia coli. J Bacteriol 161:627-635
13. Barnickel G, Naumann D, Bradaczek H (1983) Computer aided molecular modelling of the three-dimensional structure of bacterial peptidoglycan, p 61-66. In Hadenbeck R, Holtje J-V, Labischinski H (eds), The target of penicillin. Walter de Gruyter Berlin New York
14. Labischinski M, Barnickel G. Naumann G (1979) The state of order of bacterial peptidoglycan, p 49-54. In Hadenbeck R, Holtje J-V, Labischinski H (eds) The target of penicillin. Walter de Gruyter Berlin New York
15. Labischinski H, Johannsen L (1985) On the relationship between conformational and biological properties of murein, p 37-42. In Seidl PH, Schleifer KH (eds), Biological properties of peptidoglycan. Walter de Gruyter Berlin New York
16. Koch AL (1985) Inside-to-outside growth and turnover of the wall of gram-positive rods. J Bacteriol 117:137-157
17. Goodell EW (1985) Recycling of murein by Escherichia coli. J Bacteriol 163:305-310
18. Goodell EW, Schwarz U (1985) Release of cell wall peptides into culture medium by exponentially growing Escherichia coli. J Bacteriol 162:391-397
19. Dziarski R (1986) Effects of peptidoglycan on the cellular components of the immune system, p 229-247. In Seidl PH, Schleifer KH (eds), Biological properties of peptidoglycan. Walter de Gruyter Berlin New York
20. Krueger J, Walter J, Levin C (1985) Factor S and related somnogens: An immune theory for slow-wave sleep, p 253-276. In McGinty DJ (ed), Brain mechanisms of sleep. Raven Press New York

CHEMICAL NATURE AND CELLULAR LOCATION OF ADHESINS OF E. COLI

Klaus Jann, Heinz Hoschützky and Thomas Moch

Max-Planck-Institut fur Immunbiologie
D-7800 Freiburg, F.R.G.

Introduction

The term adhesion describes a specific interaction of bacterial recognition proteins (adhesins) with the carbohydrate moiety of glycoproteins or glycolipids on mammalian cells. Adhesion of pathogenic bacteria to epithelial cells initiates the infective process, enabling the bacteria to withstand the rinsing forces of body fluids or to optimize the delivery of toxins to target cells. It can also be considered as a first stage in bacterial invasion. A function hitherto less appreciated may be the acquisition of nutritional advantages on a substratum rich in nutrients.

The finding that some adhesions could be inhibited with mannose and others could not led to the designations of mannose sensitive adhesion (MS, i.e. inhibitable) and mannose resistant adhesion (MR, i.e. not inhibitable) (1). Carbohydrate specificities could later be attributed to some adhesive processes: α-Gal-(1,4)-β-Gal (P specificity) recognized by uropathogenic E. coli (2), α-NeuNAc-2,3-β-Gal (S specificity) in E. coli causing septicemia or neonatal meningitis (3), an unknown structural arrangement of galactose, neuraminic acid and serin (M specificity) in uropathogenic E. coli (4). Some enteropathogenic E. coli exhibit adhesins (CFA/I or CFA/II) (5,6) which mediate adhesion to cell surface structures containing galactose and neuraminic acid.

It was observed that adhesiveness of the bacteria is frequently correlated with the expression on their surface of filamentous surface appendages, termed fimbriae or pili. This correlation does not, however, always apply: there are unfimbriated bacteria which are, nevertheless, adhesive. Thus, adhesins can be associated with fimbriae or can be

NATO ASI Series, Vol. H24
Bacteria, Complement and the Phagocytic Cell
Edited by F. C. Cabello und C. Pruzzo
© Springer-Verlag Berlin Heidelberg 1988

expressed without the presence of fimbriae. It is noteworthy that both types of adhesins are not expressed if the bacteria are cultivated at 20°C or below. In the following these adhesins will be briefly characterized.

Fimbriae associated adhesins (FAA)

Whereas MR-fimbriae are not or only very poorly expressed during growth on agar, MR-fimbriae are well expressed on solid media. For the study of MR-fimbriae, the bacteria are therefore usually cultivated on agar.

The adhesive complex, consisting of fimbriae to which the respective adhesin is attached, can be removed from the bacteria by one of several simple procedures (agitation of the bacterial suspension with an Omnimixer, passage through a hypodermic needle or incubation in the presence of a detergent such as octyl glucoside). After removal of the bacteria, the fimbriae-adhesin complex is obtained by ultracentrifugation. The sedimented material appears as fimbriae, when inspected in the electron microscope. SDS-PAGE reveals peptide subunits (fimbrillin, pilin) with molecular weights in the range of 12-28 Kd. Fimbrial preparations from E. coli strains often exhibit several distinct subunits (7). Molecular cloning as well as analysis with monoclonal antibodies have shown that the presence of several subunits indicate the presence in the bacterial population of several distinct fimbriae, each with its own subunit. Although a large number of fimbriae have been described, the nature of the adhesin itself or its relation to the fimbriae could not be defined in any one of these reports. It was generally assumed that the adhesin is either identical with the fimbrillin, or that it is some protein of unknown nature, present on the bacterial surface. Genetic evidences, advanced in several laboratories, indicated that the characteristics of fimbriation and adhesiveness can be mutated independently (8-11). The most straightforward interpretation of these findings was that the fimbrillin and adhesin of the strains studied were different proteins, directed from different sites of the bacterial chromosome.

In collaboration with J. Hacker (Universität Würzburg, F.R.G.) we have undertaken the physical separation and characterization of fimbriae and adhesin from S-specific uropathogenic E. coli (12). The adhesive complex, obtained as described above, was dissociated into fimbriae and adhesin by incubation in PBS at 65-70°C.

Subsequent ultracentrifugation separated the fimbriae (pellet) from the adhesin (supernate). The latter was contaminated with some fimbriae and proteins of the bacterial outer membrane. It could be purified to apparent homogeneity by HPLC in the presence of chaotropes, which were then removed by dialysis. The fimbrillin of the S-specific E. coli has a molecular mass of 16 Kd and a pI of 6.6. It contains 2 cysteine units per molecule. In contrast, the adhesin has a molecular weight of 12 Kd, a pI of 4.7 and contains only one cysteine per molecule. It therefore cannot contain a disulfide loop. Also, disulfide linkage between adjacent fimbrial subunits can be excluded.

We have used the fimbriae-adhesin complex for the preparation of monoclonal antibodies (BALB/c mice, PAI myeloma cells, use of purified fimbriae and adhesin for the screening of hybridoma supernates). Four fimbriae-specific antibodies, not reactive with the adhesin, and two adhesin-specific antibodies, not reactive with the fimbriae, were obtained. Their properties are shown in Table 1. Whereas the adhesin-specific antibodies reacted with the adhesin in the native state (ELISA) and in the denatured state (SDS-PAGE/Immunoblot), the fimbriae-specific antibodies reacted only with native and not with denatured fimbriae. One of the adhesion-specific antibodies inhibited the agglutination of RBC with fimbriated E. coli (which is the usual test for adhesion) and also the adhesin of the S-specific bacteria to human kidney cells. It can thus be considered as anti-adhesive, recognizing an epitope either in the adhesive recognition site of the adhesin molecule or close to it.

The anti-adhesive monoclonal antibody was used to probe the location of the adhesin on the bacteria. Figure 1A shows that the adhesin is closely associated with the fimbriae. As seen in Figure 1B, a distal location of

<u>Table 1</u>. Properties of monoclonal antibodies (mab) against (A, Adh) or
fimbriae (F, Fim)

Mab	Ig Subclass	specificity fim	adh	state of antigen[a]	inhibition of HA[b]	agglutination of E. coli[c] Fim⁺Adh⁺	Fim⁻Adh⁺	Fim⁺Adh⁻
A1	G1	-	+	n,d	+	+	+	-
A2	M	-	+	n,d	-	+	+	-
F1	G2a	+	-	n,d	-	+	-	+
F2	G1	+	-	n	-	+	-	+
F3	M	+	-	n	-	+	-	+
F4	G1	+	-	n	-	+	-	+

[a]n, native (ELISA); d, denatured (immunoblot); [b]HA, hemagglutination;
[c]strains as described in (11) and (12).

the adhesin at the tips of the fimbriae can be assumed. Although
suggestive, this interpretation is at the moment tentative. A tip position
of the adhesin would render the individual fimbriae monovalent with
respect to adhsein.. Indeed, after ultracentrifugation, the S-fimbriae do
not agglutinate RBC, although they inhibit the haemagglutination and, as
demonstrated by immunofluorescence studies, bind to human kidney cells.
After incubation with magnesium and/or calcium ions, or on incubation at
pH 5 the fimbriae aggregate and in this state they agglutinate RBC. These
results support the notion of a tip position of the S-adhesin.

The purified S-adhesin forms aggregates of an apparent particle size of
more than 10^7 Kd, as evidence by gel permeation chromatography. In the
aggregated state it not only binds to human kidney cells, but also
agglutinates RBC. The reason why the adhesin is monomeric (or at least not
polymeric) when attached to the fimbriae and aggregates after isolation
can probably only be explained when the mode of synthesis and transport of
the adhesin across the bacterial cell wall are understood. The fact that

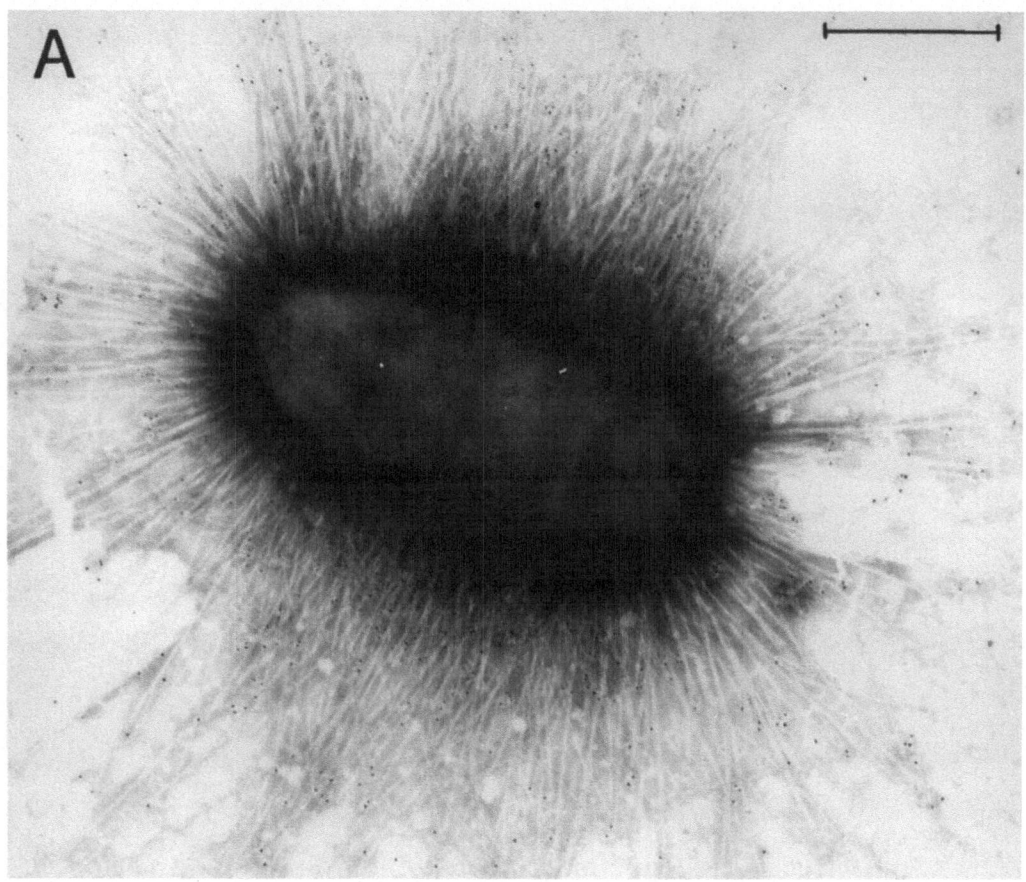

Fig. 1A

Fig. 1B

Fig. 1. Electron microscopic localization of the S-adhesin from E. coli
06:K15 with the immuno gold technique (A) and magnification (B).
Bars indicate 0.5 μm (K.D. Kröncke, M.P.I. Freiburg)

fimbriae can be released from the bacteria, together with the attached
adhesin, under mild conditions, together with the fact that the isolated
adhesive complex preferentially releases the adhesin with the fimbriae
remaining intact, can be used for a simplified model of the adhesive
complex on the bacterial cell surface. We assume that detergent treatment
breaks a non-covalent linkage of the fimbriae to an anchoring protein in
the bacterial outer membrane and that heating of the isolated adhesive
complex releases the tip protein which is the adhesin. This is indicated
in Figure 2 by arrows. Examination of the products obtained in these
reactions show that other proteins are also released, which can then be
removed from the fimbriae during the purification procedure.

One of these proteins may attach the adhesin to one (distal) end of the fimbriae and another one may attach the fimbriae at their proximal end to the anchoring protein of the bacterial outer membrane. Although the model shown in Figure 2 is in agreement with genetic data, it is at this moment speculative.

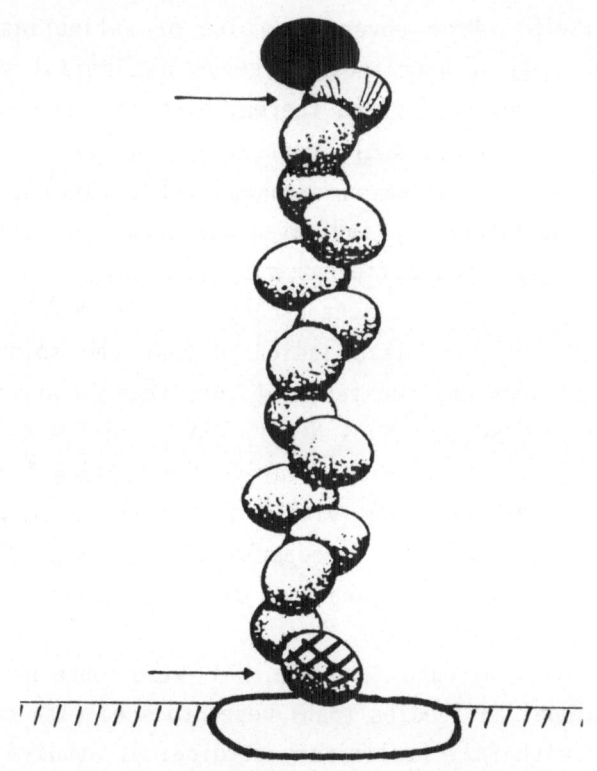

Fig. 2. Hypothetic model of adhesive frimbriae on the E. coli surface. Arrows indicate preferential breakage points.

After the isolation and characterization of the fimbriae-associated S-specific adhesin was reported as a first example (12), we have succeeded in also isolating the adhesin from P-fimbriae (Pap, and F7$_2$) and from the CFA/I-fimbriae of enteropathogenic E. coli. This prompts us to postulate that the topographic organization of adhesive complexes, in which adhesins

are attached to fimbriae, is general at least for E. coli and probably
also for related genera.

Nonfimbrial adhesins

Although unfimbriated adhesive bacteria have been described earlier
(1), studies on nonfimbrial adhesins have begun only recently in several
laboratories (13-16). From several strains of unfimbriated but adhesive
uropathogenic E. coli we have isolated seven nonfimbrial adhesins (NFA-1 -
NFA-7), using the above described thermal elution procedure. Purification
was achieved with fractional centrifugation and ammonium sulfate
precipitation, followed by anion exchange HPLC. Elution profiles during
anion exchange chromatography indicated different surface charges of the
NFA and SDS-PAGE indicated different molecular weights of their subunits.
As shown by gel permeation chromatography, all NFA had apparent molecular
weights above 10^7 daltons. This indicated that the solubilized adhesins
had complex structures and consisted of more than 10 apparently identical
subunits. Amino acid sequences of NFA-1, NFA-2 and NFA-4 (Lottspeich, MPI
Martinsried, FRG) showed as only positive compositional correlation with
other proteins a 70% homology within 20 amino terminal acids of the
fimbrial K88 antigen of calf enteropathogenic E. coli. The NFA are very
resistant to proteolysis; they are also stable to heating.

Polyclonal antisera (pab) to the NFA were obtained with purified
adhesins; monoclonal antibodies (mab) were produced with crude adhesins as
described above with FAA. Preliminary serological studies showed that all
NFA are distinct antigens, some of them cross-reacting in antisera (pab).

In haemagglutination tests it was found that, in contrast to the FAA,
all NFA reacted only with human RBC. Likewise, adhesion experiments could
only be performed with human and not with animal cells. The adhesion
specificities of the seven NFA were tested with haemagglutination and a
number of glycoproteins as putative inhibitors. Interestingly, most NFA
exhibited blood group M specificity, NFA-3 having blood group N
specificity and NFA-1 a hitherto undefined specifity. All NFA reacted with

glycophorin of the respective RBC. The receptor(s) for NFA-1, NFA-2, and NFA-4 could be isolated from RBC membranes with a procedure including affinity chromatography on immobilized NFA. First analyses indicate that the isolated material is glycophorin. Immunofluorescence studies with NFA-1, NFA-2, and NFA-4 using homologous monoclonal antibodies, indicated that human RBC, B- and T- lymphocytes, polymorphonuclear leukocytes as well as HeLa and human kidney cells, but not rat glioma C6 cells, expressed receptors. From these results we assume that the receptor epitope for these (and probably also the other) NFA, although present on glycophorin, is not characteristic of this erythrocyte glycoprotein.

To study the cellular expression of the nonfimbrial adhesins, NFA-1, NFA-2, and NFA-4 were analyzed in immuno electron microscopy using homologous antibodies. As shown in Figure 3, NFA-1 is expressed by the bacteria as a capsule-like extracellular layer.

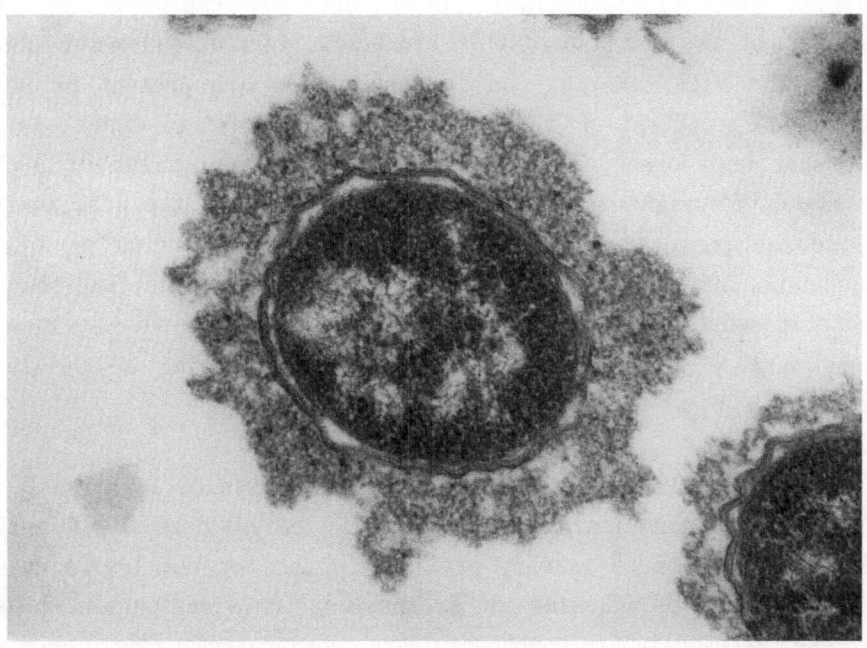

Fig. 3: Electron microscopic picture of a thin section from E. coli 083:K1 exhibiting a nonfimbrial adhesin (NFA-1). The adhesin was stabilized with a specific monoclonal antibody.

This was absent in bacteria grown at 17⁰C. When the bacteria were first grown at 17⁰C and then incubated further at 37⁰C, a slow appearance fo the extracellular material was observed, first in patches and then (after about 30 min.) full expression was seen. Such a capsule-like material was also observed with coli bacteria exhibiting NFA-4, but not with those exhibiting NFA-2. These findings indicate that topographical organization of NFA is different in different strains.

Summary and Conclusions

It is generally accepted that many bacterial infections are initiated by adhesion of the pathogenic bacteria to epithelial cells of the host. The adhesion, which can be easily monitored in vitro by agglutination of erythrocytes, is carbohydrate specific and is mediated by bacterial recognition proteins (adhesins). These are not expressed at growth temperatures of the bacteria of 20⁰C and below. In some cases the adhesins are associated with fimbriae and in others they are present as soluble extracellular proteins. A variation from one form to the other has hitherto not been observed. Both types of adhesins consist of peptides with a molecular weight in the range of 12-35 Kd. From the data available today one can postulate that FAA are exposed at the tips of fimbriae (which consist of a subunit different from the FAA peptide) and that they aggregate when isolated. In contrast, NFA normally seen to be exposed on the cell surface as aggregates, in most cases forming a capsule-like envelope.

It is important to notice that in most cases studied invasive E. coli not only exhibit adhesins but also capsular polysaccharide (K antigen) capsules. Although there is only preliminary and not conclusive evidence for a coexpression of adhesins and K capsules, it is reasonable to assume that these extracellular components are coexpressed under permissive growth conditions. If this is the case, the adhesins must penetrate the carbohydrate capsule, since E. coli are also adhesive in the presence of a polysaccharide capsule. With respect to extracellular topographical arrangements this is especially interesting with the nonfimbrial adhesins. In such cases two soluble extracellular materials may penetrate each other and form a matrix. Since adhesins as well as capsular polysaccharides have been cloned, this problem, along with others such as biosynthesis and regulation, can now be studied in detail.

References

1. Duguid, J.P. and Old, D.C. (1980) Adhesive Properties of Enterobacteriaceae. In: Bacterial Adherence, ed. Beachey, E.H. (Chapman and Hill, London) pp. 185-217.
2. Leffler, H. and Svangorg, Edén, C. (1980) Chemical identification of glycosphingolipid receptors to Escherichia coli attaching to human urinary tract epithelial cells and agglutinating human erythrocytes. FEMS Microbiol. Lett. 8, 127-134.
3. Korhonen, T.K., Vaisänen-Rhen, V., Rhen, M., Pere, A., Parkkinen, J. and Finne, J. (1984) Escherichia coli fimbriae recognizing sialyl galactosides. J. Bacteriol. 159, 762-766.
4. Vaisänen, V., Korhonen, T.K., Jokinen, M., Gahmberg, C.G. and Ehnholm, C. (1982) Blood group M specific haemagglutinin in pyelonephritogenic Escherichia coli. Lancet i:1192.
5. Evans, D.G. and Evans, D.J., Jr. (1978) New surface-associated heat labile colonization factor antigen (CFA/II) produced by enterotoxigenic Escherichia coli of serogroups 06 and 08. Infect. Immun. 21, 638-647.
6. Evans, D.G., Evans, D.J., Jr., Clegg, S. and Pauley, J. (1978) Purification and characterization of CFA/I antigens of enterotoxigenic Escherichia coli. Infect. Immun. 25, 738-748.
7. Jann, K., Jann, B. and Schmidt, G. (1981) SDS polyacrylamide gel electrophoresis and serological analysis of pili from Escherichia coli of different pathogenic origin. FEMS Microbiol. Lett. 11, 21-25.
8. Minion, F.C., Abraham, S.N., Beachey, E.H. and Goguen, J.D. (1986) The genetic determinant of adhesive function in type 1 fimbriae of Escherichia coli is distinct from the gene encoding the fimbrial subunit. J. Bacteriol. 165, 1033-1036.
9. Lund, B., Lindberg, F.P., Baga, M. and Normark, S. (1985) Globoside-specific adhesins of uropathognic Escherichia coli are encoded by similar trans-complementable gene clusters. J. Bacteriol. 162, 1293-1301.
10. van Die, I., Zuidweg, E., Hoekstra, W. and Bergmans, H. (1986) The role of fimbriae of uropathogenic Escherichia coli as carriers of the adhesin involved in mannose-resistant hemagglutination. Microb. Pathog. 1, 51-56.
11. Hacker, J., Schmidt, G., Hughes, C., Knapp, S., Marget, M. and Goebel, W. (1985) Cloning an characterization of genes involved in production of mannose-resistant, neuraminidase-susceptible (X) fimbriae from a uropathogenic 06:K15:H31 Esxherichia coli strain. Infect. Immun. 47, 434-440.
12. Moch, T., Hoschützky, H., Hacker, J., Kröncke, K.-D. and Jann, K. (1987) Isolation and characterization of the α-sialyl-β-2,3-galactosyl-specific adhesin from fimbriated Escherichia coli. Proc. Natl. Acad. Sci. USA 84, 3462-3466.

13. Forestier, C., Welinder, K.G., Darfeuille-Michaud, A. and Klemm, P. (1987) Afimbrial adhesin from Escherichia coli strain 2230: purification, characterization and partial covalent structure. FEMS Microbiol. Lett. 40, 47-50.
14. Orskov, I., Birch-Andersen, A., Duguid, J.P., Stenderup, J. and Orskov, F. (1985) An adhesive protein capsule of Escherichia coli. Infect. Immun. 47, 191-200.
15. Walz, W., Schmidt, A., Labigne-Roussel, A.F., Falkow, S. and Schoolnik, G. (1985) AFA-1, a cloned afimbrial X-type adhesin from human pyelonephritic Escherichia coli strain. Eur. J. Biochem. 152, 315-321.
16. Williams, P.H., Knutton, S., Brown, M.G.M., Candy, D.A.C. and NcNeish, A.S. (1984) Characterization of nonfimbrial mannose-resistant protein hemagglutinins of two Escherichia coli strains isolated from infants with enteritis. Infect. Immun. 44, 592-598.

GENETICS OF O ANTIGEN POLYSACCHARIDE BIOSYNTHESIS IN SHIGELLA AND VACCINE DEVELOPMENT

S. D. Mills and K. N. Timmis

Department of Medical Biochemistry
University of Geneva, Switzerland

Introduction

Lipopolysaccharide is a major structural component of the cell envelope of Gram negative bacteria and is a high molecular weight, long heteropolymer oriented from the cell surface into the external medium. The innermost portion of the molecule, lipid A, constitutes the lipid portion of the outer leaflet of the lipid bilayer of the outer membrane. The outermost portion of the molecule, the O antigen, is frequently an extremely long polymer of oligosaccharide repeat units, is a major structural component of the surface of many bacteria, and fulfills a number of important functions, such as providing resistance to host non-specific and immune defenses in the case of many bacteria pathogenic for animals. The polysaccharide component of LPS is linked to lipid A, and thus to the bacterial outer membrane, via the core, an oligosaccharide having a composition distinct from that of the O antigen.

Because of the structural and functional importance of LPS, it has been the subject of extensive genetic and structural analysis, and a reasonably comprehensive picture of the genetic organization of LPS determinants, the biosynthesis of the individual components of LPS and their assembly into the mature molecule, and the chemical structure of LPS, has emerged (25). Other aspects, however, such as the transport of LPS to the cell surface, its detailed topological structure, and the molecular mechanisms of its various functional activities, remain to be defined. Moreover, the detailed studies which have been carried out so far have been largely restricted to Salmonella and Escherichia coli and it is not certain that the picture of LPS in these organisms will be valid for others. It is nevertheless important for a number of reasons to study this component in other organisms; the O antigen, for example, is a candidate antigen for some vaccines (see below).

NATO ASI Series, Vol. H24
Bacteria, Complement and the Phagocytic Cell
Edited by F.C. Cabello und C. Pruzzo
© Springer-Verlag Berlin Heidelberg 1988

Although genetic engineering techniques have been used extensively to manipulate protein antigens, they have been hardly used at all to manipulate non-protein antigens such as LPS. Despite the greatly increased complexity of such experiments, due to the fact that genetic determinants of entire and sometimes multiple pathways are involved, the importance of such antigens as vaccines can be high. Moreover, application of molecular genetics in the development of LPS-based vaccines will undoubtedly bring important new insights into LPS biology. In this brief overview, we shall describe the cloning, manipulation for vaccine developmemt, and preliminary analysis of the O antigen determinants of Shigella dysenteriae 1, an important cause of bacillary dysentery in Developing Countries.

Shigella and Shigellosis

Shigellae and enteroinvasive strains of Escherichia coli cause bacillary dysentery, an acute but generally self-limiting ulcerative colitis which can range in severity from mild inflammation to the formation of diffuse ulcerative colonic lesions (14). The course of a typical infection involves temporary bacterial colonization of the small bowel resulting in cramps, fever and watery diarrhoea, followed by extensive colonization and invasion of the large bowel (5). The bacteria multiply within the cells of the colonic epithelium and invade the submucosa or lamina propria causing local destruction and provoking an acute inflammatory response. This later stage is characterized by acute abdominal pain and the production of mucoid bloody stools. Spontaneous recovery generally follows several days after onset of the inflammatory reaction. Invasion of deeper tissues and the development of bacteremia is rare, although complications such as convulsions in children, arthritis, purulent keratitis, leukaemoid reactions, hypoproteinaemia and haemolytic uraemic syndrom can occur (24,32).

Shigellosis is uncommon in Developed Countries, having a lower incidence than diarrhoea due to Salmonella, E. coli and Campylobacter (23). In the infections reported, S. sonnei is most frequently the causative agent, and when S. flexneri or S. dysenteriae 1 are isolated, it is usually from patients recently returned from the tropics. In countries where nutrition is poor and sanitary conditions are suboptimal, however, shigellosis can be epidemic. The severity of shigellosis varies according to the bacterial species involved, S. sonnei producing the mildest infections and S. dysenteriae 1 producing the most serious, and according to the state of health and nutrition of the infected individual. In Developing Countries within Africa, Asia and Latin America, S. flexneri and S. dysenteriae 1 predominate and the resulting infections are often characterized by elevated mortality rates and a high incidence of development of the severe complications indicated above (13,23,30). Effective clinical management of infections in these countries is frequently compromised by the multiple antibiotic resistances of the epidemic strains (13,30,43).

Only man and primates are the natural hosts and reservoirs of Shigella sp. Because direct human-human transmission via the faecal-oral route is the most common route of infection, and because of the high infectivity of shigellae (as few as 10 organisms can produce diarrhoea in an otherwise healthy adult; 22), adequate sanitation and detection and treatment of carriers is the most effective way to combat the disease. Given the difficulties in attaining these objectives, however, a safe and efficacious vaccine, particularly against S. dynsenteriae 1, is urgently needed.

In order to produce disease, dysentery-producing bacteria must be able to penetrate, multiply within and kill epithelial cells of the colon and then to spread to and kill neighbouring cells, thereby provoking an acute inflammatory response (5,14). Genetic studies have shown that these attributes are multifactorial, involving products encoded by determinants located on plasmids and on four separate regions of the bacterial chromosome. All fresh isolates of S. sonnei, S. flexneri and S. dysenteriae carry a large plasmid about 200kb in size that specifies the ability to invade and multiply within epithelial cells, and that is easily lost on serial subculture of bacteria in the laboratory (33,34,36). Loss of the plasmid is accompanied by loss of virulence, i.e. bacteria are no longer able to invade tissue culture cells nor cause keratoconjunctivitis in the guinea pig (Sereny test; 33,34).

Shigellae produce a powerful cytotoxin (31), termed Shiga toxin (S. dysenteriae 1) or Shiga-like toxin (other species); although the tissue destruction that occurs in shigellosis is consistent with a role for the cytotoxin in the infection (31), no causal relationship between toxin production and virulence has yet been established (43). A genetic locus for Shiga toxin was recently mapped to the pyrF-trp region of the chromosome of S. dysenteriae 1 (36). The xyl-mtl region of the chromosome appears to code for a product or products involved in bacterial virulence and has been associated with production of an iron-scavenging system and with the ability to persist within the intestinal mucosa (7,12,13,35). The purE region of the chromosome contains a gene designated kcp (keratoconjunctivitis production) which also encodes a virulence product necessary for a positive Sereny test (10,35). Finally, the rfb gene cluster in the his region specified biosynthesis of the bacterial O antigen (9), a critical virulence factor: rough strains of Shigella, i.e. those not producing the O antigen, are impaired in their ability to provoke a Sereny reaction (1). Although it has been suggested that the presence of the O antigen is important in establishing initial contacts with epithelial cells, it is more likely that the antigen plays a critical role in bacterial resistance to intracellular and extracellular nonspecific host defenses. The genes that determine O antigen biosynthesis and structure in S. flexneri are located in two areas of the chromosome: the his-linked rfb locus determines the group specific antigen whereas the pro-linked locus specified the type specific antigen (11). In contrast, in S. sonnei all O antigen biosynthesis genes seem to be located on the large invasion plasmid (20). Surprisingly, in S. dysenteriae 1, one or more O antigen biosynthesis genes are located on a small (9kb) plasmid (47,48) whereas the remainder are located in the chromosomal rfb region (16,40,41).

Vaccine Development

The clinical importance of shigellosis has led to several attempts to develop effective vaccines. Presumably because the infection is generally confined to the superficial layer of the bowel, parenteral vaccines, either killed bacteria or live attenuated organisms, have not proven to be protective (3,17). However, the fact that persons recovering from shigellosis are resistant to subsequent infection with strains of the same serotype suggested that oral vaccination with live organisms may prove effective. S. flexneri strains which had lost the large plasmid and were thus unable to invade the mucosa were found to be safe and provide protection in both man and monkeys, although multiple doses were required which limited their usefulness in Developing Countries (6,26). An E. coli K-12 - S. flexneri hybrid strain in which the xyl-rha region of the K-12 strain had been incorporated into S. flexneri was invasive but had reduced ability to multiply in the mucosa (7). Although a single dose of this vaccine gave protective immunity, it also caused diarrhoea in a large percentage of human volunteers (12). Reversion to virulence was also a problem with a streptomycin-dependent mutant vaccine strain (27-29).

An important finding of these studies is that protection is associated with the bacterial type specific antigen, that is the O antigen (7,12,27). Although the large number of distinct serotypes among Shigellae (greater than 30) at first sight seems to be formidable, epidemiological studies have shown that of the S. dysenteriae serotypes, type 1 predominates, and that in any one geographical area there area usually only 2-3 serotypes of S. flexneri or S. sonnei. This clearly limits the number of distinct antigens that need to be developed into vaccines. In conclusion, therefore, current information suggests that an effective Shigella vaccine candidate will be a live non-virulent organism which, when given orally, is able to penetrate the mucosal epithelial layer to a limited extent and deliver an effective dose of Shigella O antigen to the mucosal immune system (43). Although these are several options with regard to the development of such organisms, we are currently following two routes, namely to construct hybrid vaccines based on attenuated Salmonellae, that have been engineered to produce Shigella O antigens, and to develop attenuated Shigellae.

Live Antigen Delivery Systems

Several different live Salmonella vaccine candidates that stimulate the immune system of the intestinal mucosa, and that are in principle useful for delivering heterologous antigens to this system, have been developed (Table 1).

Table 1. Attenuated <u>Salmonella</u> candidate vaccine strains and antigen
delivery systems

Bacteria	Strain	0 antigen	Phenotype
<u>S</u>. <u>typhimurium</u>	SL3235	4, 5, 12	<u>leu</u> <u>mal</u> <u>cys</u> <u>his</u> <u>aro</u>*
" "	SL3237	4, 5, 12	<u>leu</u> <u>mal</u> <u>cys</u> <u>his</u> str^r
" "	SL3261	4, 5, 12	<u>his</u> <u>aro</u> <u>Fus</u>^{r}
<u>S</u>. <u>dublin</u>	SL1438	9, 12	<u>aro</u>$^-$
<u>S</u>. <u>typhi</u>	Ty21a	9, 12	<u>galE</u>

----

* <u>aro</u> mutants require the aromatic intermediates p-aminobenzoate (for folate
synthesis) and 2,3-dihydroxybenzoate (for enterochelin synthesis)

<u>Salmonella</u> <u>typhi</u> strain Ty21a is a <u>galE</u> mutant that is presently being
tested as a live vaccine against typhoid fever (15). When this bacterium
enters the host it uses galactose as a carbon source and produces smooth,
0 antigen-containing LPS, but then rapidly undergoes lysis due to the
accumulation of toxic levels of galactose-1-phosphate and UDP-galactose.
As a result, the host immune system is stimulated by the 0 antigen but the
bacterium does not persist in the host long enough to cause disease.
Extensive testing of the strain showed it to be safe and stable, and
resulted in its approval for human use by regulatory agencies. An early
trial in Alexandria, Egypt, showed that 3 doses of vaccine given to 6-7
year old school children produced a 96% efficacy over a three year period
(43). However, these findings were not confirmed in subsequent trials in
Santiago, Chile, where efficacy was closer to 50% (21).

<u>S</u>. <u>typhi</u> Ty21a strain has also been used as a carrier for the 0 antigen
of <u>S</u>. <u>sonnei</u> (8). The resulting hybrid containing the large form I plasmid
of <u>S</u>. <u>sonnei</u> produced LPS with serological specificities of both
<u>Salmonella</u> and <u>Shigella</u>, and when used for parenteral immunization of mice
protected against challenge with either organism (8). After injection into
isolated ileal loops of rabbits, mucosal IgA to both <u>S</u>. <u>typhi</u> and <u>S</u>.
<u>sonnei</u> was demonstrated (19). Doses of up to 10^9 bacteria caused no ill
effects in volunteers (44).

A second type of carrier system developed by Stocker et al (37-39) is a set of aroA mutants of S. typhimurium and S. dublin these mutants, obtained using transposon Tn10 to produce non-reverting mutations in aroA, are non-virulent due to their requirement of p-aminobenzoate (necessary for biosynthesis of the siderophore enterochelin) and 2,3-dihydroxy-benzoate (necessary for biosynthesis of folic acid). Although the efficacy of these strains as vaccines has not yet been established, they have advantages over Ty21a in that they are better characterized genetically, and thus are more readily manipulated, and are more hardy (see below).

Cloning of O antigen determinants of S. dysenteriae 1

In order to evaluate the potential utility of these various strains as carriers of the O antigen of S. dysenteriae 1, it was decided to construct a hybrid plasmid that contains all essential O antigen biosynthesis genes and that could be easily introduced into any selected host bacterium. In order to do this, it was necessary to localize and delimit the O antigen genes on the 9kb plasmid, designated pHW400, and on the chromosome. S. dysenteriae 1 derivatives lacking the pHW400 plasmid are rough and lack the O antigen (47). Introduction as well as hybrid plasmids containing cloned subfragments of pHW400, localized the essential O antigen gene, that was designated rfp, to a 2.0 kb region of pHW400 (48). The presence of this cloned segment in a pHW400-negative derivative of S. dysenteriae 1 caused the bacterium to produce typical smooth type LPS, as demonstrated by SDS gel electrophoresis of phenol-water extracted LPS, followed by silver staining. It was further shown that the product of the rfp gene is a protein of 41 kDa (48). Southern hybridization using a DNA probe derived from an internal region of the rfp gene showed that this determinant is present in all of the S. dysenteriae 1 strains tested but not in any other serotype, nor in S. sonnei, S. flexneri or E. coli (48).

Following the discovery of the plasmid-carried rfp gene, it was found that other essential O antigen biosynthesis genes (rfb) are located near the his locus on the bacterial chromosome (16). In order to clone this region, RP4::miniMu-prime hybrid plasmids containing the his region of the chromosome were generated in vivo (40). One such R-prime plasmid, R'40, was, in combination with the cloned rfp gene, able to direct in E. coli K-12 the synthesis of smooth type LPS which reacted in an immune blot with antiserum against the O antigen of S. dysenteriae 1. An 11.5 kb fragment of R'40, which was subcloned into pBR322 to produce plasmid pSS9, was found to contain all essential rfb genes, because when pSS9 and the rfp-containing hybrid plasmid pSS3 were maintained in the same E. coli K-12 host, an O antigen indistinguishable from that produced by S. dysenteriae 1 was formed (40). The rfp gene from pSS3 and rfb genes from pSS9 were then combined in a single vector to generate plasmid pSS37 (40).

The pSS37 plasmid was introduced into strains S. typhi Ty21a, S. typhimurium SL3235, SL3237, SL3261 and S. dublin SL438 (Table 1). All of the strains thereby formed were agglutinated both by homologous antisera (i.e. SL3261, SL3235 and SL3237 with anti-04,5 and S. typhi Ty21a and SL1438 with anti-09) and the heterologous anti-S. dysenteriae 1 serum, whereas the pSS37-lacking parent strains were agglutinated only by homologous sera. Solubilization of whole bacteria with detergent, followed by treatment with proteinase K, SDS-polyacrylamide gel electrophoresis, and staining of the gel with either silver stain or immune blotting revealed that all bacteria produced the heterogeneous mixture of high molecular weight LPS species typical of smooth (i.e. O antigen-linked) LPS which reacted with homologous antisera. LPS from the derivatives carrying pSS37 additionally reacted with anti-S. dysenteriae 1 serum (Fig. 1).

From the LPS banding patterns observed, it appeared that the S. dysenteriae 1 O antigen was attached to the Salmonella LPS core of all strains except S. typhi Ty21a which, although able to synthesize O antigen from Shigella and to express it on the cell surface, could not linked it to the core. It is not, however, certain that O antigen linkage to the LPS core will be essential for stimulation of an adequate anti-bacterial immune response.

Although the hybrid salmonellae produce the Shigella O antigen when cultivated under conditions (media containing chloramphenicol) selective

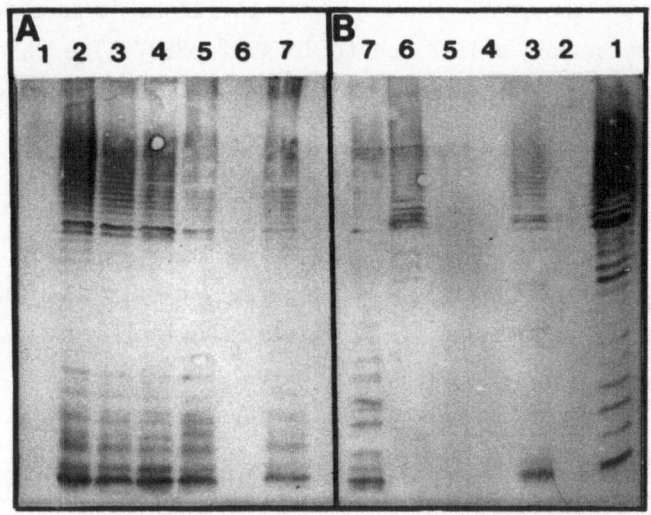

Fig. 1. Immune blots of proteinase K-digested whole cells of various strains of Salmonella. Part A: Immune blot with 04,5 antiserum. Part B: Immune blot with S. dysenteriae 1 antiserum. Lanes: 1 E. coli K-12 (pSS37); 2, SL3235; 3, SL3235 (pSS37); 4, SL3237; 5, SL3261; 6, SL3237 (pSS37); 7, SL3261 (pSS37).

Table 2. Rate of loss of pSS37 from various vaccine strains of Salmonella and E. coli OT99

Bacteria	% Loss/generation*
S. typhimurium SL3235	12
SL3237	15
SL3261	15
S. typhi Ty21a	34
S. dublin SL1438	4
E. coli K12 OT99	16

*Strains were cultivated in the absence of chloramphenicol (chloramphenicol resistance is the selection marker of pSS37) for varying numbers of generations then plated in the absence and presence of the drug to determine the number of colonies with the plasmid versus the total number of colonies.

for maintenance of the pSS37 plasmid, they failed to do so after prolonged growth in non-selective media, due to loss of pSS37. The rate of loss of pSS37 in the absence of selection pressure varied somewhat from strain to strain (Table 2).

To increase the stability of the cloned S. dysenteriae 1 LPS genes, they will be inserted into the Salmonella chromosome by means of a transposon. The LPS genes will first be inserted into the Tn5 transposon element (specified resistance to kanamycin), carried on a temperature sensitive derivative of plasmid PSC101 (specifies resistance to tetracycline; 1). Once the plasmid has been introduced by transformation into the Salmonella strains and established by the growth of transformants at 30°C, the cultivation temperature of the bacteria will be raised to 40°C and kanamycin resistant, (i.e. transposon-carrying) tetracycline sensitive (i.e. pSC101-lacking) clones will be selected that synthesize the S. dysenteriae O antigen. The Salmonella vaccine candidate strains obtained in this manner, containing the O antigen genes in their chromosomes, will be tested in appropriate animal models, for their ability to provide protection against subsequent challegne with virulent S. dysenteriae 1 bacteria.

Attenuated Shigella dysenteriae 1 vaccine candidates

A second approach to vaccine development is to generate attenuated derivatives of Shigella that are non-pathogenic but that still retain the capacity to colonize and invade the intestinal mucosa, and thus to establish contact with the mucosal immune system. A problem with attenuated derivatives, such as aroA mutants, of S. dysenteriae 1 is the production by this serotype of high levels of Shiga toxin, a property that has been implicated in several disease syndromes, including haemorrhagic colitis and haemolytic uraemic syndrome (31).

To eliminate this particular undesirable feature of such a vaccine candidate we have developed a general method for inactivating Shiga toxin production, through the targeted replacement of toxin genes by mutant alleles. Careful mapping of genes specifying the production of Shiga toxin, through conjugal and transductional transfer of S. dysenteriae 1 chromosomal determinants to E. coli K-12, has identified a locus designated stx located under pyrF that is the structural gene for Shiga toxin (Sekizaki et al., 1987). Transposon mutagenesis of an E. coli K-12 Stx$^+$ transductant, followed by co-transduction of the PyrF$^+$ marker and the antibiotic resistance marker carried by the transposon to a PyrF$^-$ recipient E. coli K-12 Hfr host, resulted in the isolation of several K-12 derivatives carrying transposon insertion mutations within the transduced stx gene. Since these derivatives can conjugally transfer chromosomal genes to other bacteria, they can be used to introduce the transposon mutant stx genes into Shiga toxin-producing bacteria. Direct selection for recipients carrying the antibiotic resistance marker of the mutant stx gene selects bacteria having replaced their endogenous Stx$^+$ gene with the mutant gene by homologous recombination, since this is the only means by which the transposon (which is deficient in transposition functions) can survive in the recipient bacterium. When such a conjugal transfer was carried out between an E. coli K-12 Hfr stx::Tn donor strain and several S. dynsenteriae 1 clinical isolates, acquisition of the transposon marker by the shigellae was, in all cases, invariably associated with loss of production of Shiga toxin. Inactivation of Shiga toxin production in aroA mutants of S. dysenteriae 1 is now in progress.

Preliminary analysis of the rfp-rfb gene clusters

The cloning of the rfp and the rfb genes from S. dysenteriae 1 has enabled not only the the construction of vaccine candidates but also the initiation of a molecular genetic study of LPS biosynthesis. Most of the earlier studies on LPS biosynthesis have been carried out with salmonellae (25). The O antigens of these organisms consist of polymers of identical repeat units consisting of between 3 and 5 sugars, some of which are usually unique to LPS. Differences in the nature of the sugars, their types of linkage, their order in the repeat unit, and in their chemical modification (O-acetylation and glycosylation), are the basis for serological typing of salmonellae (2,18). The O antigen unit is assembled on a C_{55} undecaprenol lipid phosphate carrier located within the inner membrane. Nucleotide sugars such as UDP-galactose, UDP-glucose,

UDP-N-acetylglucosamine and dTDP-rhamnose function as activated donors for transferase enzymes which successively add individual sugar moieties to the growing oligosaccharide chain until a single O antigen repeat unit is formed (25). Sugar transferases therefore recognize the activated sugar and the terminal acceptor (unsubstituted or sugar-substituted lipid carrier) and form a specific bond between the two. Once a single repeat unit has formed, it serves as an acceptor for a polymerase-mediated transfer of a growing chain of already-linked repeat units. The growing O side chain also has the potential to be transferred (by a transferase) to the core-lipid A molecule. Competition between transferase(s) and polymerase(s) for the growing O side chain is the likely origin of the enormous molecular weight heterogeneity of complete LPS molecules.

The chemical structure of the O antigen repeat unit in $\underline{S}$. dysenteriae 1 was reported to be (rha-rha-gal-GlcNAc) (4). Figure 2 shows the likely

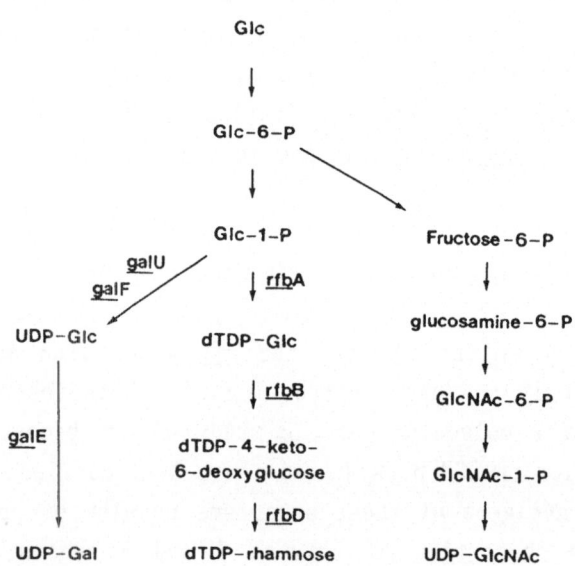

Fig. 2. Probable pathway for the biosynthesis of high energy sugar donors used to assemble the O antigen in $\underline{S}$. dysenteriae 1. Abbreviations: Glc, glucose; Gal, galactose; P, phosphate; GlcNAc, N-acetylglucosamine; dTDP, deoxythymidine diphosphate; UDP, uridine diphosphate. Genes shown are for the analogous genes in Salmonella: rfbA, TDP-glucose pyrophosphorylase; rfbB, TDP-glucose oxidoreductase; rfbD, TDP-rhamnose synthetase; galE, galactose-4-epimerase; galF, unknown; galU, galactose-1-phosphate uridyl transferase.

routes for the formation of the sugar donors required for O antigen synthesis in S. dysenteriae 1, assuming similar metabolic pathways in Salmonella and Shigella.

Since the order of the sugars in the repeat unit of the S. dysenteriae 1 O antigen (GlcNAc-rha-rha-gal) is now known (see below), it is possible to predict the nature of some of the enzymes that may be involved in O antigen assembly in this organism (Table 3).

Table 3.   Enzymes predicted to be involved in assembly of the O antigen of
           S. dysenteriae 1[1]

Enzyme	Function	Gene[2]
galactose-phosphate transferase	gal -> carrier	rfb
rhamnose transferase	rha -> gal-carrier	"
rhamnose transferase	rha -> rha-gal-carrier	"
GlcNAc transferase	GlcNAc -> rha-rha-gal-carrier	"
polymerase	linkage of oligosaccharide units	rfc (18-34 min)
O antigen transferase	O unit transfer from carrier to core	rfaL (rfa-79min) rfbT (rfb)

[1]Based upon the O antigen biosynthetic pathway of Salmonella; note that the carrier for assembly of the O repeat unit is not yet defined in S. dysenteria 1 (see text), nor is the mechanism(s) of assembly of the O polysaccharide elucidated.

[2]Gene identified in Salmonella specifying this function.

These considerations have provided an approximation of what functions are probably required for synthesis of the O antigen of S. dysenteriae 1 and therefore what genes might be located within the rfp-rfb gene clusters. Although preliminary studies have provided some confirmations, some differences in O antigen biosynthesis in Salmonella and Shigella are emerging.

E. coli K-12 produces rough lipopolysaccharide consisting only of core oligosaccharide linked to lipid A. When the rfp-containing plasmid pSS3 is introduced into E. coli K-12, core oligosaccharides having new electrophoretic mobilities are formed (41). Chemical analysis of the modified core showed that it is substituted with galactose, suggesting that the function of rfp is to link galactose to the core (41).

Insertion and deletion mutagenesis of the rfb gene cluster-containing plasmid pSS9, and the introduction of mutant plasmids into E. coli K-12 harboring the rfp plasmid (pSS3), provided 5 classes of mutants that could be distinguished on the basis of bacterial sensitivity to rough specific phages, mobility of extracted LPS on polyacrylamide gels, immunoreactivity of LPS with antisera against the O antigen of S. dysenteriae 1, and chemical composition of the polysaccharide portion of the LPS (Table 4; ref. 42). Class 1 and 2 mutants produced LPS virtually identical to that produced by E. coli (pSS3), although the mutations were located in two separate areas of the rfb region (Figure 3). Class 3 mutants produced LPS in which the core oligosaccharide was additionally substituted with a rhamnose moiety, Class 4 mutants produced core that was additionally substituted with at least two rhamnose moieties, while Class 5 mutants were characterized by a core substituted with a galactose residue, two residues of rhamnose and a residue of N-acetylglucosamine, i.e. in all likelyhood carrying a complete single O side chain repeat unit. These results confirm the chemical composition of the repeat unit determined earlier (4) and suggest the order of the sugars in the repeat as GlcNAc-rha-rha-gal, with galactose being the sugar attached to the core (42). In addition, they suggest that Class 1 and 2 mutants are unable to transfer the first rhamnose residue to the gal-core oligosaccharide and could therefore be defective in genes specifying rhamnose synthesis and/or a rhamnose transferase enzyme(s). Class 3 mutants are unable to add a second rhamnose and probably are defective for a rhamnose to rhamnose transferase. Class 4 mutants probably lacked the N-acetylglucosamine transferase and, since Class 5 mutants are unable to extend the single chain unit, they may be defective in a polymerase function.

Table 4. Structure of LPS produced by <u>E. coli</u> K-12 carrying pSS3 (rfp) and
mutant pSS9 (rfb) plasmids as suggested by chemical analysis and
electrophoretic profiles

Plasmid	Class	Components attached to core	Immunoreactivity	Probable enzyme inactivated
-	-	gal	-	
pSS9	-	complete O antigen	+	
pSS9-21	I	gal	-	rha -> gal transferase
pSS9-41	III	rha-gal	-	rha -> rha transferase
pSS9-36	IV	rha-rha-gal	+	GlcNAc transferase
pSS9-78	V	GlcNAc-rha-rha-gal	+	polymerase(s)

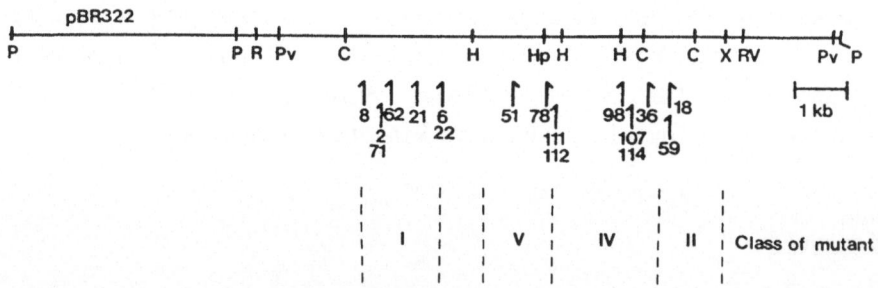

Fig. 3. Physical and functional map of the <u>rfb</u> region of plasmid pSS9.
Restriction sites for the enzymes <u>Pst</u>I (P), <u>Eco</u>RI (R), <u>Pvu</u>II
(Pv), <u>Cla</u>I (C), <u>Hind</u>III (H), <u>Hpa</u>I (Hp), <u>Xho</u>I (X), and <u>Eco</u>RV (RV)
are shown. Tn<u>1000</u> insertions which abolished biosynthesis of the
O antigen are shown by the numbered flags. The direction of the
flag shows the orientation of Tn<u>1000</u> with respect to pSS9. The
<u>rfb</u> segment could be delineated into regions of function
according to the phenotypes of the mutations.

Although the analysis of the <u>Shigella</u> <u>dysenteriae</u> 1 <u>rfp</u> and <u>rfb</u> genes and their functions is in an early phase, a number of interesting and possibly novel findings have been made. In particular, there are indications that the mechanism whereby the 0 antigen is formed in <u>S.</u> <u>dysenteriae</u> 1 may be different from that of the well characterized system of <u>Salmonella</u> (40,46). For example, the first 0 repeat unit in <u>S.</u> <u>dysenteriae</u> 1 would appear to be assembled directly on the core. Whether or not the remaining units are assembled directly on the first unit or transferred intact from a lipid carrier remains to be elucidated. Further characterization of the molecular mechanisms involved in LPS biosynthesis in this important pathogen will not only provide useful information on LPS synthesis and assembly in general, and facilitate development of efficacious anti-dysentery vaccines in particular, but may also promote progress in the design of drugs that inhibit enzymes involved in LPS biosynthesis and that may be useful alternatives to currently available antibiotics.

Acknowledgements

Work reported here from the authors' laboratory has been supported by grants from the World Health Organization Diarrhoeal Diseases Control Programme and the Swiss National Science Foundation. S.D.M. is a postdoctoral fellow of the Ontario Ministry of Health.

References

1. Binns WM, Vaughan S, Timmmis KN (1985) "0" antigens are essential virulence factors of <u>Shigella</u> <u>sonnei</u> and <u>Shigella</u> <u>dysenteriae</u> 1. Zbl Bakt Hyg I Abt Orig B 181:105-109
2. Carlin NIA, Lindberg AA, Bock K, Bundle DR (1984) The <u>Shigella</u> <u>flexneri</u> 0-antigen polysaccharide chain: nature of the biological repeating unit. Eur J Biochem 139:189-194
3. Devoino LV (1959) The specific prophylaxis of dysentery with vaccines from complete antigens. Zhur Microbiol Epidemiol Immunol 30:22-25
4. Dmitriev BA, Knirel YA, Kocketkov NK, Horman IL (1976) Somatic antigens of <u>Shigella</u>. Structural investigations on the 0-specific polysaccharide chain of <u>Shigella</u> <u>dysenteriae</u> type 1 lipopolysaccharide. Eur J Biochem 66:559-566

5. Duguid JP, Marmion BP, Surain RHA (eds) (1978) Medical Microbiology, Vol. 1, p. 323-326. Churchill Livingstone Edinburgh

6. Formal SB, LaBrec EH, Palmer A, Falkow S (1965) Protection of monkeys against experimental shigellosis with attenuated vaccines. J Bacteriol 90:63-68

7. Formal SB, LaBrec EH, Kent TH, Falkow S (1965) Abortive intestinal infection with an Escherichia coli-Shigella flexneri hybrid strain. J Bacteriol 89:1374-1382

8. Formal SB, Baron LS, Kopecko DJ, Washington O, Powell C, Life CA (1981) Construction of a potential bivalent vaccine strain: introduction of Shigella sonnei form I antigen genes into the galE Salmonella typhi Ty21a typhoid vaccine strain. Infect Immun 34:746-750

9. Formal SB, Gemski P Jr, Baron LS, LaBrec EH (1970) Genetic transfer of Shigella flexneri antigens to Escherichia coli K-12. Infect Immun 1:279-289

10. Formal SB, Gemski P Jr, Baron LS, LaBrec EH (1971) A chromosomal locus which controls the ability of Shigella flexneri to evoke keratoconjunctivitis. Infect Immun 3:73-79

11. Formal SB, Hale TH, Kapfer C, Cogan JP, Snoy PJ, Chung R, Wingfield ME; Elisberg BL, Baron LS (1984) Oral vaccination of monkeys with an invasive Escherichia coli K12 hybrid expressing Shigella flexneri 2a somatic antigen. Infect Immun 46:465-469

12. Formal SB, Kent TH, May HC, Palmer A, Falkow S, LaBrec EH (1966) Protection of monkeys against experimental shigellosis with a living attenuated oral polyvalent dysentery vaccine. J Bacteriol 92:17-22

13. Frost JA, Rowe B, Vandepitte J, Threfall EJ (1981) Plasmid characteristics in the investigation of an epidemic caused by multiply resistant Shigella dysenteriae type 1 in Central Africa. Lancet 2:1074-1076

14. Gemski P Jr, Formal SB (eds) (1975) Shigellosis: an invasive infection of the gastrointestinal tract. In D Schlessinger (ed) Microbiology-1975). Amer Soc Microbiol Washington DC, pp 165-169

15. Germanier R, Furer E (1975) Isolation and characterization of galE mutant Ty21a of Salmonella typhi: a candidate strain for a live, oral typhoid vaccine. J Infect Dis 131:533-538

16. Hale TL, Guerry P, Seid RC, Kapfer C, Wingfield ME, Reaves CB, Baron LS, Formal SB (1984) Expression of lipopolysaccharide O-antigen in Escherichia coli K-12 hybrids containing plasmid and chromosomal genes from Shigella dysenteriae 1. Infect Immun 46:470-475

17. Higgins AR, Floyd TM, Kader MA (1955) Studies in Shigellosis III. A controlled evaluation of a monovalent Shigella vaccine in a highly endemic environment. Amer J Trop Med Hyg 4:281-288

18. Kenne L, Lindberg B, Petersson K, Katzenellerbogen E and Romanowska E (1978) Structural studies of Shigella flexneri O antigens. Eur J Biochem 91:279-284

19. Keren DF, Collins HH, Baron LS, Kopecko DJ, Formal SB (1982) Intestinal immunoglobulin A responses in rabbits to a Salmonella typhi strain harboring a Shigella sonnei plasmid. Infect Immun 37:387-389

20. Kopecko DJ, Washington O, Formal SB (1980) Genetic and physical evidence for plasmid control of Shigella sonnei form 1 cell surface antigen. Infect Immun 21:207-214

21. Levine MM, Black RE, Ferreccio C, Clements ML, Lanata C, Rooney J, Germanier R (1985) The efficacy of attenuated Salmonella typhi oral vaccine strain Ty21a evaluated in controlled field trials. In Holmgren J, Lindberg A, Mollby R (eds). Development of Vaccines and Drugs against Diarrhea. 11th Nobel Conference, Stockholm, pp. 90-101

22. Levine MM, Kaper JB, Black RE, Clements ML (1983) New knowledge on pathogenesis of bacterial enteric infections as applied to vaccine development. Microbiol Rev 47:510-550

23. Levine MM (1982) Bacillary dysentery. Mechanisms and treatment . Med Clin North Amer 66:623-638

24. Koster F, Levin J, Walker L, Tung KS, Gilman RH, Rahaman MM, Majid A, Islam S, Williams RC (1978) Hemolytic uremic syndrome after shigellosis: relation to endotoxemia and circulating immune complexes. N Engl J Med 298:927-933

25. Makela PH, Stocker BAD (1984) Genetics of lipopolysaccharides. In Rietschel ET (ed). Handbook of Endotoxin, Vol 1: Chemistry of Endotoxin. Elsevier Science publishers B V, pp. 59-137

26. Meitert T, Istrati G, Sulea IT, Baron E, Andronesciu C, Gogulescu L, Templea C, Inaopol L, Galan L, Fleserice M, Onciu C, Bogos L, Lupovici R, Boghitoiu G, Onmt E, Popescu G, Mihailiuc I, Maftei S, Tapu J, Zebruniuc P (1973) Prophylaxie de la dysenterie baccillaire par un vaccin vivant anti-dysenterique dans une collectivite d'enfants neuropsychiques. Arch Roum Pathol Exp Microbio 32:35-44

27. Mel DM, Terun AL, Vuksic L (1965) Studies on vaccination against bacillary dysentry. 3. Effective oral immunization against Shigella flexneri 2a in a field trial. Bull WHO 32: 647-655.

28. Mel DM, Gangarosa EJ, Radovanic MD (1971) Studies on vaccination against bacillary dysentery. VI protection of children by oral immunization with streptomycin-dependent Shigella strains. Bull WHO 45:457-464

29. Mel DM, Papo RG, Terzin AL, Vucsis L (1965) Studies on vaccination against bacillary dysentery. 2. Safety tests and reactogenicity studies on a live dysentery vaccine intended for use in field trials. Bull WHO 32:637-645

30. Olarte J (1981) R factors present in epidemic strains of Shigella and Salmonella species found in Mecico. In Levy SB, Clowes RC, Koenig E (eds). Molecular Biology Pathogenicity and Ecology of bacterial plasmids. Plenuum Press New York, pp. 11-19

31. O'Brien A, Holmes RK (1987) Shiga and Shiga-like toxins. Microbiol Rev, in press

32. Rahaman MM, Alam J, Islam MR, Greenough WB, Lindenbaum J (1975) Shiga bacillus dysentery associated with marked leukocytosis and erythrocyte fragmentation. Johns Hopkins Med J 136:65-70

33. Sansonetti PJ, Kopecko DJ, Formal SB (1981) Shigella sonnei plasmids: evidence that a large plasmid is necessary for virulence. Infect Immun 34:75-83

34. Sansonetti PJ, Kopecko DJ, Formal SB (1981) Involvement of a plasmid in the invasive ability of Shigella flexneri. Infect Immun 35:853-860

35. Sansonetti PJ, Hale TH, Dammin GH, Kapfer C, Collins HII Jr, Formal SB (1983) Alterations in the pathogenicity of Escherichia coli K-12 after transfer of plasmid and chromosomal genes from Shigella flexneri. Infect Immun 39:1392-1402

36. Sekizaki T, Harayama S, Brazil G, Timmis KN (1987) Genetic manipulation in vivo of stx, a determinant essential for high level production of shiga toxin by Shigella dysenteriae serotype 1: Localization near pyrF and generation of Stx⁻ transposon mutants. Infect Immun (in press)

37. Smith BP, Reina-Guerra M, Stocker BAD, Hoseith SK, Johnson E (1984) Aromatic-dependant Salmonella dublin as a parenteral modified live vaccine for calves. Amer Vet Med Assoc 45:2231-2236

38. Smith BP, Reina-Guerra M, Hoseith SK, Stocker BAD, Habasha F, Johnson E, Merritt F (1984) Aromatic-dependant Salmonella typhimurium as modified live vaccine for calves. Amer Vet Med Assoc 45:59-64

39. Stocker BAD, Hoseith SK, Smith BP (1983) Aromatic dependant Salmonella sp. as a live vaccine in mice and calves. Develop Biol Standard 5:47-54

40. Sturm S, Timmis KN (1986) Cloning of the rfb gene region of Shigella dysenteriae 1 and construction of an rfb-rfp gene cassette for the development of lipopolysaccharide-based live anti-dysentery vaccines. Microb Pathog 1:289-297

41. Sturm S, Jann B, Jann K, Fortnagel P and Timmis KN (1986) Genetic and biochemical analysis of Shigella dysenteriae 1 O antigen polysaccharide biosynthesis in Escherichia coli K-12: 9kb plasmid of S. dysenteriae 1 determines addition of galactose residue to the lipopolysaccharide core. Microb Pathog 1:299-306

42. Sturm S, Jann B, Jann K, Fortnagel P, Timmis KN (1986) Genetic and biochemical analysis of Shigella dysenteriae 1 O antigen polysaccharide biosynthesis in Escherichia coli K-12: Structure and functions of the rfb gene cluster. Microb Pathog 1:307-324

43. Timmis KN, Sturm S, Watanabe H (1985) Genetic dissection of pathogenesis determinants of Shigella and enteroinvasive Escherichia coli. In Holmgren J, Lindberg A, Mollby R (eds). Development of Vaccines and Drugs against Diarrhea. 11th Nobel Conf Stockholm, pp. 107-126.

44. Tramont EC, Chung R, Berman S, Beren D, Kapfer C, Formal SB (1984) Safety and antigenicity of typhoid-Shigella sonnei vaccine strain (strain 5076-IC). J Infect Dis 149:133-136

45. Wahdan MH, Serie C, Cerisier Y, Sallam S, Germanier R (1982) A controlled field trial of live Salmonella typhi strain Ty21a oral vaccine against typhoid: three year results. J Infect Dis 145:292-296

46. Watanabe H, Nakamura A (1985) Large plasmids associated with virulence in Shigella species have a common function necessary for epithelial cell penetration. Infect Immun 48:260-262

47. Watanabe H, Timmis KN (1984) A small plasmid in Shigella dysenteriae 1 specifies one or more functions essential for O antigen production and bacterial virulence. Infect Immun 43:391-396

48. Watanabe H, Nakamura A, Timmis KN (1984) Small virulence plasmid of Shigella dysenteriae 1 strain W30864 encodes a 41,000-dalton protein involved in formation of specific lipopolysaccharide side chains of serotype 1 isolates. Infect Immun 46:55-63

CAPSULAR POLYSACCHARIDES (K ANTIGENS) OF ESCHERICHIA COLI

Klaus Jann and Barbara Jann

Max-Planck-Institut für Immunbiologie
Stübeweg 51
D-7800 Freiburg, F.R.G.

Many coli bacteria, and especially the invasive ones, are surrounded by an extracellular layer, the capsule. The electronmicroscopic appearance of a capsule is shown in Figure 1. It is easy to imagine that capsules protect the bacteria in inadvertent surroundings. They render the bacteria resistant to complement lysis and to phagocytosis and are, therefore, considered as virulence factors.

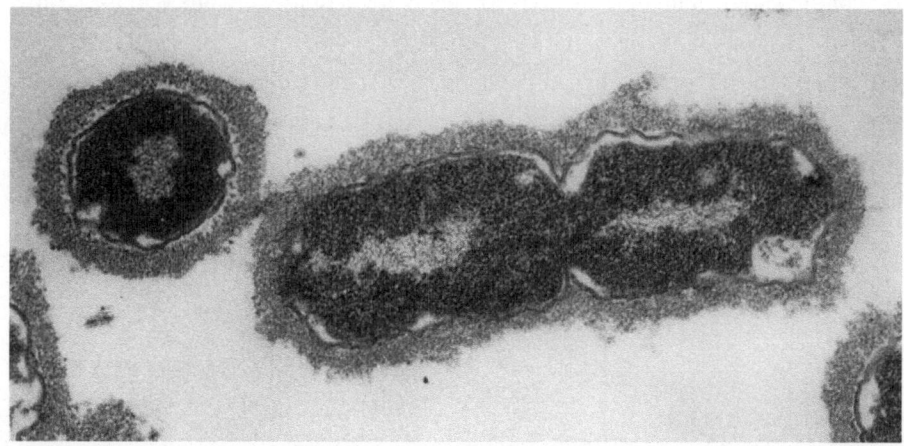

Fig. 1. Electronmicroscopic picture of a thin section from E. coli 08:K40. The capsule was stabilized with homologous antibody. (J. Golecki, MPI Freiburg)

A few capsular polysaccharides are difficult to recognize by the host, because their structures are similar to or identical with structures occurring in the higher organism. Consequently, the host cannot, or only very poorly, produce antibodies against these capsules and, thus, against the bacteria which are surrounded by them. Such bacteria have an advantage over the host and are, therefore, very virulent. Some capsular

NATO ASI Series, Vol. H24
Bacteria, Complement and the Phagocytic Cell
Edited by F. C. Cabello und C. Pruzzo
© Springer-Verlag Berlin Heidelberg 1988

polysaccharides seem to have cytotoxic effects and may, thus, play a role in pathogenic processes.

The capsules of E. coli consist of acidic polysaccharides, which are called K antigens. More than 70 distinct capsular polysaccharides (K antigens) are known today. Their structural and immunochemical characterization has been described in several articles (I. Orskov et al., 1977; K. Jann and B. Jann, 1983; K. Jann, 1983; K. Jann and B. Jann, 1985).

If one compares molecular weights, nature of the acidic components, mode of expression and stability at slightly acidic pH, one can arrange the capsular polysaccharides into two groups. The polysaccharides of group I have high molecular weights, contain predominantly hexuronic acids as acidic constituent and are stable at pH 6 at higher temperatures. They are only found in combination with a few O antigens: O8, O9 (and O20). Their biosynthesis is directed by chromosomal sites close to the his and trp operons. They are expressed at all growth temperatures. Some group I polysaccharides contain amino sugars and other do not, as shown with the K40 and K28 antigens as examples (E. Altman et al., 1985; T. Dengler et al., 1986).

K28   3)-α- Glc- (1,4)β- GlcUA- (1,4)- α- Fuc- (1,
                4   2/3
                1   OAc
            β- Gal
K40   4)-β- GlcUA- (1,4)- α- GlcNAc- (1,6)- αGlcNAc- (1,
                CO- NH- serine

The polysaccharides without amino sugars resemble the capsular antigens of Klebsiella, which also do not contain amino sugars. The K antigens which contain amino sugars are sometimes also substituted with amino acids. The polysaccharides of group II have low molecular weights and show a much greater variability in expressing their negative charge. Thus, hexuronic acids, N- acetylneuraminic acid (NeuNAc), N- acetylmannos- aminuronic acid, 2- keto- 3- deoxy mannosoctonic acid (KDO) or phosphate may be found as acidic sugar components. Many of these polysaccharides are labile in a slightly acidic milieu, especially at elevated temperature and they depolymerize when heated to 100°C at pH 6. Their expression is directed by a gene cluster close to the ser A gene. They are not expressed when the bacteria are grown at 20°C or below. Coli bacteria causing septicemia, neonatal meningitis or urinary tract infections (UTI) usually have K antigens belonging to this group. Some group I capsular polysaccharides are related to capsular polysaccharides of Klebsiella and some group II capsular polysaccharides are related to capsular polysaccharides of Neisseria meningitidis or of Haemophilus influenzae.

K2a   4)-α-Gal-(1-2)-Gro-(3-P-

K2ab   4)-α-Gal-(1-2)-Gro-(3-P-
             ⋮
        2/3-OAc

K51   3)-α-GlcNAc-(1-P-
             ⋮
      4,6-di-OAc

K52   3)-α-Gal-(1-P-
            2
            2
     β-Fru-1-OAc

K18   2)-β-Rib-(1-2)-RibOH-(5-P-

K22   2)-β-Rib-(1-2)-RibOH-(5-P-
             ⋮
       3-OAc

K100   3)-β-Rib-(1-2)-RibOH-5-P-

Fig. 2.   Repeating units of phosphate containing capsular antigens from E. coli

We have recently focused our attention on the group II capsular antigens, which can be subdivided on the basis of their acidic components. Figure 2 shows some phosphate containing polymers. They should not be termed polysaccharides, since the linkages between the repeating units are not of glycosidic but of phosphodiester nature. These polymers thus resemble the teichoic acids of Gram-positive bacteria. An interesting group of related ribosyl-ribitol phosphate polymers (M.L. Rodriguez, B. Jan and K. Jann, Carbohydr. Res. Submitted) are shown in Figure 3.

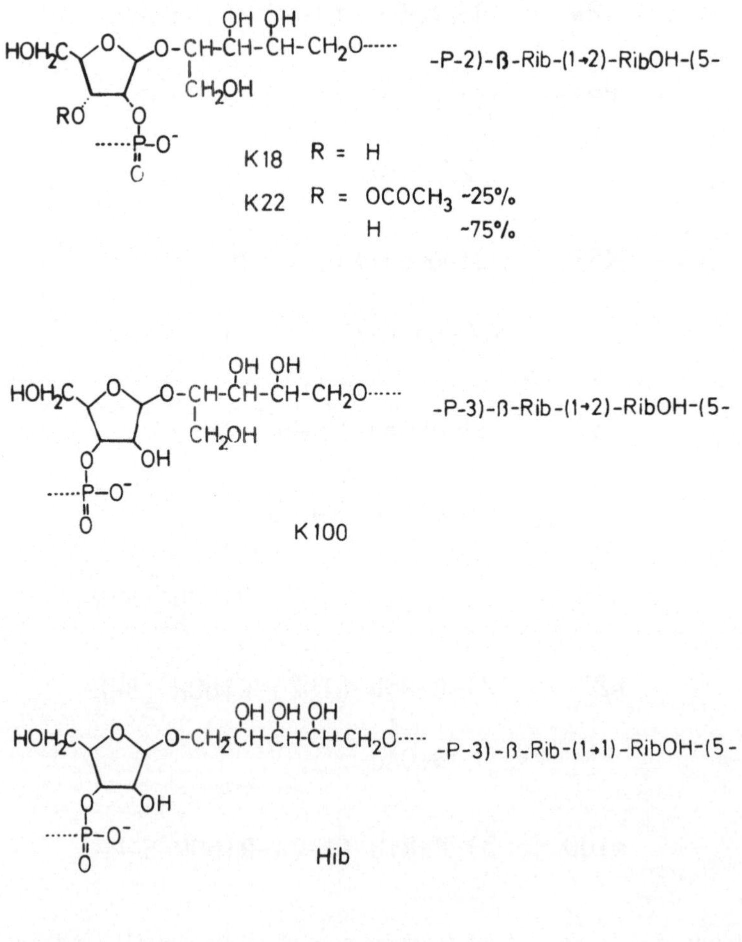

Fig. 3. Repeating units of the K18, K22, K100 antigens of E. coli and of the capsular antigen of H. influenzae b (Hib)

The K18 and K22 antigens which differ only by acetylation, and the K100 antigen have the same sequence of constituents as the capsular antigen of H. influenzae type b. There are, however, differences between these antigens with respect to the linkage of the constituents. As a result, there is a greater relatedness (serological cross reactivity of the K18 (K22) antigen with the K100 antigen than with the Hib antigen.

Until some years ago KDO has been thought to be a typical constituent of the LPS of Gram-negative bacteria, functioning as linkage region between the polysaccharide and lipid A (Rietschel et al., 1987). During the last few years we have found an increasing number of capsular polysaccharides with KDO in their repeating units. Table 1 shows the composition of twelve different capsular antigens, all of which contain KDO as a major constituent.

Table 1. Composition of KDO containing capsular polysaccharides

K antigen			composition	
K6	1 KDO	2 Rib	0.3 0-acetyl	
K12	1 KDO	2 Rha	0.5 0-acetyl	
K13	1 KDO	1 Rib	0.5 0-acetyl	
K14	1 KDO	1 GalNAc	0.5 0-acetyl	
K15	1 KDO	1 GlcNAc	0-acetyl	
K19	1 KDO	1 Rib	0.3 0-acetyl	
K20	1 KDO	1 Rib	0.5 0-acetyl	
K23	1 KDO	1 Rib	- - -	
K74	1 KDO	2 Rib	0.8 0-acetyl	
K95	1 KDO	1 Rib	0.8 0-acetyl	
K97	1 KDO	1 Rib	- - -	
K24	1 KDO	1 Gro	1 P	0.5 0-acetyl

In two cases the other component is an aminosugar; in one case it is rhamnose, but mostly ribose occurs together with KDO - either in a disaccharide or a trisaccharide repeating unit. Acetyl substitution is frequent. It is interesting to note that between 40-60% of the weight of these capsular polysaccharides is made up of KDO. Figure 4 shows the structures of the KDO containing disaccharide repeating units. With the exception of K95, KDO is linked pyranosidically, and the anomeric KDO linkage is always β. Four different linkage positions on KDO (C4-, C5-, C7- and C8) were found. Acetyl substitution, which in most cases is not statistical, is marked by an asterisk. The K13-, K20- and K23 antigens have the same primary structure and substitution. De-0-acetylation converts the K13- and the K20 antigen into the K23 antigen.

K antigen	repeating unit	KDO linkage
K19	3 – βRib – 1,4 – βKDO – 2, *	4–KDOp
K97	2 – βRib – 1,5 – βKDO – 2,	
K14	6 – βGalNAc – 1,5 – βKDO – 2, *	5–KDOp
K15	4 – βGlcNAc – 1,5 – βKDO – 2,	
K13	3 – βRib – 1,7 – βKDO – 2, *	
K20	3 – βRib – 1,7 – βKDO – 2, *	7–KDOp
K23	3 – βRib – 1,7 – βKDO – 2,	
K95	3 – βRib – 1,8 – βKDO – 2, *	8–KDOf

Fig. 4. Disaccharide repeating units of KDO containing capsular polysaccharide (K) antigens of E. coli. Asterisks indicate 0-acetylation.

As shown in Figure 5 three KDO containing trisaccharide repeating units contain either 5- or 7-linked pyranosidic KDO, α in the K6 antigen and β in the K12 antigen. The K74 antigen has a 6-linked furanosidic βKDO. Figure 6 summarizes the linkage positions and anomeric configurations of KDO in the capsular antigens hitherto found. Information from LPS are included.

K antigen	repeating unit	KDO linkage
K12	3 – αRha – 1,2 – αRha – 1,5 – ßKDOp – 2, *	5–KDOp
K6	2 – ßRib – 1,2 – ßRib – 1,7 – αKDOp – 2, *	7–KDOp
K74	3 – ßRib – 1,2 – ßRib – 1,6 – ßKDOf – 2, *	6–KDOf

Fig. 5. Trisaccharide repeating units of KDO containing capsular polysaccharide (K) antigens of E. coli. Asterisks indicate O-acetylation.

The capsular K1 and K92 antigens containing NeuNAc are both homopolymers:

K1   8)-α-NeuNAc-(2,

K92   9)-α-NeuNAc-(2,9)-α-NeuNAc-(2,

In the K1 antigen NeuNAc is α-2,8-linked and in the K92 antigens α-2,8- and α-2,9-linkages alternate (F. Ørskov et al., 1979; W. Egan et al., 1977). The K1 antigen which has the same structure as the capsular antigen of N. meningitidis b, is not or only very poorly immunogenic, whereas the K92 antigen containing 9-linked NeuNAc gives an immune response, with the antibodies being directed only against the 9-linked NeuNAc. The low immunogenicity of 8-linked NeuNAc is probably due to structural identity (Figure 7) of the K1 antigen with the carbohydrate terminal of a glycoprotein (n-CAM) of the developing brain (G.M. Edelman, 1985; J. Finne, 1985).

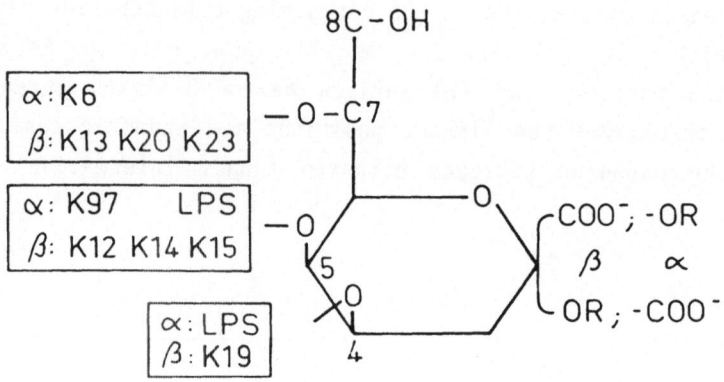

Not yet found : 8-KDOp

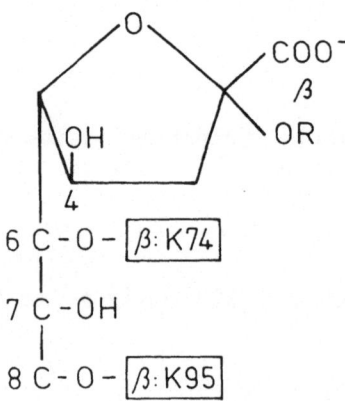

Not yet found : 4-KDOf, 7-KDOf

Fig. 6. Substitution patterns of KDO in E. coli polysaccharides

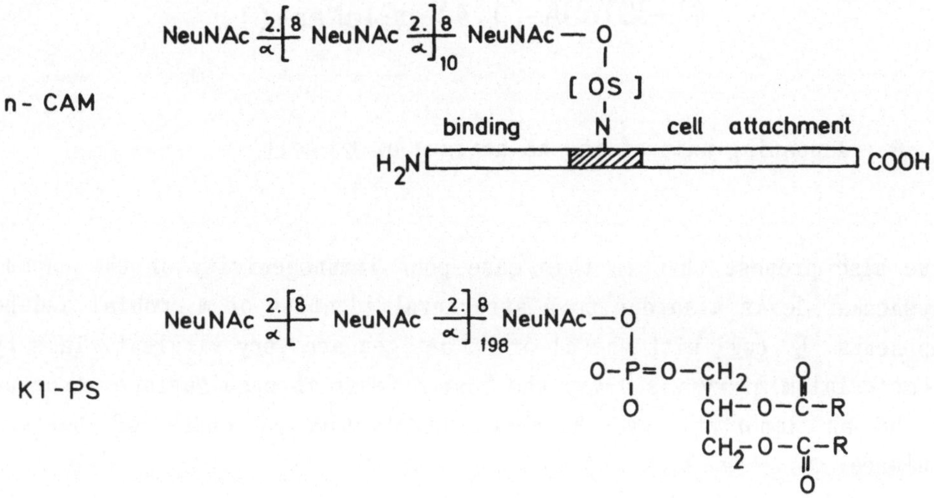

n- CAM

$$\text{NeuNAc} \xrightarrow[\alpha]{2.}\left[8 \text{ NeuNAc} \xrightarrow[\alpha]{2.}\right]_{10}^{8} \text{NeuNAc} - \text{O}$$

[OS]

binding   N   cell attachment

H_2N —————————— COOH

K1-PS

$$\text{NeuNAc} \xrightarrow[\alpha]{2.}\left[8 \text{ NeuNAc} \xrightarrow[\alpha]{2.}\right]_{198}^{8} \text{NeuNAc} - \text{O}$$

$$O-P=O-CH_2 \quad O$$
$$O \quad CH-O-C-R$$
$$CH_2-O-C-R$$
$$O$$

Fig. 7. Schematic structural representation of n-CAM and the K1 polysaccharide

In the adult brain n-CAM does not contain the polysialyl terminus any more. Immunogenicity is, therefore, probably not restricted because of the presence of an identical structure in the adult host, but because of constraints imposed by the immunological memory.

A similar situation exists with the K5 antigen. Its structure (W.F. Vann et al., 1981), which is shown in Figure 8 is identical to that of the first polymeric carbohydrate intermediate in the biosynthesis of heparin.

$$4)-\beta GlcUA-(1,4)-\alpha GlcNAc-(1,$$

Fig. 8. Repeating unit of the K5 antigen of E. coli

We also propose that in this case poor immunogenicity of the capsular polysaccharide is also due to a structural identity of microbial and host components. E. coli with the K1 or K5 antigen are very virulent. This type of bacterial mimicry vis à vis the host defense through surface structures may be an important and hitherto underestimated cause of bacterial virulence.

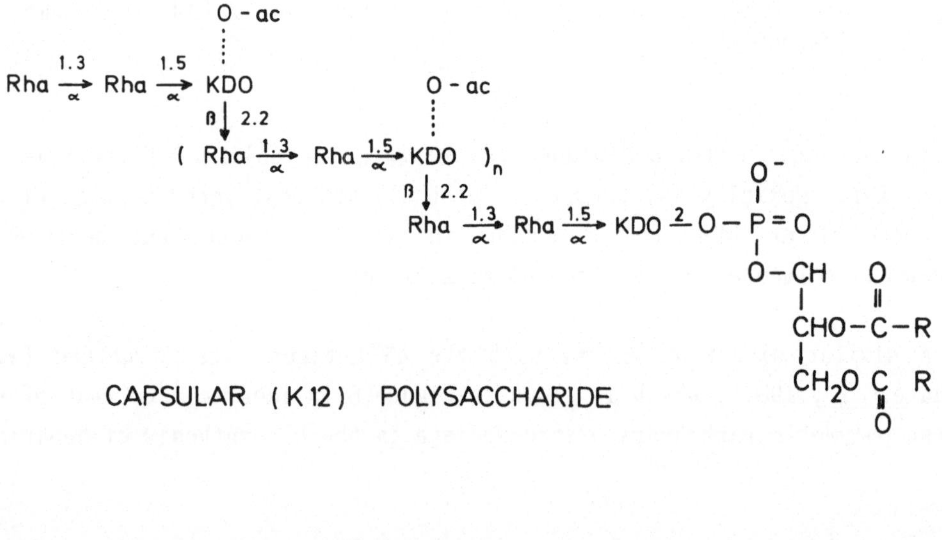

CAPSULAR (K12) POLYSACCHARIDE

Fig. 9. Structure of the K12 antigen of E. coli with the terminal substitution of phosphatidic acid

It is interesting to note how the polysaccharides, especially of group II which have low molecular weights, high charge density and which should therefore have a propensity to dissolve in the medium surrounding the bacteria, are nevertheless retained by the bacterial cells with the formation of a capsule. In this context, the following finding is important. As shown in Figure 9 with the K12 antigen as an example, the low molecular weight group II polysaccharides contain a phosphatidic acid at their reducing end (M.A.Schmidt and K. Jann, 1982). The same results were obtained by Gotschlich et al., (1981) at the Rockefeller University with polyneuraminic acid capsules of E. coli and Neisseria.

We had previously obtained evidence that the high molecular weight group I polysaccharides are linked to core lipid A. We have reinvestigated this with the K40 polysaccharide from E. coli 08:K40. It exhibited in SDS-PAGE the ladder-like pattern characteristic of LPS. In immunoblots the bands reacted with anti-K40 antiserum but not with anti-08 antiserum, which indicated that the K-40 polysaccharide is linked to core-lipid A. This interpretation was corroborated by chemical analysis.

These findings indicate that capsular antigens of E. coli are expressed as polysaccharides linked to a lipid moiety, group I capsular polysaccharides to core-lipid A and group II capsular polysacharides to a phosphatidic acid in labile linkage. As far as we can tell, only a fraction of these polysaccharides is substituted by the respective lipid, a large proportion being present as free polysaccharides. We propose that the capsule is formed and maintained around the bacterial cell by hydrophobic interaction between the lipid moiety and the cell wall as well as charge interactions via bivalent cations and by hydrogen bonds.

The analysis of the cellular expression of E. coli group II capsules has been facilitated by cloning of the relevant genes (Echarti et al., 1983, Silver et al., 1984). In collaboration with G. Boulnois, University of Leicester, we found that the DNA site governing capsule expression consists of three regions (I. Roberts et al., 1986). The region determining the biosynthesis of the capsular polysaccharides at the cytoplasmic membrane (region 2) is flanked by region 1, determining transport of the polysaccharide across the periplasm and the outer membrane, and by region 3, determining a hitherto undefined post-polymerization modification of the polysaccharide, possible the attachment of the phospholipid (see Figure 9). Regions 1 and 2 are conserved in all E. coli strains tested, independent of the structure of the respective capsular polysaccharide. Region 2, however, is characteristic of each strain in so far as it determines the structure of its capsular polysaccharide.

We study the expression of the K1, K5 and K12 capsules in the respective wild types and clones with temperature upshift experiments, where the appearance of capsules is observed after shifting the temperature from 17°C (restrictive) to 37°C (permissive). The capsules are expressed over a period of about 10 minutes, after a lag phase of about 25 minutes and the expression depends on protein biosynthesis and membrane potential. These results are in agreement with findings of C. Whitfield et al. (1984) on the capsular K1 polysaccharide. Immunoelectronmicroscopic studies (K.D. Kroncke and K. Jann, in preparation) show that the cellular distribution of capsular material in wild types and clones (predominantly external) differs from that in group I and group III mutants (only intracellular). Results similar to those with the mutants are obtained with wild types in which the extracellular expression is prevented by blocking protein biosynthesis or by reducing the transmembrane potential. These studies, together with in vitro analyses of capsular polysaccharide biosynthesis (A. Finke and K. Jann, in preparation, see also F. Troy, 1979) extend and complement our present knowledge of capsular polysaccharide structures. We are certain that the structural, genetic and biochemical studies will enable us to better understand the biological function of bacterial capsules.

References

Altman E, and Dutton GGS (1985) Chemical and structural analysis of the capsular polysaccharide from Escherichia coli 09:K28(A):H⁻ (K28 antigen). Carbohydr Res 138:293-303
Dengler T, Jann B and Jann K (1986) Structure of the serine-containing capsular polysaccharide K40 antigen from Escherichia coli 08:K40:H9. Carbohydr Res 150:233-240

Echarti CE, Hirschel B, Boulnois GJ, Varley JM, Waldvogel F, and Timmis KN (1983) Cloning and analysis of the K1 capsule biosynthesis genes of Escherichia coli: Lack of homology with Neisseria meningitidis group B DNA sequences. Infect Immun 41:54-60

Edelman GM (1985) Cell adhesion and the molecular processes of morphogenesis. Annu Rev Biochem 54:135-169

Egan W, Liu TY, Dorow D, Cohen JS, Robbins JD, Gotschlich EC, and Robbins JB (1977) Structural studies on the sialic acid polysaccharide antigen of Escherichia coli strain Bos-12. Biochemistry 16:3687-3692

Finne J (1985) Polysialic acid - glycoprotein carbohydrate involved in neural adhesion and bacterial meningitis. Trends Biochem. Sci. 10:129-132

Gotschlich EC, Fraser BA, Nishimura O, Robbins JB, and Liu T-Y (1981) Lipid on capsular polysaccharides of Gram negative bacteria. J. Biol. Chem. 256:8915-8921.

Jann K (1983) Capsular polysaccharides of pathogenic Escherichia coli. In: Bacterial lipopolysaccharides: structure, synthesis and biological activities (ACS Symposium Series No 231), eds. Anderson, L. and Unger, F. (Am. Chem. Soc.) pp. 171-191

Jann K and Jann B (1983) The K antigens of Escherichia coli. In: Progress in Allergy, eds. Kallos, P. et al. (S. Karger, Basel) pp. 53-79

Jann K and Jann B (1985) Cell surface components and virulence: Escherichia coli O and K antigens in relation to virulence and pathogenicity. In: The virulence of Escherichia coli, ed. M. Sussman (Academic Press, New York) pp. 156-176

Ørskov I, Ørskov F, Jann B and Jann K (1977) Serology, chemistry and genetics of O and K antigens of Escherichia coli. Bacteriol. Rev. 41:667-710

Ørskov F, Ørskov I, Sutton N, Schneerson R, Lin W, Egan W, Hoff, GE and Robbins JB (1979) Form variation in Escherichia coli K1: determined by O-acetylation of the capsular polysaccharide. J. Exp. Med. 149:669-685

Rietschel ET, Brade H, Brade L, Brandenburg K, Schade U, Seydel U, Zähringer U, Galanos C, Lüderitz O, Westphal O, Labischinski H, Kusumoto S, and Shiba, T (1987) Lipid A, the endotoxic center of bacterial lipopolysaccharides: Relation of chemical structure to biological activity. Prog. Clin. Biol. Res. 231:25-53

Roberts I, Mountford R, High N, Bitter-Suermann D, Jann K, Timmis K and Boulnois G (1986) Molecular cloning and analysis of genes for production of K5, K7, K12 and K92 capsular polysaccharides in Escherichia coli. J. Bacteriol. 168:1228-1233

Silver RP, Vann WF and Aaronson W (1984) Genetic and molecular analyses of the Escherichia coli K1 antigen genes. J. Bacteriol. 157:568-575

Schmidt MA and Jann K (1982) Phospholipid substitution of capsular (K) polysaccharide antigens from Escherichia coli causing extraintestinal infections. FEMS Microbiol. Lett. 14:69-74

Troy FA (1979) The chemistry and biosynthesis of selected bacterial capsular polymers. Annu. Rev. Microbiol. 33:519-560

Vann WF, Schmidt MA, Jann B and Jann K (1981) The structure of the capsular polysaccharide (K5 antigen) of urinary-tract-infective Escherichia coli O10:K5:H4. Eur. J. Biochem. 116:359-364

Whitfield C, Vimr ER, Costerton JW and Troy FA (1984) Protein synthesis is required for in vivo activation of polysialic acid capsule synthesis in Escherichia coli K1. J. Bacteriol. 159:321-328

BIOGENESIS OF Galα(1-4)Gal BINDING P-PILI OF UROPATHOGENIC E. COLI

Monica Båga, Kristina Forsman, Mikael Göransson, Frederik Lindberg,
Björn Lund, Britt-Inger Marklund, Mari Norgren, Staffan Normark,
Jan Tennent and Bernt Eric Uhlin

Department of Microbiology
University of Umeå
S-901 87 Umeå
Sweden

Introduction

Escherichia coli is the most common cause of urinary tract infections (UTI). Depending on the virulence of the organism and the efficiency of the host defense system, these UTI appear either as asymptomatic bacteriuria, acute cystitis or acute pyelonephritis. A number of putative virulence properties have been recognized in E. coli isolates associated with acute pyelonephritis in otherwise uncompromised children. One such property is the ability to express mannose-resistant (MR) hemagglutinins. Most E. coli are capable of forming type 1 pili that carry a mannose-specific adhesin. Pyelonephritogenic E. coli additionally carry one or more chromosomal gene clusters that encode pili associated with MR hemagglutination. Of the several MR hemagglutinins that are known today, the best characterized is the Galα(1-4)Gal-specific adhesin associated with P-pili. The name P-pili refers to the observation that this adhesin acts as a hemagglutinin for erythrocytes which express any of the P blood group antigens.

P-pili are heteropolymeric structures

Our current knowledge of gene functions required for the biogenesis of Galα(1-4)Gal-binding P-pili is summarized in the figure. The P-pili gene cluster, pap, was originally cloned from the UTI strain J96 (0:4) by Hull et al. (1). We have recently demonstrated that P-pili are heteropolymers incorporating at least three proteins (PapE, PapF, PapG) at the tip of the

NATO ASI Series, Vol. H24
Bacteria, Complement and the Phagocytic Cell
Edited by F.C. Cabello und C. Pruzzo
© Springer-Verlag Berlin Heidelberg 1988

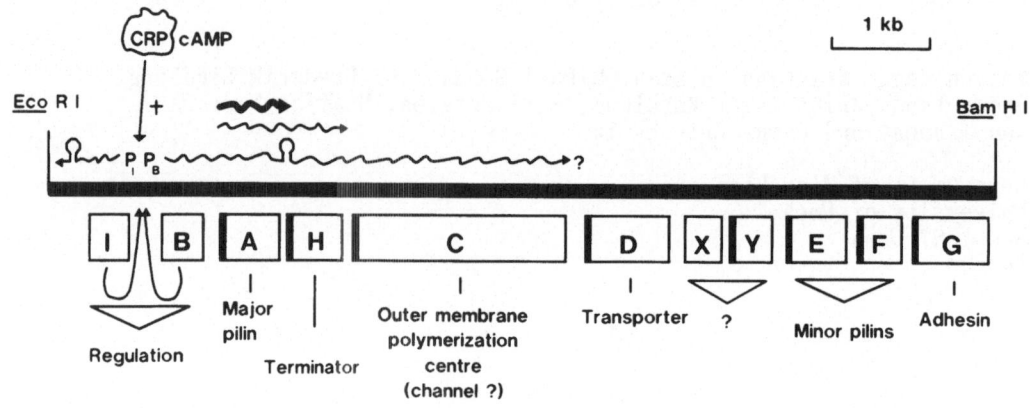

Summary of the structure, function and regulation of the P-pilus gene cluster

The pap gene cluster from the UTI E. coli isolate J96 has been physically mapped on the 9.7 kb EcoRI-BamHI fragment of plasmid pPAP5 (6). We have indicated the established and postulated functions of the various Pap proteins A-H. Two open reading frames, X and Y, have been identified by DNA sequence analysis; as yet, no function has been assigned to the possible products of these genes. The positions of the papI and papB promotors, designated P_I and P_B, respectively, together with the proposed regulatory targets of the PapI, PapB and CRP proteins, are shown. Messenger RNA transcripts emanating from P_I and P_B are depicted as wavy lines, while postulated transcriptional terminators are shown as stem and loop structures (14). Map units are in kilobase pairs (kb).

pilus (2). The pilus shaft is composed of a repeated pilin subunit, PapA, that represents the major pilus antigen. It is proposed that the fully developed P-pilus contains a minor subunit, PapH, at its base. One role of PapH is to firmly associate the pilus to the outer membrane. Recent data also suggests that the stoichiometric ratio between PapA and PapH regulates the length of the pilus (3); on this basis, we call PapH the terminator protein. Like PapH, the tip proteins PapE and PapF share several structural characteristics with PapA and other known E. coli pilins (4). PapG (35.5 kd) is about twice the size of PapA (19.5 kd) and its carboxyl terminal region has some similarities to that of pilin proteins. The PapE, PapF and PapH proteins are referred to as minor pilins, whereas PapG, as will be shown below, is the adhesin.

Prerequisites for pilus biogenesis and binding

Antibodies directed against the tip proteins PapE, PapF and PapG are present in antisera obtained by immunization with whole wild-type pili. Specific absorbed antisera were tested in a radio-immuno assay using intact bacteria harbouring various mutated pap plasmids as solid phase antigen. In papA⁻ cells, PapE, PapF and PapG are expressed on the cell surface, whereas in a papA⁻ papE⁻ strain PapF and PapG no longer reach the cell surface (5). Furthermore, P-pili purified from a papE⁻ strain can only poorly agglutinate P₁-erythrocytes (6). In contrast, whole piliated cells of the same strain are able to express Galα(1-4)Gal-specific binding. These data implied that PapE was not adhesin but rather a minor pilin protein essential for associating the adhesin to the pilus. Immunogold electron microscopy has established that PapF and PapG are still found at the pilus tip in a papE⁻ strain (2).

Mutations in either papF or papG completely abolish pilus attachment. In addition, a mutation in papF causes a marked reduction in piliation. Apparently, PapF does not affect the surface localization of PapG and vice versa. Thus, PapG and PapF are found at the pilus tip in papF⁻ and papG⁻ strains, respectively. Our conclusion is that both PapF and PapG must be accessible at the tip of the pilus for bacteria to express Galα(1-4)Gal-specific binding. We believe that each pilus tip contains very few copies of these two proteins. In contrast, PapE is more abundant; we suggest that PapE subunits at the pilus tip may form one turn of a helix that is in contact with both PapF and PapG and/or PapA. Such a PapE disc may still be formed at the cell surface in association with the PapF and PapG proteins in papA⁻ strains.

The biogenesis of P-pili absolutely requires one periplasmic protein, PapD, and one large molecular weight outer membrane protein, PapC. The

PapD protein has been purified to homogenicity from the periplasmic space (Lindberg, unpublished). In the absence of PapD, which is a basic 28.5 kd protein, both the major PapA subunit and the minor pilins are rapidly broken down. PapD has also been purified in complex with pilin proteins providing evidence that this protein transports pilin proteins through the periplasmic space. The PapC protein is an 88.3 kd outer membrane protein that we believe forms a channel and polymerization centre for P-pili. No pili are formed in the absence of PapC and antisera specific for major and minor subunits do not react with intact cells. However, the total amount of PapA antigen is not decreased in cell extracts of a papC⁻ strain (7).

At present, we only have indirect evidence that PapC forms an outer membrane channel. The primary sequence of PapC, as deduced from the nucleotide sequence, contains a large hydrophilic sequence of amino acids that could form an aqueous pore. Moreover, overproduction of PapC renders the cells lysozyme sensitive and leaky for periplasmic enzymes such as β-lactamase. Constructs have been made that allowed us to compare the effect of a two-fold reduction in PapC production without affecting the expression of other Pap proteins. In such cells we observed a roughly two-fold decrease in the number of pili/cell (7). Thus, it is conceivable that pilus polymerization actually occurs within a PapC pore. We anticipate that P-pili, like type 1 pili (8), grow from the base. In our model, the tip proteins PapE, PapF and PapG are thought to form an initiator complex for pilus formation. Pilus polymerization would then proceed until the minor pilin PapH enters the polymerization complex. Compared with other pilins, PapH has a proline rich hydrophobic N-terminal extension that we believe anchors the P-pilus to the outer membrane. In the absence of PapH, roughly 70% of the total pilus antigen is found released from the cell (3).

Related pili with different binding specificity

We now know that J96 contains a second chromosomal gene cluster that is structurally and functionally related to pap. This second gene cluster was, therefore, denoted prs (pap related sequence). E. coli J96 displays MR agglutination of both human and sheep erythrocytes. The cloned pap gene cluster only caused MR agglutination of human erythrocytes, whereas a clone carrying prs caused the agglutination of sheep erythrocytes but not of human erythrocytes (5). Thus, E. coli J96 is able to express two distinguishable adhesins from highly related gene clusters. In order to elucidate which protein(s) in the respective gene clusters mediated the receptor binding specificity, we performed a series of trans complementation experiments between specific pap and prs genes. We found that mutations in either papF or papG resulted in the loss of the ability to hemagglutinate human erythrocytes, while inactivation of either prsF or prsG abolished sheep-specific hemagglutination. It was possible to complement a mutation in a papF with an intact prsF gene and vice versa without affecting the specificity of hemagglutination. In contrast, complementation of a papG mutation with the wild-type prsG resulted in bacteria which expressed the sheep-specific hemagglutinin. The reverse combination (prsG$^-$/papG$^+$) gave rise to human-specific hemagglutination. Hence, the binding specificity of the pap and prs gene clusters is mediated by the papG and prsG gene products, respectively. Comparative restriction endonuclease mapping of pap and prs revealed no differences between the two gene clusters except for within the regions encoding papG and prsG (9). In fact, an internal papG probe did not hybridize to prs DNA. PrsA, the major subunit of pili expressed by the prs clone, is identical in size to PapA. We do not yet know how similar the PrsA and PapA pilins are at the amino acid level. We conclude that E. coli J96 can express two structurally related pili which bind to two different receptors. Since pap and prs are so similar, it is possible that in wild type J96 cells Pap pili may contain minor Prs proteins and vice versa. The trans complementation data suggest that such heterologous pili may be formed.

The exact nature of the receptor for the sheep hemagglutinin is not yet known. The fact that E. coli expressing Prs pili bind to purified Forsmann antigen, a known major component of the sheep erythrocyte membrane, suggests that the binding epitope is within this structure. The Forsmann antigen contains Galα(1-4)Gal. However, unlike the Pap adhesin, Prs pili do not bind to certain Galα(1-4)Gal containing glycolipids (N. Strömberg, personal communication). Therefore, we believe that the Prs-binding epitope on the Forsmann antigen is distinct from Galα(1-4)Gal. Since the Forsmann antigen has been isolated from the renal pelvis (10), it is exciting to speculate that the Prs adhesin may also contribute to the uropathogenicity of E. coli J96.

Variation of the P-pilus tip proteins

Antigenic variation is a common mechanism by which extracellular parasites avoid the host immune response. A number of serological variants of P-pili have been identified (11,12). The variation in antigenicity has been correlated with structural differences in the major pilin subunit of the variants. Since the minor tip proteins PapE, PapF and PapG are either close to or interact directly with the host receptor, antibodies directed against them might be more protective than antibodies directed against the major subunit. The pap clone E. coli J96 (0:4) expresses P-pili of serotype F13. PapE-, PapF- and PapG-specific antisera prepared from this strain were tested against HB101 expressing P-pili of serotype F7$_2$ or F11. The PapE-antiserum crossreacted with F11 pili but not with F7$_2$ pili, whereas the PapF specific antiserum crossreacted with pili of both F7$_2$ and F11 serotype. However, no crossreaction was observed using the PapG specific antiserum. Therefore, it is clear that the tip proteins may also vary antigenically. Nucleotide sequence information revealed that papF was the most conserved gene of the three. The sequence of the papG adhesin gene from the F7$_2$ and F11 pili gene clusters was highly conserved but quite different from that of papG encoded by the F13 gene cluster. Of 24 UTI E. coli isolates expressing the Galα(1-4)Gal-specific adhesin, 23 hybridized to an internal papG probe from the F11 pili gene cluster (9). None of these UTI strains (except J96) hybridized to a corresponding probe prepared from the PapG gene of the F13 gene cluster. Therefore, we believe that papG may be more conserved than was at first apparent by comparing only three clones.

Regulatory features of pilus biogenesis

The pap gene cluster contains at least 9 closely linked genes, of which four have been shown to be part of an operon (7). Transcription of this operon is initiated from a promoter which lies proximal to the papB gene. The transcriptional activity from this papB promoter is positively affected by the environmental glucose concentration via the CAP protein and cAMP (13). Both PapB and PapI (expressed from its own promoter) appear to act as activators of pap transcription. Most transcripts initiated at the papB promoter are terminated at the papA terminator positioned in the intercistronic region between papA and the terminator pilin gene papH. Readthrough transcription is responsible for the low transcriptional activity over papH and papC. An abundance of the major pilin PapA relative to PapH and PapC is ensured by this terminator. It was recently shown that increasing the expression of PapH while keeping the PapA level constant resulted in cells with shortened pili, whereas unusually long pili were observed when PapA was overproduced relative to the level of PapH (3). The dominance of PapA in the pilus is presumably further ensured by differential stability of the pap messenger RNA. Recent findings suggest that the primary transcript initiated at the papB promoter is cleaved in the intercistronic region between papB and papA (14). The processed transcript which encodes the PapA subunit shows an unusually long half life. As yet there is no detailed information concerning the regulation of papD, which encodes the pilin transport protein, or the distal pap genes encoding the tip proteins. It is possible that these genes are also transcribed from the papB promoter.

Although the transcriptional features of the pap gene cluster described above ensure the preferential expression of PapA, we believe that various protein-protein interactions dictate the ordered polymerization of the P-pilus heteropolymer. Specific interactions between PapD and various pilin proteins, homologous and heterologous interactions between pilin subunits, and possible interactions between PapD and PapC are likely to be decisive factors in P-pilus biogenesis. It is probable that these interactions place constraints on the antigenic variation that the various subunits can display. An understanding of the polymerization rules for P-pili will probably require purification of each of the Pap proteins and the design of an in vitro system for pilus biogenesis.

References

1. Hull RA, Gill RE, Hsu P, Minshew BH, Falkow S (1981) Construction and expression of recombinant plasmids encoding type 1 or D-mannose-resistant pili from a urinary tract infection Escherichia coli isolate. Infect Immun 33:933-938.

2. Lindberg F, Lund B, Johansson L, Normark S (1987) The receptor-binding adhesin is located at the tip of the bacterial pilus. Nature, accepted for publication.

3. Båga M, Norgren M, Normark, S (1987) Biogenesis of E. coli Pap pili: PapH, a minor pilin subunit involved in cell anchoring and length modulation. Cell 49: in press.

4. Lindberg F, Lund B, Normark S (1986) Gene products specifying adhesion of uropathogenic Escherichia coli are minor components of pili. Proc Natl Acad Sci USA 83:1891-1895.

5. Lund B, Lindberg F, Marklund BI, Normark S (1987) PapG is the Galα1-4Gal-specific adhesin of uropathogenic E. coli. Proc Natl Acad Sci USA, accepted for publication.

6. Lindberg FP, Lund B, Normark S (1984) Genes of pyelonephritogenic E. coli required for digalactoside-specific agglutination of human cells. EMBO J 3:1167-1173.

7. Norgren M, Båga M, Tennent JM, Normark S (1987) Nucleotide sequence, regulation, and functional analysis of the papC gene required for cell surface localization of Pap pili of uropathogenic Escherichia coli. Submitted for publication.

8. Lowe MA, Holt SC, Eisenstein BI (1987) Immunoelectron microscopic analysis of elongation of type 1 fimbriae in Escherichia coli. J Bacteriol 169:157-163.

9. Lund B (1987) Umeå University Medical Dissertations No. 194, Department of Microbiology, University of Umeå, Umeå, Sweden.

10. Breimer ME (1985) Chemical and immunological identification of the Forssman pentaglycosylceramide in human kidneys. Glycoconjugate J 2:375-385.

11. Ørskov I, Ørskov F (1983) Serology of Escherichia coli fimbriae. Prog Allergy 33:80-105.

12. Abe C, Schmitz S, Moser I, Boulnois G, High NJ, Orskov I, Jann B, Jann K (1987) Monoclonal antibodies with fimbrial FIC, F12, F13, and F14 specificities obtained with fimbriae from E. coli 04:K12:H⁻. Microbial Pathogenesis 2:71-77.

13. Båga M, Göransson M, Normark S, Uhlin BE (1985) Transcriptional activation of a Pap pilus virulence operon from uropathogenic Escherichia coli. EMBO J 4:3887-3893.

14. Båga M, Göransson M, Normark S, Uhlin BE (1987) Processed mRNA with differential stability in the regulation of E. coli pilin gene expression. Submitted for publication.

THE EVOLUTION OF M PROTEINS OF GROUP A STREPTOCOCCI

June R. Scott[1], Susan K. Hollingshead[1], and Vincent A. Fischetti[2]

[1]Department of Microbiology and Immunology
Emory University
Atlanta, Georgia 30322 USA

[2]The Rockefeller University
1230 York Ave.
New York, N.Y. 10021 USA

Introduction

Protection of the human host against infection by the Group A streptococcus is mediated by antibodies directed against the M protein, a molecule found on the bacterial surface (Lancefield, 1962; Swanson et al, 1969). This protein is a dimeric coiled-coil (Phillips et al, 1981) attached to the bacterial cell by its carboxy terminus (Fischetti et al, 1985) (Fig. 1). So far, over 80 different antigenic types of M protein have been identified with specifically absorbed antisera, and additional strains have been isolated for which no specific antisera have been developed. The M proteins appear to be the most important virulence determinants for the group A streptococci because they protect the organism against phagocytosis. In a polyclonal antiserum directed against one streptococcal strain, some anti-M antibodies are opsonic and thus overcome the protection that the M protein provides (Lancefield, 1962). The opsonic antibodies are thought to be largely type-specific, because little cross-opsonization is observed, at least with the available standardized typing antisera.

Epidemiologically, the group A streptococci are classified on the basis of their M type using sera developed for this purpose. Typing sera are produced by immunizing rabbits with killed whole streptococci and absorbing the resulting antisera with a series of other streptococcal strains. Antisera that show strong precipitin reactions with M

NATO ASI Series, Vol. H24
Bacteria, Complement and the Phagocytic Cell
Edited by F.C. Cabello und C. Pruzzo
© Springer-Verlag Berlin Heidelberg 1988

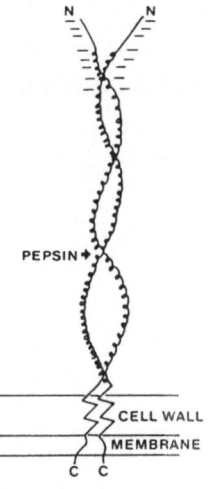

Fig. 1. <u>Cartoon representing the structure of the M protein dimer on the streptococcal surface</u>. The negative charges concentrated at the amino terminus and the predominant pepsin-sensitive site are indicated. The molecule is a predominantly alph-helical coiled-coil dimer with a carboxy- terminal beta pleated sheet region that is probably buried in the cell wall.

protein extracted from the immunizing organism are used to define serological types (Lancefield, 1928; Rotta et al, 1971). Therefore, typing sera from two different laboratories will not be identical. It appears, from our recent data, that there may be a continuum of different antigenic determinants among streptococcal strains and that the classification into specific M types may be an oversimplification.

Many pathogens have a mechanism for alteration of surface antigens which generate production of protective antibodies. This allows the pathogen to evade the host's immune response by changing its immunogenic protein. The existence of a large number of different antigenic types of M protein suggests that a mechanism exists by which the streptococcus can alter this surface protein. Our studies have led to a greater understanding of one possible mechanism for this type of antigenic variation in the streptococcus.

The possibility of development of an anti-streptococcal vaccine is complicated by two factors: the antigenic diversity of M types and the cross-reactivity of certain anti-M protein antibodies with human heart tissue (Dale and Beachey, 1982; Dale and Beachey 1985, a,b). Identification of an opsonogenic determinant common to all M types that does not generate anticardiac cross-reactive antibody would be important for a successful vaccine. We are addressing the question of whether such epitopes exist.

Intact emm6 Cloned Into Escherichia coli

We isolated DNA from the M6 group A streptococcal strain D471, digested it partially with Sau3a, and ligated DNA fragments into the BamHI site of a cosmid vector. Using a colony immunoblot screening technique, we succeeded in isolating an E. coli clone that produced M protein (Scott and Fischetti, 1983). The protein appears to be the same as that produced in the streptococcus because it gives rise to a line of identity with streptococcal-derived M protein in Ouchterlony double diffusion against M6-specific antiserum, because it has the same N-terminal amino acid sequence, and because it reacts with a monoclonal antibody directed to the streptococcal M protein. However, since the M6 protein from E. coli migrates slightly slower on SDS-PAGE than streptococcal-derived M protein, we were concerned that the chimeric cosmid might contain an extra piece of DNA. Alternatively, the extraction of M protein from the streptococcus (with phage-derived lysin which solubilizes the cell wall) might not release the intact molecule from the gram positive cell wall. Our more recent work has shown that the clone contains the entire structural gene for M6 (emm6): the DNA sequence of the cloned DNA corresponds exactly with the sequence derived from the mRNA of the streptococcus (Hollingshead et al, 1986).

When the cloned emm6 gene was transferred to an M$^-$ streptococcal strain, all the known M$^+$ phenotypes were restored (Scott et al, 1986a). The recipient strain we used for this experiment was derived from an M type 28 streptococcus, but appeared from DNA hybridization experiments to be deleted for the M protein structural gene (Scott et al, 1985; see below). The reconstituted M6 strain showed resistance to phagocytosis in human blood, and was opsonized by anti-M6 hyperimmune serum but not by anti-M28 serum. That the M molecule is located on the bacterial surface was demonstrated by both immunofluorescence and ELISA. Furthermore, the reconstituted M6 strain was able to remove specific M6 opsonic antibodies from hyperimmune rabbit serum, and to generate opsonic anti-M6 antibodies. These results indicate that the cloned DNA contains all the information necessary to produce the expected M6 phenotype.

Because the M protein is found on the surface of the streptococcal cells, it must contain signals for protein transport across the cytoplasmic membrane. Fractionation of the E. coli cells containing the emm6 gene indicated that, as expected, the M protein was transported across the cytoplasmic membrane and was located in the periplasm (Fischetti et al, 1984). Comparison of the translated emm6 DNA sequence with the aminoterminal amino acid sequence of mature M6 protein revealed the presence of a long (42 amino acid) signal peptide, typical of transported proteins in gram positive organisms (Hollingshead et al, 1986).

Identification of a Common Region Among all M Types Tested

From subcloning experiments and analysis of the DNA sequence, we localized the emm6 gene within the cloned fragment (Scott et al, 1985). As a hybridization probe, we then selected an intragenic NciI/PvuII fragment which contained all but about 35 bases from each end of the sequence encoding the mature M6 protein (Fig. 2). DNA extracted from strains representing 56 different M types and 4 for which no typing sera are currently available was hybridized to this emm6 probe on a solid support. Under conditions permitting about 23% mismatch, DNA from all 59 group A strains hybridized with our probe (Scott et al, 1985). DNA from the gram positive bacteria Bacillus subtilis, Staphylococcus aureus and Streptococcus pneumoniae served as negative controls. Of the other streptococcal groups, only C and G, which are also beta-hemolytic human pathogens (Mohr et al, 1979) and have been reported to have M-like proteins (Woolcock 1974; Bryans and Moore 1972) hybridized with the emm6 probe.

DNA from 2 of 3 functionally M⁻ group A strains also hybridized with the M6 probe (Scott et al, 1985). This means that the M⁻ strains contain at least part of the structural gene for M protein. Their functional defect may result from a lower level of M protein expression, from a missense mutation, or from a small deletion within the emm gene. The one M⁻ strain that showed no homology with the emm6 probe was derived from the same clinical M28 isolate as a strain that did hybridize, suggesting that this M⁻ strain has an extensive deletion in its M protein gene. This strain was used as the recipient for transfer of emm6 from E. coli in the experiment described above. The homology with emm6 displayed by all group A streptococci tested suggests the possibility that all M proteins share a common region.

Translation of the DNA sequence reveals that the carboxy-terminus of M6 has a hydrophobic region typical of protein domains thought to be associated with membranes (Hollingshead et al, 1986). Immediately adjacent to the membrane anchor region is a segment rich in proline and glycine, which would be expected to form a beta-pleated sheet structure and possibly to be invovlved in attachment to the peptidoglycan layer of the streptococcal cell wall. Protein A of S. aureus, a surface protein of another gram positive organism, has a carboxyl sequence that is both analogous and homologous to the M6 regions we believe are involved in wall attachment and membrane anchoring (Guss et al, 1984). In addition, the gram negative organism E. coli has a homologous amino terminal region in the H protein, which is thought to be responsible for attaching pap pili to the cell wall (Baga et al, 1987).

We anticipate that this carboxy-terminal region might be conserved in all M proteins, since it might be involved in attachment of the protein to the streptococcal cell. On the other hand, the amino-terminal portion of the M molecule, which is distal to the bacterial cell surface, should be readily available for interaction with the host's immune system and therefore should be more sensitive to selective pressure. Thus, we anticipated that among different M proteins, the greatest structural diversity would occur in this part of the protein.

To determine which regions of the molecule were most conserved among unrelated streptococcal M proteins, smaller intragenic DNA probes were used in Southern blot hybridization experiments (Scott et al, 1986b). The results (represented in Fig. 2) indicate that the greatest conservation occurs at the carboxy-terminal portion and the greatest variation at the amino-terminal region of the molecule. Location of the epitopes of 4 different monoclonal antibodies which have different degrees of cross-reactivity with M proteins from streptococci of different M types is

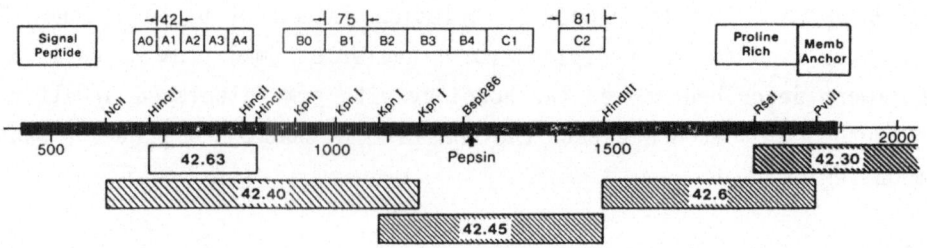

Fig. 2. Diagram representing the major features of the sequence of emm6 (Hollingshead et al, 1986; Scott et al, 1986). Relevant restriction enzyme sites are indicated above the line, and the scale below is in kilobases. The black box represents the M6 coding region. The predominant site for pepsin cleavage is indicated by the bold arrow below the map. The boxes immediately below the line indicate the locations of the signal peptide, proline-rich region and membrane anchor. The three regions of directly reiterated DNA sequence are shown as boxes A, B, and C and the number of base pairs in each is indicated above it. The boxes below the line indicate the probes used for hybridization with DNA of strains of different M type. The proportion of strains hybridizing is indicated by the shading within the box: the probes that hybridize with more strains are more completely shaded.

in agreement with this assignment (Jones et al, 1986; Jones et al, unpub.). The most carboxy-teminal of the epitopes cross-reacts with the largest number of different M proteins, and the most amino-terminal epitope with the smallest number. Comparison of the RNA sequence of about the 3' half of the message from M5, M19, M24, M30, M55, and M6 also indicates extensive conservation among these genes in this region of the molecule (Hollingshead and Scott, in prep).

Size Variation Among M Proteins

In addition to the well-known antigenic diversity of streptococcal M proteins, we have recently discovered that they are very diverse in apparent size (Fischetti, 1985). SDS-PAGE analysis reveals differences of 40,000 in apparent molecular weight for M proteins isolated from streptococcal strains of different types and of almost that much among M proteins of type 6. Such size heterogeneity strongly suggests that the actual amino acid sequence of most of the molecule cannot be critical for the antiphagocytic function of the M molecule.

We believe that evolutionary selection may lead to conservation of only three aspects of the molecule. First, a particular amino acid sequence of the carboxyl-terminal region may be needed for anchoring the molecule to the bacterial surface. Second, antiphagocytic function may require the presence of negative charge at the amino terminal region. Such amino-terminal charged domains have been identified in both M5 (Fischetti, 1983; Manjula et al, 1985) and M6 (Fischetti et al, in prep). And third, the alpha-helical coiled-coil structure may also be required for M protein function. This structure results from a seven amino acid periodicity in which hydrophobic residues occur at positions 1 and 4 and charged residues at position 7 (Manjula and Fischetti, 1980). Such a coiled-coil structure can be maintained with divergent amino acid sequences as long as the periodicity of hydrophobic amino acids is conserved. We anticipate the existence of limits on the length of the coiled-coil region which will dictate limits on the total permissible molecular weights of M proteins.

DNA sequence analysis of emm6 revealed an unusual structure that might generate size diversity in the molecule (Hollingshead et al, 1986). The sequence contains three regions of long direct repeats (Fig. 2). Region A contains five tandem 42 base reiterations, region B contains 5 tandem 75 base repeats, and region C consists of two 81 base repeats separated by 45 bases. Such extensive reiterations would be expected to serve as substrates for homologous recombination which would generate both duplications and deletions of the repeat blocks.

To test the hypothesis that homologous recombination produces size differences among M proteins of different streptococcal strains we had to isolate a size variant mutant from a strain whose M protein we had studied. By screening pools of D471 colonies by Western blots, we obtained three independent mutants with deletions in their M protein and one additional mutant derived in a second event from an initial deletion (Fischetti et al, 1986). Such deletion mutants arise in laboratory-grown cultures at a frequency of about 1 in 2,000 colony forming units. Reconstruction experiments demonstrated that, for our technique, detection of a mutant with a smaller M protein requires that the mutant constitute about 5% of the population being analyzed. Similar reconstruction experiments indicated that mutants with larger M proteins would have to constitute about 50% of the population before we could detect them. This probably accounts for our inability to isolate such mutants.

The DNA sequence of the emm6 gene of each mutant that produces a smaller M protein was obtained from the mRNA (Hollingshead et al, 1987). Comparison of these sequences verified that each could have arisen by a homologous recombination event between repeated sequences. There was no tendency, in this small sample, to recombine more frequently in any particular region.

To assess the possible importance of homologous recombination in the generation of M protein size heterogeneity in nature, we performed similar sequence analyses on two groups of strains that displayed size alterations in their M6 proteins (Hollingshead et al, 1987). One group, obtained before penicillin treatment was available, consisted of sequential isolates from a patient in an epidemic of streptococcal pharyngitis and the other was from different patients in a recent outbreak of group A streptococcal pharyngitis in a day care center. In both cases, the sequences of the emm6 alleles were consistent with the size differences having arisen by homologous recombination.

Generation of Antigenic Diversity

There are two other well-studied cases in which a bacterial pathogen has a mechanism for alteration of a surface antigen to which the host responds by producing protective antibody. The antigenic variation of the major pilin subunit of Neisseria gonorrhoeae appears to occur as a result of recombination between the expressed gene copy and an extragenic partially homologous silent copy of a pilin gene (So, 1986). The variable major protein on the surface of the Borrelia seems to be altered as a result of a similar homologous recombination event between an expressed gene copy and a silent copy (Meier et al, 1985). In this case, however, the copies are apparently present on linear plasmid DNA. To determine whether there are partially homologous copies of emm6 in the streptococcal genome from which this gene was cloned, we used our NciI/PvuII fragment as a probe for Southern hybridization to D471 DNA. Even under conditions permitting about 27% mismatch, no emm6-homologous regions outside of the structural gene itself were detectable (Scott et al, 1985). Thus, antigenic variation of the M protein gene does not occur by the same mechanism that generates diversity for gonococcal pilin or Borrelia surface protein.

However, homologous recombination between intragenic repeat regions, which we showed generates size diversity among M molecules (Hollingshead et al, 1987), will also generate antigenic changes. In both the A and B repeat regions the terminal repeat blocks diverge slightly from the consensus sequence of the three internal blocks. Homologous recombination between one of these terminal blocks and any other block might, then, generate a new sequence. One of the deletion mutants isolated from D471 apparently resulted from a recombination between divergent A repeat blocks which led to an amino acid sequence that differs from D471 in this region (Fig. 3). If such a new sequence occurred in an immunogenic determinant, it would lead to an antigenic change. We know that both the A and B repeats of M6 contain immunologically recognized epitopes because we have isolated monoclonal antibodies that react with these regions (Jones et al, 1985; Jones et al, 1986). Furthermore, because the A and B repeats constitute such a large proportion of the total M protein sequence, we expect them to be immunologically dominant, as found for repeats in other surface proteins (Nardin et al, 1982; Ballou et al, 1985; Gysin et al, 1984). We have not yet determined whether the repeat regions contain epitopes that are opsonogenic, but studies to determine this are currently in progress. It seems apparent that antigenic diversity can arise by intragenic homologous recombination and it is possible that such changes might alter the M protein "type".

In addition to direct production of altered sequence by homology-promoted deletions, intragenic homologous recombination should be able to amplify random mutational changes that occur in the emm gene. Once a mutation arises in a repeat region, it can be duplicated by homologous recombination, and the old version of the repeat block can be deleted by the same type of process. Thus, recombination, which is a very frequent event in most organisms, would be expected to accelerate evolution of the emm gene.

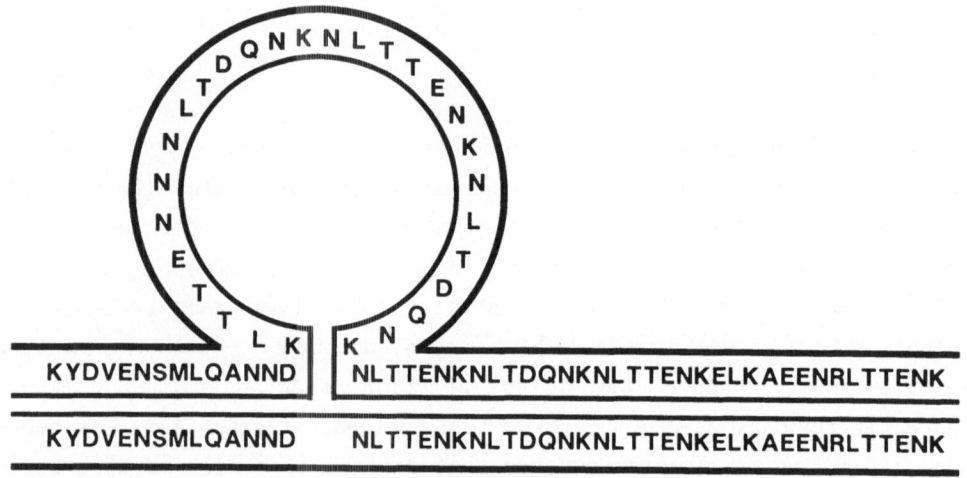

Fig. 3. Diagram showing the amino acid sequence of part of the A repeat blocks in strain D471 (top line) and the deletion mutant del113 (bottom line) derived from it (Hollingshead et al, 1987). The deletion in strain del113, which apparently arose by homologous recombination, generates a new sequence.

Homologous recombination events are therefore expected to lead to the accumulation of changes in the M molecule. Eventually, the protein should be altered sufficiently to be unrecognizable by the antiserum defining the original M type, type 6 for example. However, before that time, a series of changes will accumulate so that a range of molecules with different sequences will be found among those defined as type 6. That such a range of different molecules occurs within on type has been shown by analysis of both amino acid and DNA sequence (Hollingshead et al, 1986; Hollingshead et al, 1987; Manjula et al, 1984; Seyer et al, 1980).

As indicated above, because typing sera are polyclonal and gain their specificity by multiple absorptions with strains of different M types, these sera are artificial and irreproducible constructs. The antigenic alteration that is important for the streptococcus is one that leads to its escape from recognition by opsonic antibody previously produced in the host and to the generation of new opsonogenic determinants. We have recently obtained evidence that variation of opsonogenic determinants occurs within one M type. Antibodies prepared against a peptide consisting of the amino-terminal 20 amino acids of one M6 molecule opsonize the strain from which the sequence was derived more efficiently than some other strains of M6 streptococci (Fischetti et al, in prep). As described above, we expect this type of antigenic variation of the streptococcal M protein to occur by intragenic homologous recombination.

In summary, we are left with a picture of the evolution of M proteins in which the more carboxy-terminal unique regions of the molecule are largely conserved. At the same time, the repeated regions of the molecule, which are probably of primary immunological importance, are rapidly diverging. This process should result in a continuum of antigenically distinguishable M protein molecules and in the evolution of sequences selected by the host's immune surveillance system.

Acknowledgements

We wish to thank Mary Windels, Clara Eastby, Christina Gotschlich, Lynn Malone, William Pulliam, Sharon Coyle, Patty Guenthner, Andrew David, and Jane Rhatigan for their excellent technical assistance during the course of some of these studies, and Dr. Kevin Jones for the monoclonal antibodies. The studies were supported by NIH grants AI 25219 (to June R. Scott and Vincent A. Fischetti) and AI 11822 to Vincent A. Fischetti.

References

Baga M, Norgren M, Normark S (1987) Biogenesis of E. coli Pap-pili: PapH, a minor pilin subunit involved in cell anchoring and length modulation. Cell 49:241-251.

Ballou WR, Rothbard J, Wirtz RA, Gordon DM, Williams JS, Gore RW, Schneider I, Hollingdale MR, Beaudoin RL, Maloy WL, Miller LH, Hockmeyer WT (1985) Immunogenicity of synthetic peptides from circumsporozoite protein of Plasmodium falciparum. Science 228:996-999

Bryans JT, Moore BO (1972) Group C streptococcal infections of the horse. In: Streptococci and Streptococcal Diseases, LW Wannamaker, and JM Matsen, Eds, Academic Press, New York pp 327-338

Dale JB, Beachey EH (1982) Protective antigenic determinant of streptococcal M protein shared with sarcolemmal membrane protein of human heart. J Exp Med 156:1165-1176

Dale JB, Beachey EH (1985a) Epitopes of streptococcal M protein shared with cardiac myosin. J Exp Med 162:583-591

Dale JB, Beachey EH (1985b) Multiple heart-cross-reactive epitopes of streptococcal M proteins. J Exp Med 161:113-122

Fishcetti VA (1983) Requirements for the opsonic activity of human IgG directed to type 6 group A streptococci net basic charge and intact Fc region. J Immunol 130:896-902

Fischetti VA, Jarymowycz M, Jones KF, Scott JR (1986) Streptococcal M protein size mutants occur at high frequency in a single strain. J Exp Med 164:971-980

Fischetti VA, Jones KF, Majala BN, Scott JR (1984) Streptococcal M6 protein expressed in Escherichia coli. Localization, purification and comparison with streptococcal-derived M protein. J Exp Med 159:1083-1095

Fischetti VA, Jones KF, Scott JR (1985) Size variation of the M protein in group A steptococci. J Exp Med 161:1384-1401

Guss B, Uhlen M, Nilsson B, Lindberg M, Sjoquist J, Sjodaht J (1984) Region X, the cell-wall-attachment part of staphylococcal protein A. Eur J Biochem 138:413-420

Gysin J, Barnwell J, Schlesinger DH, Nussenzweig V, Nussenzweig RS (1984) Neutralization of the infectivity of sporozoites of Plasmodium knowlesi by antibodies to a synthetic peptide. J Exp Med 160:935-940

Hollingshead SK, Fishcetti VA, Scott JR (1986) Complete nucleotide sequence of type 6 M protein of the group A streptococcus: Repetitive structure and membrane anchor. J Biol Chem 162:1677-1686

Hollingshead SK, Fishcetti VA, Scott JR (1987) Size variation in group A streptococcal M protein is generated by homologous recombination between intragenic repeats. Molec Gen Genet 207:196-203

Jones KF, Khan SA, Erickson BW, Hollingshead SK, Scott JR, Fischetti VA (1986) Immunochemical localization and amino acid sequences of cross-reactive epitopes within the group A streptococcal M6 protein. J Exp Med 164:1226-1238

Jones KF, Manjula BN, Johnston KH, Hollingshead SK, Scott JR, Fischetti VA (1985) The location of variable and conserved epitopes among the multiple serotypes of streptococcal M-protein. J Exp Med 161:623-628

Lancefield RC (1928) The antigenic complex of Streptococcus haemolyticus. II. chemical and immunological properties of the protein fractions. J Exp Med 47:469-480

Lancefield RC (1962) Current knowledge of type-specific M antigens of group A streptococci. J Immunol 89:307-313

Manjula BN, Acharya AS, Mische SM, Fairwell T, Fischetti VA (1984) The complete amino acid sequence of a biologically active 197-residue fragment of M protein from type 5 group A streptococci. J Biol Chem 259:3686-3693

Manjula BN, Fischetti VA (1980) Tropomyosin-like seven residue periodicity in three immunologically distinct streptococcal M proteins and its implications for the antiphagocytic property of the molecule. J Exp Med 151:695-708

Manjula BN, Trus BL, Fischetti VA (1985) Presence of two distinct regions in the coiled-coil structure of the streptococcal PepM5 protein: Relationship to mammalian coiled-coil proteins and implications to its biological properties. Proc Natl Acad Sci 82:1064-1068

Meier JT, Simon MI, Barbour AG (1985) Antigenic variation is associated with DNA rearrangements in a relapsing fever borrelia. Cell 41:403-409

Mohr DN, Feist DJ, Washington JA, Hermans PE (1979) Infections due to group C streptococci in man. Am J Med 66:450-456

Nardin EH, Nussenzweig V, Nussenzweig RS, Collins WE, Harinasuta KT, Tapchairsi P, Chomcharn Y (1982) Circumsporozoite proteins of human malaria parasites Plasmodium falciparum and Plasmodium vivax. J Exp Med 156:20-30

Phillips GN, Jr, Flicker PF, Cohen C, Manjula BN, Fischetti VA (1981) Streptococcal M Protein: Alpha-helical coiled-coil structure and arrangement on the cell surface. Proc Natl Acad Sci USA 78:4689-4693

Rotta J, Krause RM, Lancefield RC, Everly W, Lackland H (1971) New approaches for the laboratory recognition of M types of group A streptococci. J Exp Med 134:1298-1315

Scott JR, Fischetti VA (1983) Expression of streptococcal M protein in Escherichia coli. Science 221:758-760

Scott JR, Guenthner PC, Malone LM, Fischetti VA (1986a) Conversion of an M⁻ group A streptococcus to M⁺ by transfer of a plasmid containing an M6 gene. J Exp Med 164:1641-1651

Scott JR, Hollingshead SK, Fishcetti VA (1986b) Homologous regions within M protein genes in group A streptococci of different serotypes. Infect & Immun 52:609-612

Scott JR, Pulliam WM, Hollingshead SK, Fishcetti VA (1985) Relationship of M protein genes in group A streptococci. Proc Natl Acad Sci 82:1822-1826

Seyer JM, Kang AH, Beachey EH (1980) Primary structural similarities between types 5 and 25 M proteins of Streptococcus pyogenes. Biochem Biophys Res Comm 92:546-553

So M (1987) The pilus of Neisseria gonorrhoeae: phase and antigenic variation. In: "Bacterial Outer Membranes as Model Systems", I Inouye (ed) John Wiley and Sons, NY pp. 401-417

Swanson J, Hsu KC, Gotschlich EC (1969) Electron microscope studies on streptococci. I. M Antigen. J Exp Med 130:1063-1075

Woolcock JB (1974) Purification and antigenicity of an M-like protein of Streptococcus equi. Infect Immun 10:116-122

A NEW FAMILY OF PORE FORMING PROTEINS, THE PERFORIN FAMILY, IN IMMUNE CYTOLYSIS BY COMPLEMENT AND BY LYMPHOCYTES

Eckhard R. Podack

Department of Microbiology and Immunology
New York Medical College
Valhalla, N.Y. 10595

Introduction

Immune cytolysis is effected by the complement proteins and by cytolytic lymphocytes. Target destruction by these humoral and cellular immune defense systems is mediated, in part, by pore forming proteins assembling transmembrane tubules on target membranes (for review see 1-5). Perforation of the membrane results in loss of membrane potential, dissipation of transmembrane ionic gradients, Ca-ion entry into target cells, loss of macromolecular cellular compounds and the eventual death of the target cell.

The effector membrane attack complex (MAC) of complement is formed from five proteins C5b, C6, C7, C8 and C9 that assemble to a macromolecular structure with the probable subunit formula $C5b-8_1$, $C9_{10-16}$ (6). Although C6b-8 has already transmembrane channel activity, polymerized C9 (poly C9), which is tightly associated with C5b-8 in the MAC (6), is believed to mediate the main pore forming activity of complement giving rise to a transmembrane tubule with an internal diameter of 100 Å. Isolated poly C9, in fact, resembles structurally the membrane lesion of the complete MAC (7,8).

Lymphocytic killer cells also cause formation of transmembrane tubules of approximately 160 Å diameter (1-12). Recently, the pore forming protein (perforin 1, P1) of cytolytic T cells (CTL) and NK-cells was isolated and identified as a 70-75000 dalton protein with molecular properties similar to C9. Isolated P1 assembles to polymeric complexes (Poly P1) in the presence of Ca ions and efficiently lyses tumor cells under these

NATO ASI Series, Vol. H24
Bacteria, Complement and the Phagocytic Cell
Edited by F.C. Cabello und C. Pruzzo
© Springer-Verlag Berlin Heidelberg 1988

conditions (13,14). It has been suggested previously that C9 and perforin are homologous proteins (12,15). Figure 1 shows the assembly of C9 and Perforin 1 to transmembrane tubules in a biochemically similar polymerization reaction.

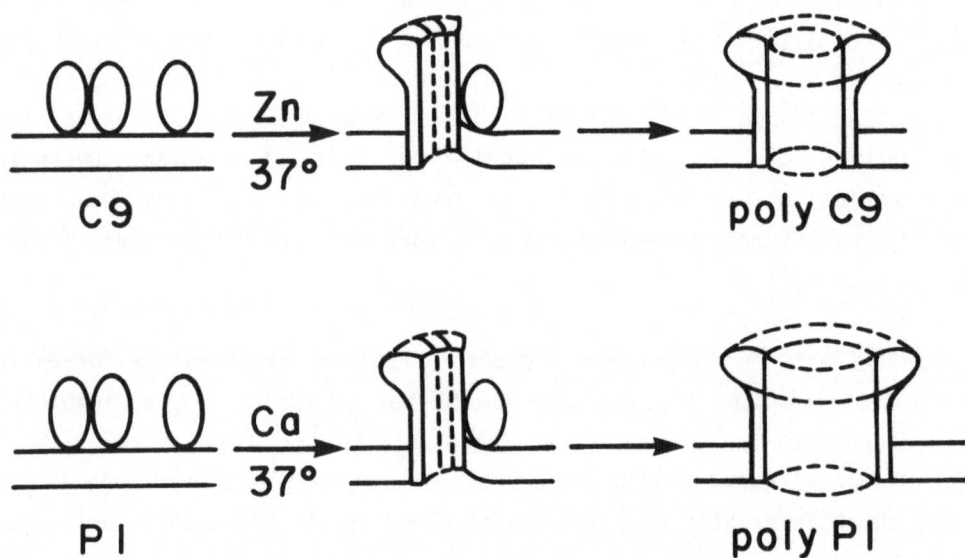

Fig. 1.   Schematic representation of polymerization and membrane insertion of C9 and Perforin during cytolysis.

Immunological cross reactivity of antibodies to C9 with perforin 1

To directly address the question of the putative homology of C9 with P1, polyclonal antibodies to C9 were generated by immunizing rabbits with all forms of native and denatured human C9. For this purpose, native C9, SDS denatured C9, SDS denatured, reduced and alkylated C9 and glutaraldehyde crosslinked C9 was injected into rabbits at multiple sites. After boosting, the resultant antiserum (anti-whole C9) was tested for reactivity with P1 by Western blot analysis and compared with an antiserum raised against native, murine P1 (anti-P1) (Figure 2). Anti-P1 (raised against unreduced P1), as expected, stains P1 in its nonreduced and reduced form on Western blots. Anti-P1 does not recognize C9 in either reduced or unreduced form.

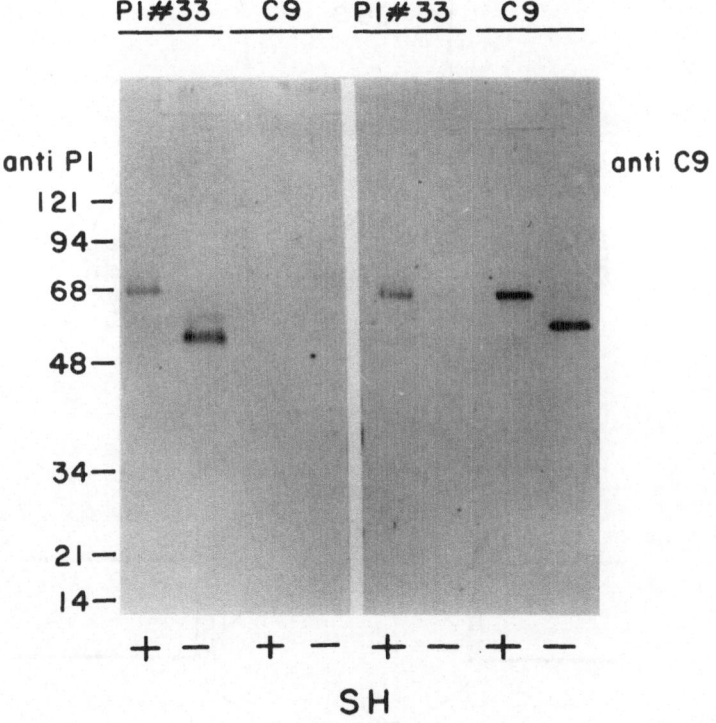

Fig. 2. Cross reaction of anti-whole C9 with reduced perforin 1 (P1). 1
μg human C9 (16) or murine perforin 1 (13) as indicated were
separated on 7.5 - 15% SDS polyacrylamide gradient gels under
reducing (+) or nonreducing (-) conditions on Western Blots.

In contrast, anti-whole C9 does recognize reduced P1 in addition to reduced and unreduced C9. Unreduced P1 is not recognized by anti-whole C9. Antibodies raised against native C9 or SDS-denatured, but unreduced C9, did not react with P1 in either form even though they stained both reduced and unreduced C9 (not shown). This finding suggests that antigenic determinants of murine P1 and human C9 are shared but are accessible only after reduction of P1 (17). It is of significance to note that both C9 and P1 undergo a large shift to an apparent higher molecular weight upon reduction and SDS polyacrylamide-gel electrophoresis indicating significant intrachain disulfide cross-linking of the peptide chains.

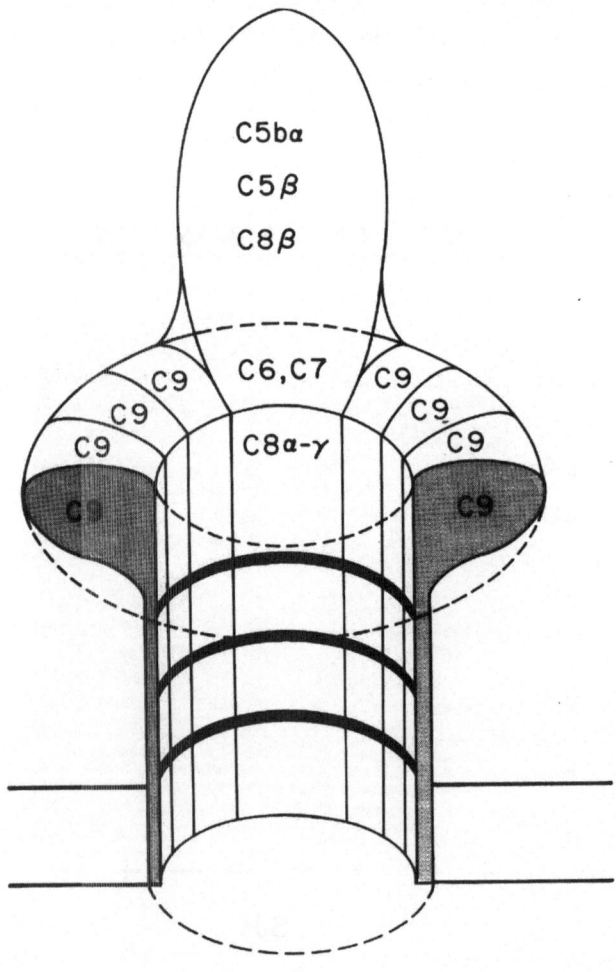

Fig. 3. Hypothetical subunit arrangement in the MAC. Note that the transmembrane channel is formed by a heteropolymer of the composition $C6_1$, $C7_1$, $C8\alpha_1$, $C9_{10-14}$.

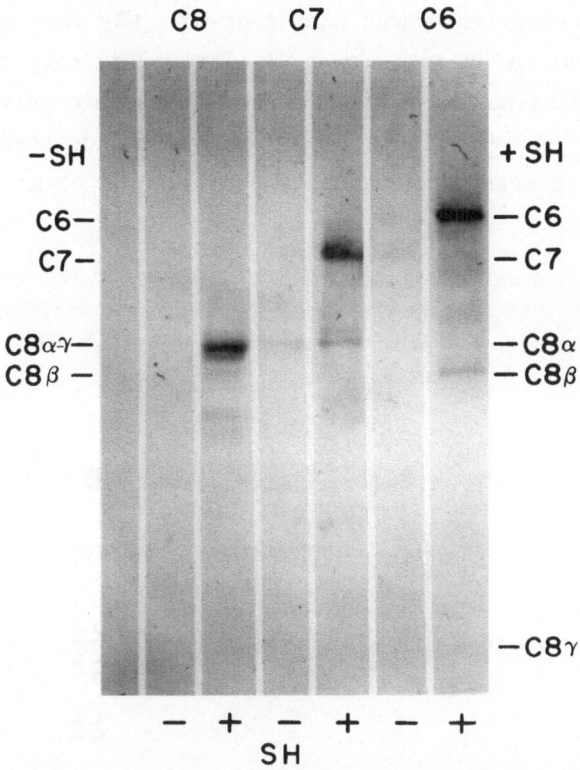

Fig. 4. Anti-whole C9 detects reduced C6, C7 and C8α but not C5 and C8β.
1 μg C5 (18), C6, C7 (19), and C8 (20) were separated by SDS
polyacrylamide electrophoresis with (+) or without reduction (-)
and immunostained with anti-whole C9 described in Fig. 1. Note
the reactivity of reduced alkylated C6, C7 and C8α with anti
whole C9. The unreduced proteins are not detected nor are the
reduced C5 chains (not shown) and reduced C8β (see Fig. 3 for
migration of reduced and unreduced proteins).

Cross reaction of anti-C9 with reduced C6, C7 and C8α

Because in the MAC polymerized C9 is intimately associated in a
heteropolymeric transmembrane tubule with C6, C7 and C8α suggesting
homology of these proteins (6) (Fig. 3), the antiserum to whole C9 was
tested with these components and with C5 and albumin as control. As shown
in Fig. 4 anti-whole C9 in Western blots stains reduced, alkylated C6,
reduced, alkylated C7 and the reduced and alkylated C8α chain. The
unreduced peptide chains are not recognized by the antiserum.

Nor are the reduced or unreduced chains of C5α, C5β (not shown), C8β, C8γ or albumin recognized by anti-whole C9. The specificity for the reduced peptide chains again points to determinants accessible only after complete unfolding of the peptide chain subsequent to disulfide bond reduction. The same reactivity is seen after affinity purification of anti C9 on reduced C7-Sepharose 4B.

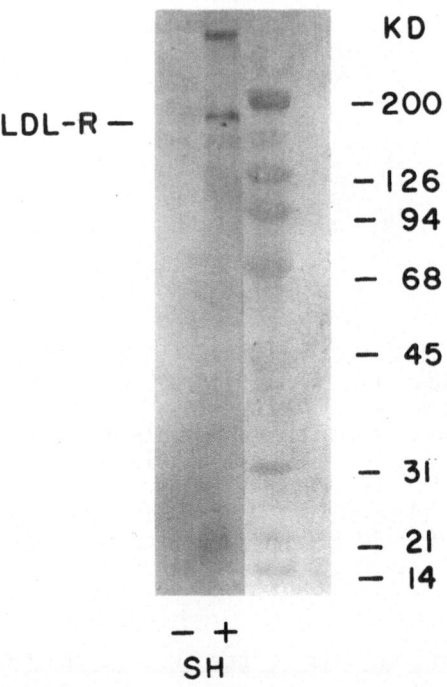

Fig. 5. Detection of the reduced LDL-receptor by anti-whole C9 which was used after affinity purification on reduced C7-Sepharose. 0.3 μg purified LDL-receptor (21) with (+) or without (-) reduction was applied to SDS polyacrylamide gels and, after electrophoretic transfer to nitrocellulose, immunostained with anti-whole C9 affinity purified on reduced C7-Sepharose. Reduced LDL-receptor is detected with an apparent molecular weight of 160,000 after reduction and as aggregated material at the interface of stacking and separating gel. Non-reduced LDL-receptor is not recognized by the antibody. The track on the right gives the position of molecular weight proteins (in kilo Dalton) run on the same gel and stained with Ponceau red.

Reactivity of anti whole C9 with the reduced LDL-receptor

Since C9 and the LDL-receptor share two cysteine rich domains (approximately 38% homology) (22-25) the anti whole-C9 antiserum after affinity purification on reduced C7-Sepharose was tested with the purified LDL-receptor in a similar assay (Fig. 5). Anti-C9 stains the receptor, again only in its reduced forms.

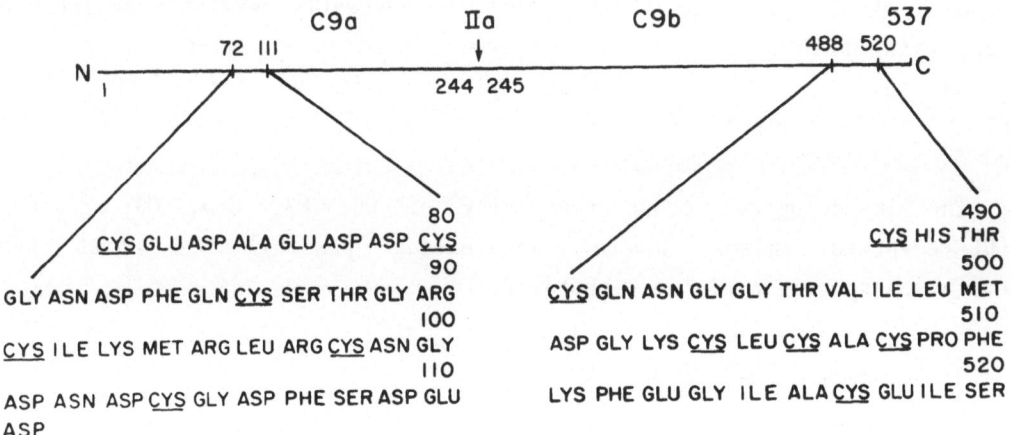

Fig. 6.   Cysteine rich domains in C9a and C9b as deduced from nucleotide sequencing of C9 cDNA (22,23). Both domains are homologous to similar domains in the LDL-receptor. IIa (244-245) thrombin cleavage site.

Summary and Conclusions

Figure 6 shows the cysteine rich domains in the sequence of C9 that are most likely the target of immunopurified anti-whole C9 antibodies. This conclusion is derived, first, from the fact that these are the only two domains shared between C9 and the LDL-receptor; second, determinants in this region are likely to become exposed upon reduction of the peptide chains. After affinity purification of anti-whole C9 on reduced C7-Sepharose or C6-Sepharose, the antibody recognized both thrombin fragments of C9 i.e., reduced C9a and C9b (not shown). It would seem likely, therefore, that C7, C6 and possibly the outer proteins contain both cysteine rich domains of C9. The presence of the cysteine rich domains of C9a in P1 and C8α was also independently suggested by Tschopp (26) with antibodies to a synthetic peptide of this region. Our studies are in agreement with this finding and extend the homology of C9 to C8α, C7, C6 and Perforin 1.

The immunological cross reactivity of C6, C7, C8α, C9, P1 and LDL-receptors raises several interesting possibilities about the evolutionary origin of these proteins.

a) Homology of Perforin 1 and C9

The immunological cross reactivity of C9 and P1 strongly supports previous suggestions of a common origin of these proteins. In light of their similar function, structure and polymerization to transmembrane tubules, the immunological cross reactivity is not a surprise and has been anticipated (12,15,27).

b) Homology of C6, C7, C8α and C9

These four polypeptides form the transmembrane tubule of the MAC are intimately associated in an SDS-resistant complex as previously shown (6). They apparently form a copolymer of the stoichiometry $C6_1$, $C7_1$, $C8\alpha-\gamma_1$, $C9_{10-16}$. The C8β-chain and the C5b subunits are not part of the tubule because they are dissociated by boiling of the MAC in SDS (6).

Based on this finding, a structural homology of these four peptides has been predicted (6) and is strongly supported by the current analysis. All four peptides (C6 to a lesser extent) are labeled with membrane-restricted photoreactive probes in the membrane-bound MAC further supporting their structural and functional relationships (28-30). It is likely that upon complex assembly the four peptide chains undergo a restricted unfolding and membrane insertion of their hydrophobic domains similar to the reaction of C9 upon polymerization or C7 upon dimerization or reaction with C5b-6 (8,31,32). Fig. 7 summarizes in a schematic drawing the putative unfolding and membrane insertion of the complement proteins and of P1 upon complex assembly and membrane attack. The putative evolutionary relationship of C6, C7, C8 and C9 is consistent with the phylogeny of complement membrane attack proteins as described in shark, trout and mammals (for discussion see ref. 1 and refs. 33,34) and with the genetic linkage of C6 and C7 in humans (35).

Table 1

Perforin Protein Family

Name	Effector System	Molecular Weight	Function
C6	Complement	120,000	Membrane insertion, tubule formation
C7	Complement	110,000	Membrane insertion, tubule formation
C8α	Complement	73,000	C9 polymerization, tubule formation
C9	Complement	73,000	Transmembrane tubule formation
P-1	Cytotoxic Lymphocytes	70-75,000	Transmembrane tubule formation

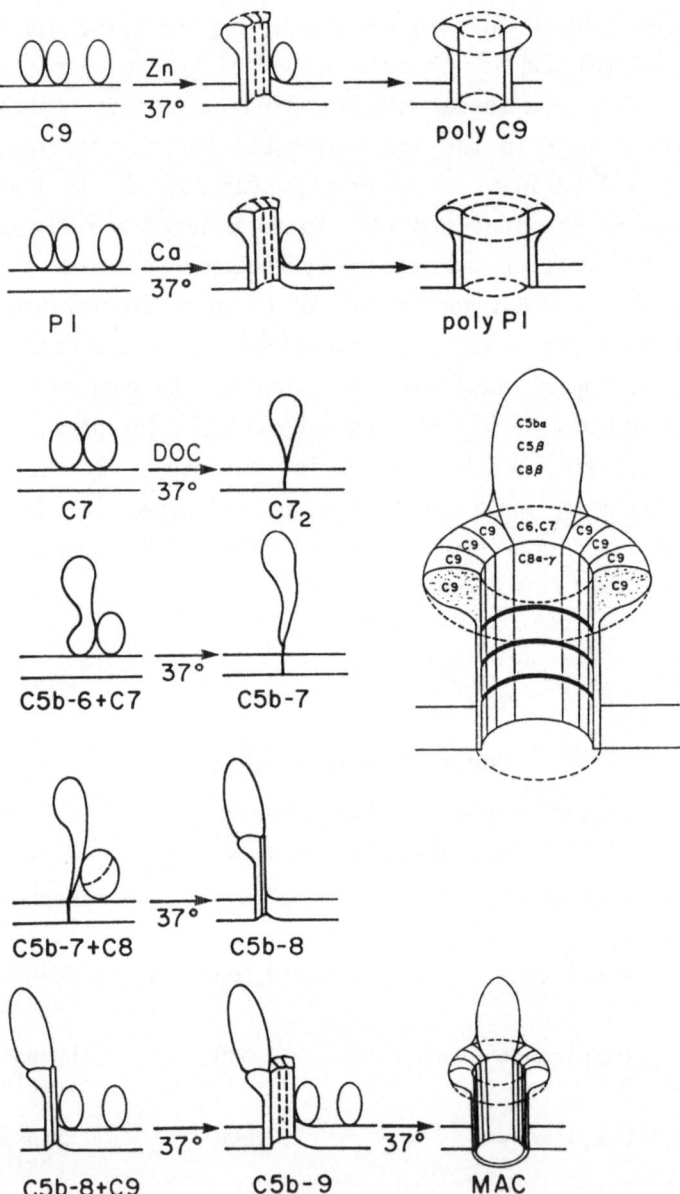

Fig. 7. Hypothetical scheme for the unfolding and membrane insertion of complement proteins and perforin 1 during complex assembly. The scheme is based on a number of previous studies with C9, C7, P1, C5b-7 and MAC (6,7,8,12,31,32). The proposed subunit arrangement in the MAC is shown in the inset on the right.

The homology of peptide chains of C5b-8 with C9 also offers a plausible mechanism for the rapid polymerization of C9 mediated by C5b-8. It is highly probable that C5b-8 by virtue of C9-homologous regions in its peptide chains imitates an activated C9-like site, thus enabling it to bind and unfold the first C9 molecule. Subsequent C9 molecules then interact with C9 until the last C9 molecule in the complex makes contact with C5b-8 to close the tubular structure (Fig. 6). This view is supported by the compositions of the MAC tubule of a heteropolymer of C6, C7, C8α-γ and C9 (6). Table 1 summarizes some of the properties of the members of the perforin-protein family.

c) <u>C9 and the LDL-receptor</u>

The homology of the cysteine-rich domains of C9 and the LDL-receptor have been described previously (23-25). Due to the scarcity of purified LDL-receptor it is not feasible to prepare an immunoabsorbent column with this protein for immunopurification of anti-whole C9. Therefore, the positive identification of homologous domains of C6, C7 and C8α with the cysteine rich domains of C9 and the LDL-receptor must await further structural studies. It is, nevertheless, possible to suggest a plausible explanation for the importance of the cysteine-rich C9a domain in these functionally different proteins. In the LDL-receptor the C9a like domain occurs in a seven fold repeat (24) and is believed to form an oligomeric domain for the binding of LDL. It is possible that the cysteine-rich C9a domain is involved in the polymeric assembly and unfolding of peptide chains of C6, C7, C8α and C9 in the MAC. Although the C9a sequence is present in C9 only in one copy it occurs ten to sixteen times in the fully assembled MAC.

<u>Summary</u>

The human complement proteins C6, C7 C8α and C9 and the pore-forming protein of murine cytolytic lymphocytes, perforin 1 (P1), as well as the low density lipoprotein-receptor show immunological cross reactivity, in their reduced form only, with an antibody produced against reduced human C9. The antigenic cross reactivity suggests shared immunological determinants in these proteins that become exposed upon reduction of disulfide bonds and that may be homologous to the known cysteine-rich domains of C9 and the LDL-receptor. It is proposed that the complement proteins C6, C7, C8α and C9 and the cytolytic lymphocyte protein perforin 1 are members of an evolutionarily related protein family, the perforin family, whose function is to form transmembrane tubules in immune cytolysis.

Acknowledgements

This work was supported by USPHS grants A1 21999 and CA 39102, by a grant of the American Cancer Society (IM 396) and funds from the Dr. I Fund Foundation. The excellent assistance of Richard Manganiello and Kristin Penichet is gratefully acknowledged as well as stimulating discussion with Dr. Jurg Tschopp, University of Lausanne, Switzerland.

References

1. Podack, E.R., and Tschopp, J. Mol. Immunol. 21, 603-689 (1984).
2. Podack, E.R., J. Cell. Biochem. 30:133-170 (1986).
3. Bhakdi, S. and Tranum-Jensen, J. Biochem. Biophys. Acta 737, 343-372 (1983).
4. Podack, E.R. Immunol. Today 6, 21-27 (1985).
5. Henkart, P.A. Annual. Rev. Immunol. 3, 31-58 (1985).
6. Podack, E.R. J. Biol. Chem. 259, 8461-8467 (1984).
7. Podack, E.R. and Tschopp, J. Proc. Natl. Acad. Sci. USA 79, 574-578 (1982).
8. Tschopp, J., Muller-Eberhard, H.J. and Podack, E.R. Nature (London) 298, 534-538 (1982).
9. Dourmashkin, R.R., Deteix, P., Simone, C.B. and Henkart, P.A. Clin. Exp. Immun. 42, 554-560 (1980).
10. Henkart, P.A., Millard, P.J., Reynolds, C.W. and Henkart, M.P. J. Exp. Med. 160, 75-93. (1984).
11. Podack, E.R. and Konigsberg, P.J. J. Exp. Med. 160, 695-710 (1984).
12. Podack, E.R. and Dennert, G. Nature (London) 302, 442-445 (1983).
13. Podack, E.R., Young, D.E. and Cohn, Z.A. Proc. Natl. Acad. Sci. 82, 8629-8633 (1985).
14. Masson, D. and Tschopp, J. J. Biol. Chem. 260, 9069-9072 (1985).
15. Lachmann, P.J. Nature (London) 305, 473-474 (1983).
16. Biesecker, G. and Muller-Eberhard, H.J. J. Immunol. 124, 1291-1296 (1980).
17. Podack, E.R., Young, J-D.E., Weeks-Levy, C., Lowrey, D. and Cohn, Z.A. Complement 2, 63-64 (1985).
18. Hammer, C.H., Wirtz, G.H., Renfer, L., Greshan, H.D. and Tack, B.F. J. Biol. Chem. 256, 3995-4004 (1981).

19. Podack, E.R., Kolb, W.P., Esser, A.F. and Muller-Eberhard, H.J. J. Immunol. 123, 1071-1077 (1979).
20. Kolb, W.P. and Muller-Eberhard, H.J. J. Exp. Med. 143, 1131-1139 (1976).
21. Schneider, W.J., Beisiegel, U.V., Goldstein, J.L. and Brown, M.S. J. Biol. Chem. 257, 2664-2673 (1981).
22. DiScipio, R.G., Gehring, M.R., Podack, E.R., Kan, C.C., Hugli, T.E. and Fey, G.H. Proc. Natl. Acad. Sci. 81, 7278-7302 (1984).
23. Stanley, K.K., Kocher, H.P., Luzio, P., Jackson, P. and Tschopp, J. EMBO J. 4, 375-382 (1985).
24. Yamamoto, T., Davis, C.G., Brown, M.S., Schneider, W.J., Casey, M.L., Goldstein, J.L. and Russel, D.W. Cell 39, 27-28 (1984).
25. Doolittle, R.F. TIBS 10, 233-237 (1985).
26. Tschopp, J. and Mollnes, T.E. Complement 2, 79-80 (1985).
27. Sundsmo, J.S. and Muller-Eberhard, H.J. J. Immunol. 122, 2371-2378 (1979).
28. Podack, E.R., Stoffel, W., Esser, A.F. and Muller-Eberhard, H.J. Proc. Natl. Acad. Sci. USA 78, 4544-4548 (1981).
29. Hu, V.W., Esser, A.F., Podack, E.R. and Wisnieski, B.J. J. Immunol. 127, 380-386 (1981).
30. Steckel, E.W., Welbaum, B.E. and Sodetz, J.M. J. Biol. Chem. 258, 4318-4324 (1983).
31. Preissner, K.T., Podack, E.R. and Muller-Eberhard, H.J. J. Immunol. 135, 445-451 (1985).
32. Preissner, K.T., Podack, E.R. and Muller-Eberhard, H.J. J. Immunol. 135, 452-458 (1985).
33. Jensen, J.A., Festa, E., Smith, D.S. and Cayer, M. Science 213, 566-569 (1981).
34. Nonaka, M., Yamagndi, N., Natsiume-Sakai, S., Takahashi, M., J. Immunol. 126, 12489-1494 (1981).
35. Hobart, M.J., Joysey, V. and Lachmann, P.J. J. Immunogenetics 5, 157-163 (1978).

THE MOLECULAR BASIS OF THE INTERACTION OF IMMUNOGLOBULIN G WITH COMPLEMENT AND PHAGOCYTIC CELLS

D.R. Burton

Department of Biochemistry
University of Sheffield
Sheffield S10 2TN
United Kingdom

Introduction

The elimination of bacterial cells mediated by antibodies involves the antibody molecule in a dual recognition process. First, the antibody recognizes foreign antigenic determinants on the bacterial cell surface. Second, the antibody-coated bacterial cell is recognized by an effector molecule, such as complement C1q or cellular Fc receptor, triggering effector mechanisms, such as complement or phagocytosis, leading to elimination of the bacterial cell. The molecular basis of the interaction of antibody and antigen is reasonably well understood. Recently, the crystallographic structure of the complex of the Fab fragment of an anti-lysozyme antibody with lysozyme has been described (Amit et al, 1986). In contrast, the basis for the interaction of antibody and effector molecules is only now being tackled at the molecular level.

The two recognition processes occur in different regions of the antibody molecule, most conveniently seen by reference to the schematic diagram of immunoglobulin G (IgG) seen in Fig. 1. IgG is the principal antibody in the serum and is the only one that will be considered here.

The molecule is a four-chain structure organized into domains. Antigen binding is a combined property of the V_H (variable-heavy) and V_L (variable-light) domains at the extremities of the Fab arms of the molecule. Effector binding is primarily a property of the stem (Fc) part of the molecule and, therefore, the C_H2 and/or C_H3 domains. The Fab arms are linked to the Fc via an extended segment of polypeptide chain known as

NATO ASI Series, Vol. H24
Bacteria, Complement and the Phagocytic Cell
Edited by F.C. Cabello und C. Pruzzo
© Springer-Verlag Berlin Heidelberg 1988

the hinge. The Fc is, thus, of prime interest in this review and will now be considered in some detail.

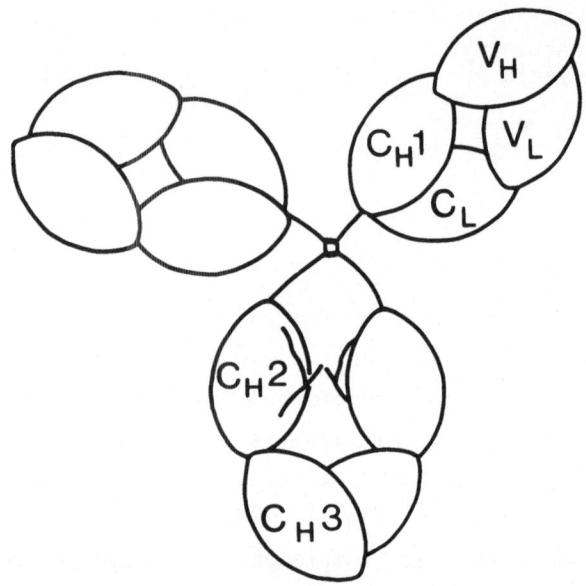

Fig. 1.   A schematic view of the domain structure of human IgG1. The Fab arms of the molecule are connected to the Fc part via a flexible hinge region. The length and number of interheavy chain disulphide bonds in the hinge differs between IgG subclasses: human IgG1 possesses two interheavy bonds as shown. The C_H2 domains, unlike the other domains of IgG, are not closely paired but have two N-linked branched carbohydrate chains interposed.

Structure of Fc

The structure of human Fc has been solved to 2.9A (Deisenhofer, 1981) and is presented in Fig. 2.

Each domain has a common pattern of polypeptide chain folding known as the "immunoglobulin fold". It consists of two twisted stacked beta-sheets enclosing an internal volume of tightly packed hydrophobic residues. The arrangement is stabilized by an internal disulphide bond linking the two sheets in a central position. One sheet has 4 and the other 3 antiparallel beta-strands. These strans are joined by bends or loops which generally

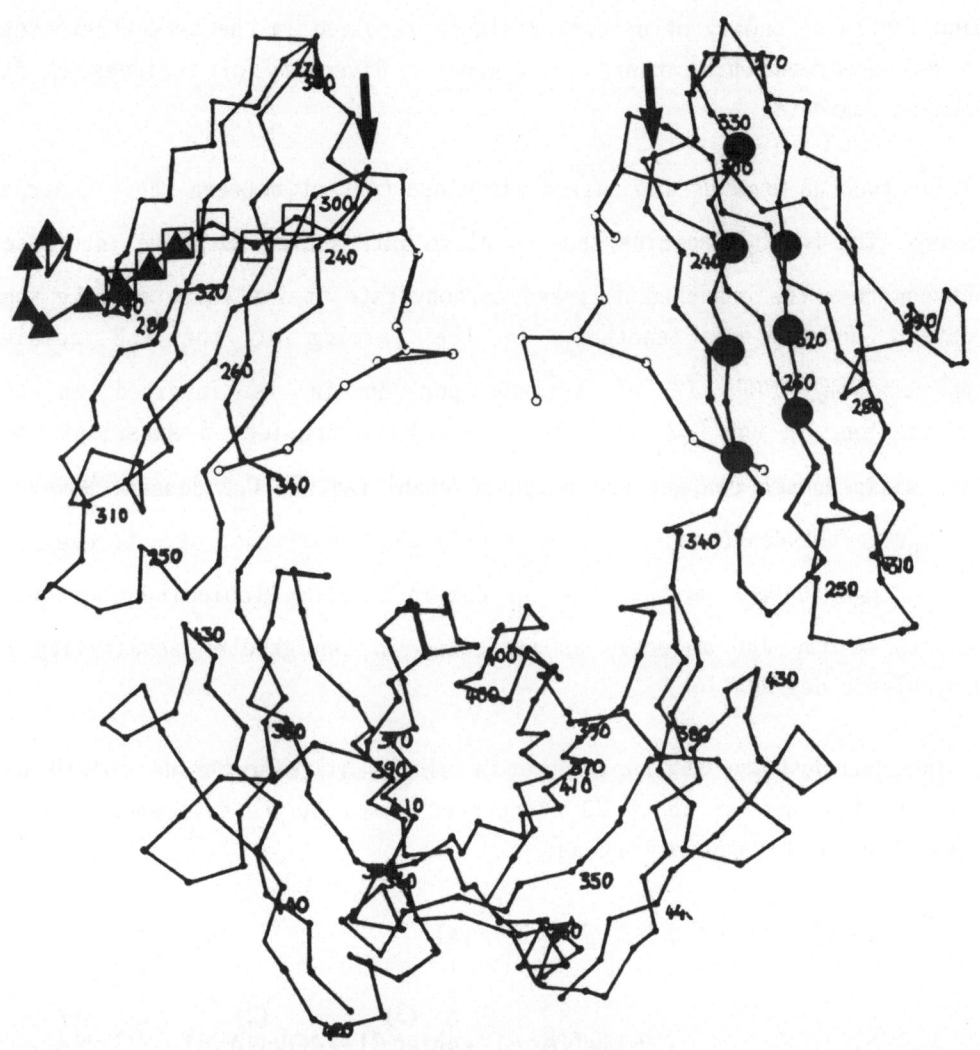

Fig. 2. The structure of Fc fragment from pooled human IgG. (·), alpha carbon positions; (o), approximate centres of carbohydrate hexose units. Co-ordinates were obtained from the Brookhaven Data Bank (after Deisenhofer, 1981). The pairing of C_H3 domains and the position of carbohydrate between the C_H2 domains is clearly seen in this view. The arrows indicate where the heavy chains are first located in the electron density map (Pro 238). The region from the interheavy disulphides in the hinge to this point is disordered, possibly reflecting flexibility. The possible complement C1q interaction sites proposed by Brunhouse and Cebra (1979) (□ ; residues 290-295), Lukas et al (1981) (▲ ; 285-292) and Burton et al (1980) (● ; 318,320,322,331,333,335,337) are represented and discussed in the text.

show little secondary structure. Residues involved in the beta-sheets tend to be conserved while there is a greater diversity of residues in the joining segments.

The two C_H3 domains are paired via close contact between their 4-strand layers. The two C_H2 domains show no close interaction but have interposed between them two branched N-linked carbohydrate chains which do make some contact between one another. In the pairing of the C_H3 domains approximately 1000 $Å^2$ of surface per domain is involved in the interaction. In the C_H2 case the carbohydrate provides a substitute for the domain-domain contact and helps to stabilize the C_H2 domain. However, the C_H2-carbohydrate contact area is only about half that of, for example, the C_H3-C_H3 contact so that the C_H2 domain is less stable than the other domains of the IgG molecule as reflected in its greater sensitivity to proteolytic degradation.

The carbohydrate chains are not a single oligosaccharide moiety but consist of a set of about 20 structures based on a mannosyl-chitobiose core which can be represented as:

$$
\begin{array}{l}
\pm \quad \pm (6') \quad\quad (5') \quad\quad\quad (4') \\
\mathrm{Sia}\alpha2 \rightarrow 6\mathrm{Gal}\beta1 \rightarrow 4\mathrm{GlcNAc}\beta1 \rightarrow 2\mathrm{Man}\alpha1 \quad\quad\quad\quad\quad \pm \mathrm{Fuc}\alpha1 \\
\hspace{8.2cm}\downarrow \hspace{3.3cm}\downarrow \\
\hspace{5.5cm} 6 \;\;(3) \hspace{1.2cm} (2) \hspace{1cm} 6 \;\;(1) \\
\hspace{3.5cm} \pm \mathrm{GlcNAc}\beta1 \rightarrow 4\mathrm{Man}\beta1 \rightarrow 4\mathrm{GlcNAc}\beta1 \rightarrow 4\mathrm{GlcNAc_{OT}} \\
\hspace{5.8cm} 3 \\
\pm \quad \pm (6) \quad\quad (5) \quad\quad\quad (4) \uparrow \\
\mathrm{Sia}\alpha2 \rightarrow 6\mathrm{Gal}\beta1 \rightarrow 4\mathrm{GlcNAc}\beta1 \rightarrow 2\mathrm{Man}\alpha1
\end{array}
$$

As shown, four types of mannosyl-chitobiose cores are found (± 'bisecting' N-acetylglucosamine/ fucose) and outer-chain variants include the presence or absence of galactose and sialic acid. The Fc carbohydrate chains have been suggested to adopt both structural and functional roles (Rademacher and Dwek, 1984; Burton, 1985). Structurally, they are important in resistance to proteolysis, and possibly in maintenance of C_H2-C_H2 domain orientation and assembly.

Structure of subclasses of IgG

As will be discussed below, the interaction of IgG with effector molecules, although involving binding sites on Fc, can be influenced by the conformation of the Fab arms. In some contexts, therefore, one needs to consider the structure of the whole IgG molecule rather than Fc alone. Crystallographic structures are available for mutant hinge-deleted IgG proteins (Silverston et al, 1977; Sarma and Laudin, 1982; Rajan et al, 1983), but these molecules do not generally trigger effector functions (Burton, 1985). Unfortunately, crystal diffraction patterns of whole intact IgG have been characterized by a lack of electron density associated with part of the hinge and the whole Fc, a phenomenon which has been related to hinge flexibility. Therefore, one needs to turn to lower resolution methods to describe the conformation of whole IgG molecules. Figure 3 shows models proposed for the solution conformations of the human IgG subclasses based on sedimentation and small angle X-ray scattering data (Gregory et al, 1987).

It is seen that the subclasses show marked differences in the relative conformation of Fab and Fc, reflecting differences in their hinge region. Thus, for example, IgG3 has an extended hinge of about 70 residues whereas IgG2 and IgG4 have only about 12 residues (Burton et al, 1986). These structural differences have important functional correlates as described below.

Interaction of IgG with Complement

The classical pathway of complement is a cascade system generating a variety of potent biological molecules including anaphylatoxins and chemoattractants and leading ultimately to lysis of antibody-coated cells (Reid and Porter, 1981; Reid, 1983). The pathway is triggered by the interaction of the first complement component, C1, with IgG in an associated state, i.e. coating a target cell or aggregated by antigen in an immune complex. The pathway is clearly not triggered by monomeric IgG which is at high concentration in the serum.

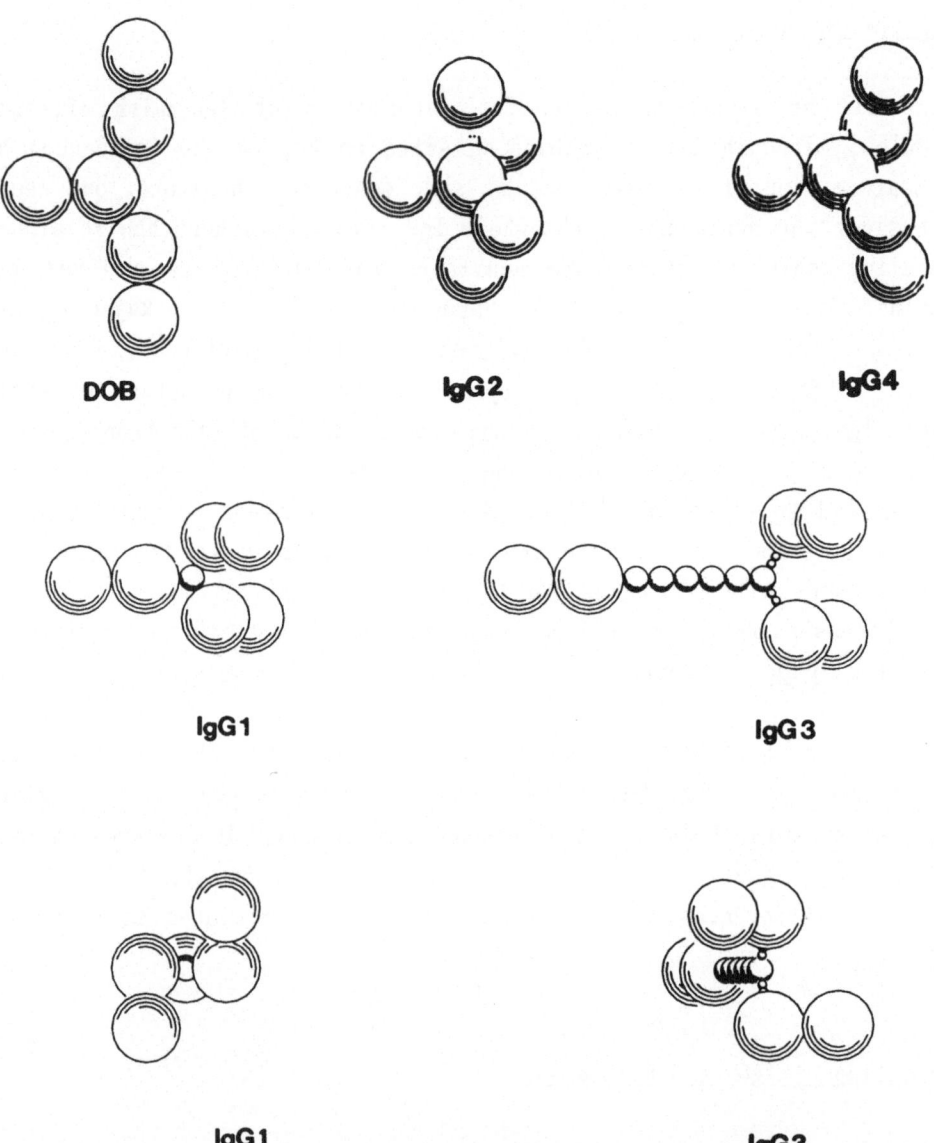

Fig. 3. <u>Models proposed for the solution conformations of human IgG
subclasses</u>. The models are proposed on the basis of sedimentation
data with supporting small-angle X-ray scattering data (Gregory
<u>et al</u>, 1987). The hinge deleted IgG1 Dob protein is used in
reference. IgG2 and IgG4 are shown to resemble Dob in the close
approach of Fab and Fc; for IgG2 the Fab arms are suggested to
fold back. In IgG1 the hinge is apparent and the Fab arms are
non-colinear as illustrated. In IgG3 the central or middle hinge
is about 90 Å long, in agreement with EM studies (Pumphrey,
1986).

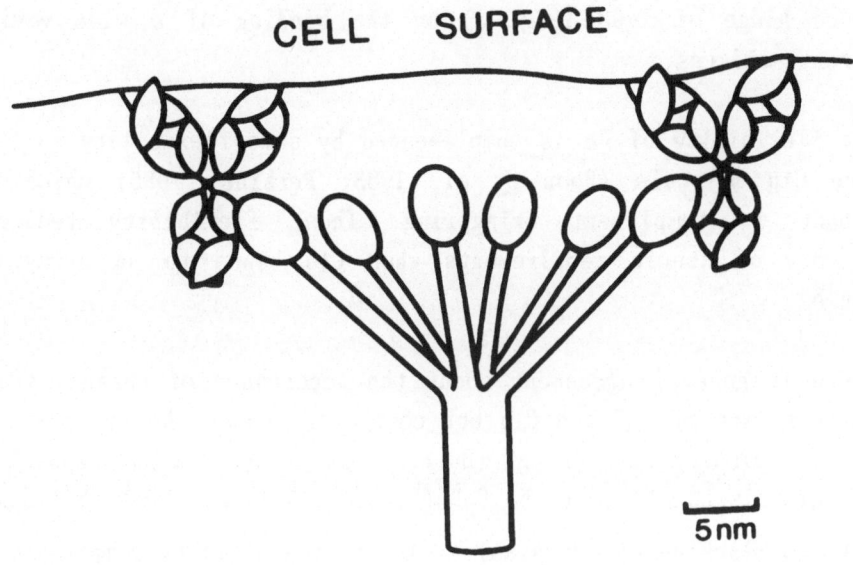

Fig. 4.   A schematic view of the binding of complement C1q to two IgG
molecules on a cell surface.

C1 is a complex of the complement components C1q, C1r and C1s. It is
the subcomponent C1q which interacts with the C_H2 domain of IgG to
initiate the enzymatic process of the pathway as shown in Figure 4.

C1q is a molecule having the appearance of a "bunch of tulips" (Reid
and Porter, 1981) and it is multivalent in its binding to IgG. This
multivalency is probably the key to why complement is only triggered by
IgG in an associated form. Binding of C1q to monomeric IgG is only weak
($K \sim 10^4 M^{-1}$) whereas binding to associated IgG and the consequent use of
two or more tulip heads makes binding much tighter ($K \sim 10^8 M^{-1}$) and allows
the activation process to proceed (Burton, 1985). Theories of activation
which involve binding of antigen to the Fab arms of IgG and the induction
of conformational changes which are passed down the molecule to Fc, thus
affecting the interaction with C1q, are now rejected by most workers,
(Metzger, 1978; Burton, 1985). Certainly, it appears very difficult to
visualize a common conformational change being transmitted through the

extended hinge of IgG3 (Fig. 3) by the binding of a wide variety of different antigens.

The flexibility of Fc is complemented by some flexibility in the arms of the C1q molecule (Poon et al, 1983; Perkins, 1985) which may be important in complement triggering. Thus, flexibility reduces the stringency of steric requirements when C1q binds to an array of IgG molecules.

There is general agreement about the importance of changed groups in the interaction of IgG and C1q but controversy over the precise location of the C1q binding site on IgG (Burton, 1985). The domain responsible for C1q binding is concluded to be C_H2 (Burton, 1985) as the Facb fragment of rabbit IgG (lacking C_H2 domains) binds C1 with affinity comparable to IgG and isolated C_H2 domains bind C1 with an affinity comparable to Fc. Early interest in the C_H2 domain centred on Trp 277 and nearby residues but this was quenched by the demonstration that Trp 277 is buried in the crystal structure of Fc. Three C1q binding sites have more recently been proposed. The first (Brunhouse and Cebra, 1978) involves residues in an extended chain region Lys 290 - Gln 295, in human IgG1, the second (Lukas et al, 1981) involves residues in the overlapping extended chain region His 285 - Arg 292 and the third (Burton et al, 1980) involves residues on the last two anti-paralellel beta-strands of the C_H2 domain (Gln 318, Lys 320, Lys 322, Pro 331, Gln 333, Thr 335, Ser 337). IgGs of different subclass and species (isotype) show differing affinity for C1q and the above proposals attempt to account for this to some extent in terms of sequence differences between isotypes. However, there is an added complication in that it appears that proximity of Fab arms can modulate the expression of the C1q binding site on Fc (Burton, 1985). Thus, hinge-deleted IgG1 does not bind C1q (Klein et al, 1981). Further and more strikingly IgG4 does not bind C1, whereas its Fc fragment (Fc4) does (Isenman et al, 1975). This is consistent with the close approach of Fc and Fab in IgG4 visualized in Figure 3.

In comparing isotype behavior with respect to complement it is important to distinguish C1q binding, C1 binding, C1 activation and whole complement activation. The most used assay in the measurement of the end-product of the whole complement cascade, i.e. cell lysis. The inability of an IgG to promote efficient lysis does not necessarily indicate an inability to bind C1q. A later stage may be implicated. For example, it appears that C1q binding is not always directly related to C1 activation (Folkerd et al, 1980; Circolo et al, 1985) and, furthermore, later components of complement, e.g. C4b, C3b also interact with IgG. A further complication is that a small change in C1q binding affinity may, through the amplification nature of the complement cascade process, produce a large change in whole complement activation measured as cell lysis.

Comparison of the human IgG subclasses provides the following view (Burton et al, 1986). All the subclasses in a monomeric state bind C1q with measurable affinity with the order of binding constants IgG3 > IgG1 > IgG2 > IgG4. IgG3 and IgG1 activate C1 and whole complement efficiently. IgG2 is less efficient in complement activation. IgG4 does not appear to bind C1 and does not activate complement. In other species mouse IgG2a and IgG2b, guinea pig IgG2, rabbit IgG, rat IgG2b, 1 and 2a bind C1q and activate complement. Mouse IgG1, guinea pig IgG1 and rat IgG2c bind C1q weakly if at all and do not significanly activate complement (Burton, 1985; Medgyesi et al, 1978; Hughes-Jones and Gornick, 1982).

Whilst it is clear that C1 interacts with associated IgG primarily through C1q there have been suggestions that C1r and/or C1s may also interact weakly with sites on IgG (e.g. Hughes-Jones and Gornick, 1982) but this has not been clearly demonstrated (Burton, 1985). Activated forms of C3 and C4 also bind to IgG (Reid, 1983) via covalent interaction with residues in the heavy chain of Fab. The former interaction is important not only in the classical pathway but also for IgGs such as rabbit IgG which activate the alternate pathway.

Interaction of IgG with cell Fc receptors

Receptors for the Fc region of IgG are found on a number of cell types and are associated with a variety of functions including phagocytosis (monocytes, macrophages, neutrophils), antibody dependent cellular cytotoxicity (monocytes, macrophages, lymphocytes), maternofoetal transport (trophoblasts) and possibly immunodulation (lymphocytes). "Fc receptor" is an operational term and does not imply the same molecular species is found on the different cell types. Indeed, there is good evidence that a number of molecular species are involved.

Human leukocyte IgG Fc receptors fall into three categories (Anderson and Looney, 1986) as defined by a number of criteria but most especially by reactivity with specific monoclonal antibodies. FcRI is a 72,000 mol. wt. receptor found on monocytes and binding monomer IgG with high affinity ($5.10^8 M^{-1}$). FcRII is a 40,000 mol. wt. receptor found on monocytes, granulocytes, plateletes and B-cells binding associated IgG but monomer IgG only very weakly ($<10^6 M^{-1}$). FcR_{10} is a 50-70,00 mol. wt. receptor found on granulocytes, macrophages, K and NK cells again binding associated but not monomer IgG. FcRI binds the human IgG subclasses with the range order IgG1 = IgG3 > IgG4. The subclasses are shown to bind to the same receptor by competition experiments. IgG2 does not bind. The subclass specificity of FcRII and FcR_{10} is not definitively demonstrated at this stage.

Until recently, the conventional wisdom was that complement interacts with the C_H2 domain and cell Fc receptor with the C_H3 domain of IgG. This has been challenged by a number of workers (Burton, 1985). The binding of IgG to the human monocyte Fc receptor (FcRI) will be briefly discussed.

The C_H3 domain was originally favoured owing to the (weak) ability of pFc' fragment of IgG (C_H3 domain dimer) to inhibit the interaction of IgG and monocyte Fc receptor. Using domain-specific anti-human IgG monoclonal antibodies, this has been shown to arise from contamination of pFc' preparations by small amounts of parent IgG (Woof et al, 1984). In contrast, C_H2 domain involvement in receptor binding is favoured by the loss of binding associated with hinge deletion (Woof et al, 1984) and aglycosylation of IgG (Leatherbarrow et al, 1985). Further, anti-human IgG monoclonal antibodies specific for the C_H3 domain and the C_H2/C_H3 domain interface do not inhibit the Fc receptor-IgG interaction and are still

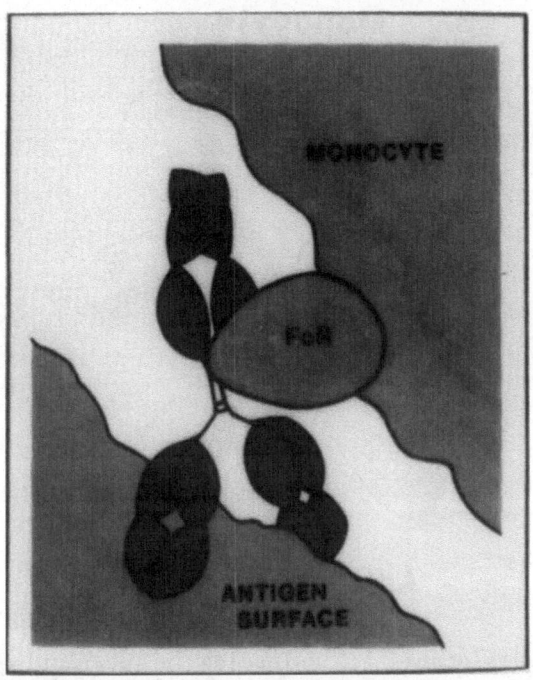

Fig. 5. <u>A model proposed for the interaction of IgG and human monocyte Fc</u>
<u>receptor (FcRI)</u>. The model proposed primarily on the basis of a
comparison of sequences of IgG of different species and subclass
showing differing affinity for monocyte FcRI (Woof <u>et al</u>, 1986).
The region of interaction on Fc is suggested to involve that
between the interheavy disulphides and the folded C_H2 domain. The

monocyte Fc receptor has a molecular weight of approximately
70,000 and is represented as a globular protein of appropriate
size. An antigenic surface is included in the diagram to show the
close approach of foreign cell and monocyte required by the
interaction although most studies have been carried out using
unliganded monomeric IgG.

able to bind to receptor-bound IgG (Partridge <u>et al</u>, 1986). Antibodies
specific for a C_H2 epitope are inhibitory and do not bind to
receptor-bound IgG. It has therefore been suggested that C_H2 is the
critical domain for monocyte receptor binding (Partridge <u>et al</u>, 1986; Woof
<u>et al</u>, 1986). A similar conclusion is reached for mouse IgG2b binding to
mouse macrophage Fc receptor based on the use of deleted mutant proteins
(Diamond <u>et al</u>, 1979). There is still controversy, however, in this area
which is complicated by the existence of the different types of Fc
receptor (Burton, 1985; Anderson and Looney, 1986). Based on a sequence
comparison of IgGs showing differing affinity for the monocyte Fc

Monocyte

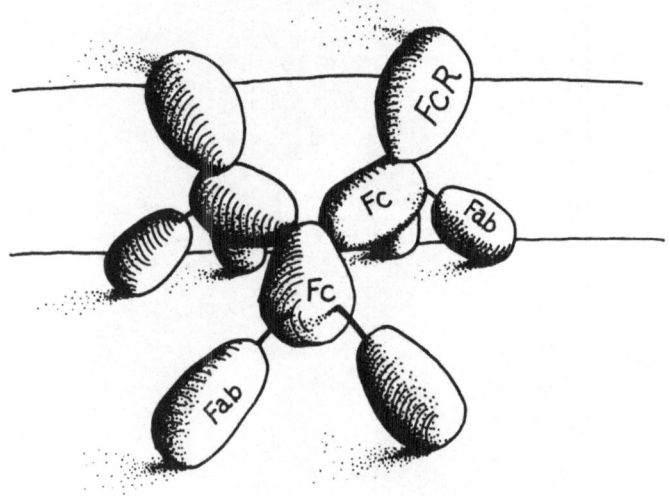

Antigen surface

Fig. 6. <u>Model of dislocated IgG molecules in interaction with antigen and</u> <u>Fc receptor</u>. The C_H3 domains of three dislocated IgG molecules are shown interacting to produce a staple-like arrangement. The monocyte FcR interacts with the region of Fc described in Figure 5 but not represented in a generalized form. The choice of three IgG molecules is arbitrary. This is or a similar arrangement could be propagated in three dimensions to maintain a parallel configuration of monocyte and foreign cell surfaces.

receptor, the model shown in Figure 5 has been proposed (Woof <u>et al</u>, 1986). The IgG binding site is suggested to comprise residues of the lower hinge, <u>i.e.</u> those between the hinge disulphides and the folded C_H2 domain (Leu 234-Pro 238) and possibly those on a nearby bend (Leu 328-Pro 331).

<u>The dislocated model of antibody function</u>

From the above, the C_H2 domain is implicated in two major effector interactions: complement triggering and Fc receptor binding. If the C_H3 domain is not involved in either of these interactions (and this remains to be shown definitively for all types of Fc receptors) one can ask the question "what is the function of C_H3?" A major function may be to maintain the unusual disposition of the C_H2 domains which serve as the critical binding units for effector molecules (Burton, 1985). Certainly, removal of the C_H3 domains would be expected to destabilize the C_H2 domains, which are already known to be relatively unstable. Another possible function is in the assembly of IgG, driving heavy chain dimerization.

However, Figure 5 suggests that the C_H3 domains may be involved in other function when IgG is bound to antigen and this brings out striking paralells with IgM (Burton, 1986). Thus, for the monocyte Fc receptor to interact most readily with the lower hinge region of IgG, it would seem opportune to take advantage of hinge flexibility and present the IgG molecule in a dislocated form (Figure 6). This would leave the C_H3 domains free to interact with anti-C_H3 domain monoclonal antibodies (as above in the absence of antigen) or possibly with C_H3 domains on a neighbouring IgG molecule. Hinge flexibility would allow the Fc to rotate perpendicular to the plane of the Fab arms to facilitate interaction. There is a large hydrophobic area at the "bottom" of the C_H3 domain which could be used in Fc-Fc interaction (Burton, 1985). Such interactions may be important in immune precipitation (Steensgard, 1984) and the model proposed would allow this to happen without obstruction of monocyte Fc receptor or (probably) C1q binding.

Perhaps the most striking feature of Figure 6 is the potential resemblance to electron micrographs of IgM bound to an antigen in the "staple" form. The uncomplexed IgM molecule, by electron microscopy, adopts a roughly star-shaped planar conformation (Figure 7) (Feinstein and Richardson, 1981; Feinstein et al, 1986), but on binding to bacterial flagella, the $F(ab')_2$ arms are dislocated out of the plane of the $(Fc)_5$ disc to give a molecule resembling a "staple". This star-to-staple transition appears sufficient to allow complement C1 binding and activation of IgM. The mechanism appears to be distortional (e.g., exposure of sites in the staple form) rather than allosteric, i.e.

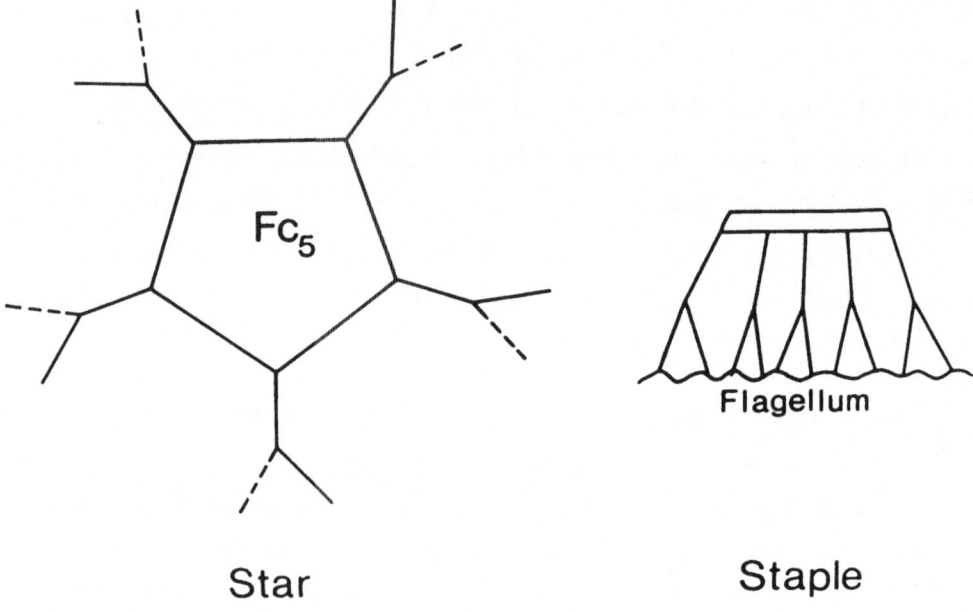

Star   Staple

Fig. 7. <u>Representation of the star and staple conformation of IgM</u> deduced
<u>from electron microscopy.</u>

involving conformation changes passed through the molecule to Fc on
antigen binding.

I have suggested that IgG molecules may also be in a staple-like
configuration on C1 binding and activation (Burton, 1986) as represented
in Figure 8. This staple may resemble only superficially that found with
IgM and may simply represent a convenient method of bringing Fc regions
into close proximity to allow multivalent binding of C1q. The flexibility
of C1q arms supports this view. Alternatively, a closer relationship may
exist between the two staples, dictated by the requirements of C1q
binding, C1 activation or possibly of the binding of a later component of
complement, <u>i.e.</u> C3b or C4b. It is interesting that complement C1 binding
and activation seem to be distinct as discussed earlier and that
activation is far more dependent on the spacing and distribution of
antigen on a cell surface than is binding.

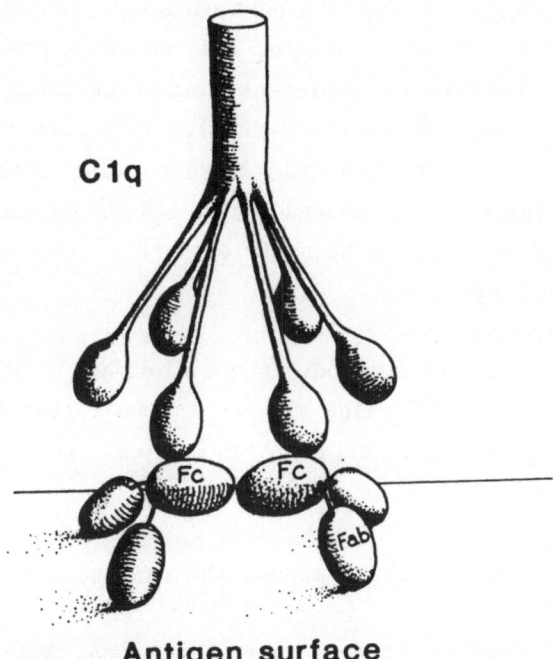

C1q

Antigen surface

Fig. 8. <u>Model of dislocated IgG molecules in interaction with antigen and complement C1q (compare Figure 4)</u>.

A possible argument against the role for the C_H3 domain postulated here is the ability of rabbit Facb (a fragment produced by plasmin digestion lacking the C_H3 domains) to activate complement in immuneprecipitates (Colomb and Porter, 1975). However, the generation of Facb requires exposure of IgG to pH 2.5, and its ability and that of an acid-treated control IgG to activate complement is considerably reduced compared with untreated IgG. Also, the details of complement activation on an immune precipitate and at an antibody-coated target cell membrane, for example, may differ.

I have therefore suggested that the physiological problem of activating complement when in contact with antigen, but not when free in serum, is solved for IgG and IgM by two different routes arriving at a similar destination (Burton, 1986). Monomeric serum IgG molecules rely on antigen to bring Fc regions together, the relationship being cemented by Fc-Fc interaction. In pentameric star-shaped serum IgM Fc regions are already close, but in an inactive form. Binding to antigen dislocates the molecule to produce an active form.

Finally, the occurrence of antibody in a dislocated Y-shape rather than the conventional Y-shape or T-shape may be a common feature of antibody interaction with antigen and effector molecules. Two recent studies support this general notion.

First, Holowka and Baird (1983) mapped the distances between sites on IgE and the membrane surface in an IgE-Fc receptor complex. Using fluorescence energy transfer, these workers have shown that "the $(C_H3C_H4)_2$ domains of IgE cannot be fully extended and oriented perpendicular to the membrane surface with the C-terminal end at the surface". They conclude that the type of dislocated model for bound Ig suggested above is fully consistent with that available for IgE bound to its Fc receptor.

Second, Greenspan et al (1987) have demonstrated that mouse IgG3 antibodies specific for streptococcal group A carbohydrate (GAC) can enhance the binding of radiolabeled prototype monoclonal IgG3 anti-GAC antibody to group A vaccine. This enhancement is not exhibited by $F(ab')_2$ fragments from IgG3 anti-GAC antibody implying the involvement of Fc-Fc interaction in the antibody interaction with this antigen.

Acknowledgements

The author is a Jenner Fellow of the Lister Institute of Preventive Medicine. The financial support of the MRC, SERC and the Yorkshire Cancer Research Campaign is acknowledged.

References

Amit, A.G., Mariuzza, R.A., Phillips, S.E.V. and Poljak, R.J. Science 233, 747-753 (1986).

Anderson, C.L. and Looney, R.J. Immunol. Today 7, 264-266 (1986).

Brunhouse, R. and Cebra, J.J. Molec. Immun 16, 907-911 (1979).

Burton, D.R. Molec. Immun. 22, 161-206 (1985).

Burton, D.R. Immunol. Today 7, 165-167 (1986).

Burton, D.R., Boyd, J., Brampton, A., Easterbrook-Smith, S.B., Emmanuel, E.J., Novotny, J., Rademacher, T.W., van Schravendijk, M.R., Sternberg, M.J.E. and Dwek, R.A. Nature (Lond.), 288-338-344 (1980).

Burton, D.R., Gregory, L. and Jeffris, R. Monogr. Allergy 19, 7-35 (1986).

Circolo, A., Battisto, P. and Boros, T. Molec. Immun. 22, 207-214 (1985).

Colomb, M. and Porter, R.R. Biochem. J. 145, 177-183 (1975).

Deisenhofer, J. Biochemistry 20, 2361-2370 (1981).

Diamond, B., Birshtein, B.K. and Scharff, M.D. J. Exp. Med. 150, 721-726 (1979).

Feinstein, A. and Richardson, N.E. Monogr. Allergy 17, 28-47 (1981).

Feinstein, A., Richardson, N.E. and Taussig, M.J. Immunol. Today 7, 169-174 (1986).

Folkerd, E.J., Gardner, B. and Hughes-Jones, N.C. Immunology 41,179-185 (1980).

Greenspan, N.S., Monafo, M.J. and Davie, J.M. J. Immunol. 138, 285-292 (1987).

Gregory, L., Davis, K.G., Sheth, B., Boyd, J., Jefferis, R., Nave, C. and Burton, D.R. Molec. Immun. 24, 821-829 (1987).

Holowka, D. and Baird, B. Biochemistry 22, 3475-3484 (1983).

Hughes-Jones, N.C. and Gorick, B.D. Molec. Immun. 19, 1105-1112 (1982).

Isenman, D.E., Dorrington, K.J. and Painter, R.H. J. Immun. 114, 1726-1729 (1975).

Klein, M., Haeffner-Cavaillon, N., Isenman, D.E., Rivat, C., Navia, M.A., Davies, D.R. and Dorrington, K.J. Proc. Natl. Acad. Sci. U.S.A. 78, 524-528 (1981).

Leatherbarrow, R.J., Rademacher, T.W., Dwek, R.A., Woof, J.M., Clark, A., Burton, D.R., Richardson, N.E. and Feinstein, A. Molec. Immun. 22, 407-415 (1985).

Lukas, T.J., Munoz, H. and Erikson, R.W. J. Immun. 127, 2555-2560 (1981).

Medgyesi, G.A., Fust, G., Gergely, J. and Bazin, H. Immunochemistry 15, 125-129 (1978).

Partridge, L.J., Woof, J.M., Jefferis, R. and Burton, D.R. Molec. Immun. 23, 1365-1372 (1986).

Perkins, S.J. Biochem. J. 228, 13-26 (1985).

Poon, P.H., Schumaker, V.N., Phillips, M.K. and Strang, C.J. J. Molec. Biol. 168, 563-577 (1983).

Pumphrey, R.S.H. Immunol. Today 7, 174-178 (1986).

Rademacher, T.W. and Dwek, R.A. Prog. Immun. 5, 95-112 (1984).

Rajan, S.S., Ely, K.R., Abola, E.E., Wood, M.K., Colman, P.M., Athay, R.J. and Edmunson, A.B. Molec. Immun. 20, 787-799 (1983).

Reid, K.B.M. Biochem. Soc. Trans. 11, 1-12 (1983).

Reid, K.B.M. and Porter, R.R. Ann. Rev. Biochem. 50, 433-464 (1981).

Sarma, R. and Laudin, A.G. J. Appl. Cryst. 15, 476-481 (1982).

Silverton, E.W., Navia, M.A. and Davies, D.R. Proc. Natl. Acad. Sci. U.S.A. 74, 5140-5144 (1977).

Steensgaard, J. Immunol. Today 5, 7-10 (1984).

Woof, J.M., Nik Jaafar, M., Jefferis, R. and Burton, D.R.   Molec. Immun.
   21, 523-527 (1984).
Woof, J.M., Partridge, L.J., Jefferis, R. and Burton, D.R.   Molec. Immun.
   23, 319-330 (1986).

C3 BINDING TO BACTERIAL SURFACES

K. Lynn Cates and R. Paul Levine

Department of Genetics
Washington University School of Medicine
Box 8031
4566 Scott Avenue
St. Louis, Mo 63110 USA

Introduction

The activation of complement and binding of C3 to the surface of pathogenic organisms are prerequisites for the biologic role of the complement system in host defense, and the virulence of many successful pathogens is the result of their ability to resist complement-mediated opsonophagocytosis and/or killing. Studies of complement binding have provided a foundation for evaluating the mechanisms of virulence of several important pathogens. More complete understanding of these organisms' resistance to complement eventually may provide means of attenuating their virulence in the clinical setting.

In this paper we will review C3 structure and the biochemistry of C3 binding, some mechanisms of bacterial resistance to complement and, finally, factors to be considered in evaluating interactions between complement proteins and pathogenic organisms in vitro.

C3 Structure and the Biochemistry of Covalent C3b Binding via its Reactive Thioester

In its native state, C3 is composed of a 115,000 dalton α and a 75,000 dalton β chain connected by disulfide bonds (1). C3 is activated and converted to C3b by either the alternative or classical pathway C3 convertases, $\overline{C3bBb}$ and $\overline{C4b2a}$, respectively. Activation by these enzymes results in cleavage of an arginine-serine bond (Arg 77 - Ser 78) and the release of C3a, a 77 residue peptide with a molecular weight of about 9,000 daltons, from the amino terminal end of the α chain. C3b is

NATO ASI Series, Vol. H24
Bacteria, Complement and the Phagocytic Cell
Edited by F. C. Cabello und C. Pruzzo
© Springer-Verlag Berlin Heidelberg 1988

comprised of the remainder of the α chain, now designated C3α', and the attached, intact β chain. Proteolytic cleavage of C3 to form C3b results in a change in the conformation (2,3) of the molecule exposing an internal thioester bond between cysteinyl and glutamyl residues in the amino acid sequence, cysteine-glycine-glutamine-glutamine, in the C3d region of C3α' (4-6). C3 activation increases this thioester bond's susceptibility to nucleophilic attack. The second glutamyl residue donates its carbonyl group to form a covalent acyl ester or amide bond with a hydroxyl or amino group on a receptive molecule or cell surface (5,7). Covalent bond formation apparently is nonspecific since C3b can bind to a variety of molecules in the fluid-phase (8,9), to immune complexes, and to the surfaces of many kinds of cells (4,7,10-13). Under certain conditions, covalent binding of C3b to receptive surfaces renders it biologically active to function either as an opsonic agent or in the enzymic cleavage of C5 in the steps leading to the formation of the membrane attack complex that causes cytolysis.

Each C3b molecule produced has the potential to bind to factor B and, in the presence of factor D and Mg^{++}, to form the C3 convertase $\overline{C3bBb}$. This enzyme cleaves C3 to produce more C3b and more enzyme. Thus, the alternative pathway can be continuously activated by positive feedback. This feedback loop is known as the alternative pathway amplification loop.

C3b inactivation, as defined by its loss of the ability to bind covalently can result from hydrolysis of the thioester bond (-OH binds to C=O, and -H to -S), or from nucleophilic inactivation by such readily available nucleophiles as methylamine (7).

C3b also can be inactivated to iC3b upon cleavage of the C3α chain at two or more sites by factor I with factor H (6,14) or CR1 (the C3b receptor on many cell surfaces) serving as cofactors. Preferential binding of factor H over factor B leads to inactivation of C3b and interruption of the alternative pathway amplification loop.

Covalent bond formation by C3b has been studied most extensively in the fluid phase using relatively small receptive molecules (3,8,9). Most fluid-phase binding studies have been carried out under strictly defined chemical conditions employing trypsin as the C3 activating enzyme (8,9,15). It should be noted that, whereas, C3 cleavage by the C3 convertases is limited to the Arg 77 - Ser 78 bond of the α chain, trypsin not only is capable of cleaving C3 at that location, but also at other sites (3,6). Thus, trypsin can inactivate as well as activate C3 if experimental conditions are not carefully controlled.

Under well-controlled conditions, Law et al. (8) examined fluid-phase binding of radiolabeled small molecules to trypsin-activated C3b. Since the thioester bond of preformed C3b is not available for binding, preformed C3b served as a control to assess nonspecific binding to sites other than the reactive thioester on the C3b molecule. They demonstrated that a variety of small molecules including amino acids and mono-, di-, and trisaccharides bound specifically to trypsin-activated C3b. The different binding efficiencies of these molecules reflected that there is an order of preference in C3b binding.

Capel et al (9) found that on a molar basis, in general, there was decreasing affinity for active C3b from tetra- to tri- to di- to monosaccharides, suggesting that molecular size influenced the ability to bind to C3b. In contrast, subsequently, Levine et al. (11) demonstrated that there was very little difference among the C3b binding efficiencies for alcohols of different sizes such as methanol, ethanol, 1-propanol, and 1-butanol. However, their binding efficiencies were much higher than those for 2-propanol, 2-butanol, and 2-methyl-2-propanol, indicating that the length of the alkyl chain does not affect binding of the hydroxyl group with the thioester, but that the hydroxyl groups on bulkier molecules may not be as accessible for binding to C3b. These findings suggest that a molecule's conformation, as well as its size, may play a role in determining its binding efficiency to C3.

Nucleophilicity of the receptive molecule or surface also is important in determining its ability to bind to the reactive thioester of C3b. Many potentially important biological receptive molecules and surfaces are only weakly nucleophilic under physiologic conditions. However, recently, Venkatesh and Levine (submitted for publication) and Dodds and Law (Complement 4: 151, 1987, Abstract) have proposed a mechanism whereby ordinarily weak nucleophiles with available hydroxyl or amino groups are made more nucleophilic and, therefore, more available for binding with the reactive thioester of C3b. Their work suggests that during activation of C3, a catalytic group in its unprotonated form near the reactive thioester on the C3d region of the α chain abstracts protons from hydroxyl (R-OH) or amino (R-NH$_2$) groups of receptive molecules and, potentially, receptive surfaces, making the oxygen or nitrogen nucleophilic and more likely to be able to form acyl ester or amide bonds with C3b. Mechanisms such as this may be critical to expanding the range of acceptor molecules available for covalent binding to C3b under physiologic conditions.

Law et al. (8) have investigated the nature of the covalent bond formed between the reactive thioester of C3b and some radiolabeled small molecules by quantitating the amount of radioactivity released from C3b-small molecule complexes after treatment with hydroxylamine under alkaline conditons. As could be anticipated, those small molecules such as glucose and glycerol with only hydroxyl and no amino groups available for binding, formed mostly hydroxylamine-sensitive, therefore, presumably acyl ester bonds, with the reactive thioester of C3b. However, molecules such as lysine, serine, and threonine with both hydroxyl and amino groups, formed mainly hydroxylamine-resistant, presumably amide bonds, with C3b.

The relative efficiency of acyl ester versus amide bond formation by trypsin-activated C3b in the fluid phase was investigated further by comparing the ability of serine and its derivatives, serine with an acetylated α-amino group, (N-acetlyserine) and serine with a methylated hydroxyl group (0-methylserine) to inhibit binding of [^{3}H]glycerol to C3b (15). At physiologic pH, serine and N-acetylserine, but not 0-methylserine inhibited glycerol binding well. These findings suggested that there is more efficient C3b binding with the hydroxyl than with the amino terminal end of the serine molecule. The fact that most of the covalent bonds formed between serine and C3b are hydroxylamine-resistant may result from an intramolecular rearrangement after the initial acyl ester bond is formed.

Law et al. (15) also have examined the formation of acyl ester and amide bonds with the reactive thioester of trypsin-activated C3b in the fluid phase as a function of the pH of the reaction mixture. They employed [^{3}H]glycerol and [^{3}H]putrescine as receptive molecules with available hydroxyl and amino groups, respectively. The binding efficiency of glycerol to activated C3 was relatively constant over a pH range from 5.5 to 9, with only a slight decrease from pH 9 to pH 10. In contrast, the binding efficiency of putrescine was low at physiologic pH, and only began to rise and approach that of glycerol at approximately pH 9. Its highest binding efficiency was at pH >10. These findings probably reflect that the nucleophilicity of hydroxyl groups is relatively independent of pH, but the nucleophilicity of amino groups rises with pH, particularly at pH >9, since a higher proportion of amino groups are deprotonated at high pH.

C3 Activation and Binding in Host Defense Against Pathogenic Bacteria

Covalent binding to receptive surfaces may render C3 biologically active to function as a lytic or opsonic agent. C3 activation by the classical pathway ordinarily requires antibody, whereas, C3 can be activated by the alternative pathway by many substances in the absence of antibody. In the nonimmune host, since no specific antibody is available, complement usually is activated _via_ the alternative pathway by surface molecules of the invading organism itself. Although many bacterial surface molecules can activate the alternative pathway (14), the virulence of several important bacterial pathogens containing these activating substances has been demonstrated to correlate with their ability to resist being opsonized and/or killed by complement (13,16-25), particularly as mediated by the alternative pathway. Disease caused by these organisms is especially prevalent among infants and children, since they have less specific antibody and, thus, are less able to activate complement by the classical pathway than adults.

There are several requirements for successful complement-mediated opsonic and/or bactericidal activity. First, sufficient numbers of C3b molecules must bind to the bacterial surface (13,26). It should be noted that even though C3 may be activated by an organism as measured by the formation of C3 degradation products or the loss of hemolytic activity, it is not functional unless it is bound to the cell surface (4,13) and, even then, it is not always effective in host defense (12,16). Next, for continued complement activation by the alternative pathway amplification loop, C3b must be available for binding to factor B for formation of the alternative pathway C3 convertase $\overline{C3Bb}$. At the same time, it must be protected from inactivation by regulatory factors (e.g., factors H and I). For formation of the membrane attack complex needed for lytic activity, C3b must be accessible for binding to terminal complement proteins. Similarly, in order to function as an opsonin, C3b must have access to phagocyte complement receptors (16). With these requirements in mind, the mechanisms employed by a few selected bacterial pathogens to resist complement will be reviewed here. Mechanisms of complement resistance of some other organisms will be reviewed elsewhere in this publication.

Streptococcus pneumoniae

The pneumococcus employs several means to resist complement and successfully evade being efficiently opsonized for uptake and killing by phagocytes. Pneumococcal cell wall constituents such as teichoic acid (27), and cell membrane lipoteichoic acid (28), unlike most types of pneumococcal capsular polysaccharide (29), can activate the alternative pathway of complement in the absence of specific antibody. Although C3b can bind to the cell wall of encapsulated pneumococci, most C3b probably binds on or near the capsular surface of the organism. Hostetter (13) demonstrated that even when human agammaglobulinemic serum is employed as a complement source to insure the absence of specific antibody, most C3b can be recovered along with the capsule when the organism is treated with a decapsulating enzyme. The C3b covalent bonds with the capsular polysaccharides are assumed to be acyl ester in nature since the polysaccharides lack amino groups. Bonds with the cell wall probably are amide in nature since less than 10 percent of C3b molecules bound to an unencapsulated strain of pneumococcus were released after treatment with hydroxylamine (13).

Hostetter (13) has reported that the number of molecules of C3b covalently bound to three of four types of encapsulated pneumococci tested (serotypes 3,4, and 6A, but not type 14) was approximately two-fold or more higher after opsonization with immune serum than with nonimmune serum. Earlier, Brown et al. (30), had shown that C3b binds to both heavy and light chains of antipneumococcal antibody eluted from pneumococci. Only about half of the C3b could be released from IgG or IgM antipneumococcal antibody by hydroxylamine, suggesting that it was bound by both amide and ester bonds. Thus, in addition to serving as an opsonin, anticapsular antibody may enhance host defense against pneumococci by providing binding sites on itself for C3b. Clearance of pneumococci by phagocytes most likely is enhanced by the organisms' binding to both antiobdy and complement since complement provides signals for attachment of the organisms to phagocytes and antibody provides signals for their ingestion.

The location of C3b on the surface of the pneumococcus is important in determining its functional activity. Brown et al. (17) found that more C3 was depleted from guinea pig serum and, correspondingly, more radiolabeled C3 was bound to encapsulated pneumococci in the presence of IgG anticell wall antibody or IgM anticapsular antibody than by IgG anticapsular antibody. Both anticapsular antibodies, however, but not the anticell wall antibody, supported opsonization of the pneumococci for phagocytosis by polymorphonuclear leukocytes (PMN) in vitro, and clearance of pneumococci in a guinea pig model of pneumococcal bacteremia (17). Furthermore, when pneumococci were sensitized with type-specific anticapsular antibody or IgG anti-cell wall antibody in quantities sufficient to allow similar numbers of C3 molecules to bind to each organism, those organisms treated with IgG or IgM anticapsular antibody, but not those treated with anti-cell wall IgG, were able to bind to the human erythrocyte C3b receptor, CR1, by the immune adherence reaction (i.e., binding of the immune complexes, antibody-sensitized pneumococci, via C3b to membrane proteins of the erythrocytes).

Electron microscopy employing avidin-ferritin to localize biotinylated C3 or antibody on the surface of the pneumococcus has revealed that anticapsular antibody, or C3b fixed by anticapsular antibody by the classical pathway, is deposited in clusters at a distance from the cell wall (16). In contrast, anti-cell wall antibody or C3 fixed by anti-cell wall antibody is not localized by avidin-ferritin on encapsulated organisms, even though the biotinylated component can be detected by radiolabeling it and C3 can be demonstrated to be bound to cell wall components of the organism by Ouchterlony analysis (16). Based on these observations, it appears that, just as the capsule interferes with avidin-ferritin localization of cell wall-bound biotinylated C3, it also interferes with the recognition of cell wall-bound C3b by complement receptors on phagocytes, and by CR1 on erythrocytes. In contrast, C3b fixed to the capsule is readily accessible for binding.

Some organisms' resistance to complement results from their ability to bind to factor H preferentially over factor B (31). Whereas, factor B binding to C3b permits formation of the C3 convertase C3bBb and favors continued activation of C3b by the alternative pathway amplification loop, preferential binding of factor H favors inactivation of C3b by factor I and leads to termination of the amplification loop. Thus, continued complement activation on the surface of an organism by the alternative pathway depends on C3b deposition in a site accessible to factor B but "protected" from regulatory proteins. Some investigators have classified organisms and other receptive surfaces as being either complement "activators" or "nonactivators". Nonactivators may fail to activate because they cannot bind C3b to begin with, because they bind it in a location or a manner not suited for continued activation, or because they are "regulatory" surfaces that favor inactivation of C3b by regulatory proteins (14).

Brown et al. (32) examined the ability of pneumococci to bind factors H and B. They found that the number of C3b molecules binding factor H, as measured by the conversion of bound C3b to iC3b, was comparable on unencapsulated and type 7 encapsulated pneumococci. However, C3b on the "nonactivator" surface of sheep erythrocytes bound factor H with higher affinity than did C3b on either organism. Also, most of the C3b on either pneumococcal capsules or cell walls bound H with a lower affinity than did the C3b on sheep erythrocytes. Thus, both capsular- and cell wall-bound C3b acted as if they were in "protected" sites, resisting degradation by the regulatory proteins, factors H and I. In contrast, whereas, C3b molecules on sheep erythrocytes or pneumococcal cell walls bound factor B with comparable affinity, most C3b molecules fixed to pneumococcal capsules bound factor B with only 1/30 as high affinity.

In summary, encapsulated pneumococci successfully evade complement in several ways. First, the nonactivating capsular material interferes with components. Second, even when C3b is deposited on the cell wall, it is not recognized by phagocyte receptors. Finally, the poor affinity of capsule-bound C3b for factor B prevents efficient continuing fixation of complement in a site where it can be recognized by phagocytic receptors.

Group A Streptococci

Isogenic mutants of group A beta-hemolytic streptococci differing only in the presence of absence of surface M protein differ in their virulence and resistance to phagocytosis (26). Peterson et al. (33) found that an M-strain could be efficiently opsonized via the alternative pathway in the relative absence of immunoglobulin, but that an isogenic M+ strain was poorly opsonized by all sera tested. Weis et al. (12) demonstrated the importance of M protein in preventing complement-dependent adherence of group A streptococci to PMN. In the presence of normal human serum or normal human serum chelated with MgEGTA to remove classical, but not alternative pathway activity, C3 promoted adherence of M+ streptococci to only about 8 to 10 percent of human PMN, but adherence of M- streptococci to almost 90 percent of PMN.

In another study, Jacks-Weis et al. (26) employed a quantitative fluorometric immunoassay to demonstrate that less C3 was fixed by M+ than by M- group A streptococcal cells. Although M+ organisms bound C3 in quantities that would have been sufficient for phagocytosis and killing of M- strains by PMN, the M+ strains were not efficiently ingested and killed. The amount of normal human C3 fixed by an M+ strain, M49, could be increased to a level comparable to that of its M⁻ isogenic mutant only in the presence of specific rabbit anti-M49 antibody, not in the presence of normal, nonimmune rabbit serum, or antiobdy to another type of M protein (anti-M12). This finding suggested that the specific antibody either provided sites on itself for C3b binding, as is the case with antipneumococcal antibody, or it allowed C3b binding elsewhere on the surface of the organism.

Interestingly, most of the radiolabeled C3b bound to intact M⁺ group A streptococci could be removed by treatment with sodium dodecyl sulfate (SDS) (12). Less than half of the C3b bound to M⁻ cells was removed by SDS, but most was extracted by treatment with SDS-methylamine to release acyl ester bonds. Of the C3b removed from M⁺ bacterial cells by SDS, most was further released from bacterial surface molecules by incubation at pH 11. Therefore, it appeared that the C3b extracted from M⁺ cells by SDS had been covalently bound to surface molecules of the organism, but that those molecules themselves had been extracted by SDS. The more detergent-extractable nature of the bacterial surface molecules fixing C3b on M⁺ versus M⁻ cells suggested that C3b probably bound to different sites on the surface of M⁺ than on the surface of M⁻ streptococci. Both M⁺ and M⁻ strains bound the two opsonic forms of C3, i.e., C3b and iC3b, as was demonstrated by SDS-polyacrylamide gel electrophoresis (SDS-PAGE) of the C3 eluted from the bacteria with chaotropic and hydrolytic agents.

Thus, M protein contributes to complement-resistance of M⁺ group A streptococci in several ways. It leads to decreased complement activation, decreased C3b binding and, probably, a difference in location of C3b binding on M⁺ versus M⁻ strains. As a result of one or a combination of these factors, even though both C3b and iC3b are present on the M⁺ cell surface, they do not promote attachment of the organisms to phagocytic cells. It is possible that C3b is bound to a structure that is only loosely attached to M⁺ cells and cannot mediate attachment to phagocytes, or that the constraints established by the rigid alpha-helix of M protein directly block C3b and/or iC3b binding to phagocyte receptors.

Bacteria with Sialic Acid-Containing Capsules

Cell surfaces rich in sialic acid often have the ability to regulate complement activation by the alternative pathway because they favor binding of factor H over factor B (34). Sialic acid is considered to be an important virulence factor and it is found in an unusually high frequency as a component of the capsular polysaccharides of pathogenic organisms. Most nonpathogenic bacteria can activate the alternative pathway of complement in the nonimmune host (i.e., in the absence of specific antibody), but several very important human pathogens having capsules containing sialic acid are resistant to the alternative pathway of complement (18-21, 35,36). This resistance is overcome, at least to some extent, with anticapsular antibody, or by removing the capsule or altering its sialic acid (18-21).

The capsule of K1⁺ E. coli is simply a homopolymer of sialic acid. Van Dijk et al., (18) found that several K1⁺ strains of E. coli required higher concentrations of human serum to be opsonized for uptake by human PMN than K1⁻ strains. The differences in uptake were even more marked when the serum source employed as the opsonin was chelated with MgEGTA leaving only the alternative pathway of complement intact. The poorer uptake of K1⁺ strains correlated with the finding that they consumed less C3 and that they consumed it more slowly than K1⁻ strains, especially by the alternative pathway.

Pluschke et al., (36) studied O18:K1 E. coli, a common cause of severe neonatal infections. (The role of O antigen in resistance to complement is discussed elsewhere in this publication.) They found that this organism was resistant to classical pathway-mediated serum bactericidal activity in the absence of specific antibody, and to alternative pathway-mediated bactericidal activity even in the presence of antibody. In contrast, K1⁻ mutants of the same strain were sensitive to classical or alternative pathway killing with or without antibody. Susceptibility to serum bactericidal activity corresponded to the organism's ability to fix complement.

Similar findings have been reported for the type III, group B Streptococcus (19,20) which, as is the case with K1⁺ E. coli, is a leading cause of serious neonatal infections. The type III capsule has a terminal sialic acid residue that can be reduced to 5-acetoamido-3, 5-dideoxynonulose or removed enzymatically by neuraminidase treatment. Edwards et al., (19,20) found that growing type III, group B streptococci in the presence of neuraminidase or opsonizing them in serum with high concentrations of specific antibody enhanced alternative pathway-mediated opsonophagocytosis of this organism (19). Furthermore, C3 and factor B consumption was greater in neuraminidase-treated or reduced type III, group B streptococcal particles than in control, fully-sialated organisms, whether the complement source was C2-deficient human serum or unchelated or MgEGTA-chelated antibody-deficient human serum (20). The fully-sialated organisms' resistance to alternative pathway-mediated opsonophagocytosis probably is the result of the sialic acid's providing a regulatory surface. The means by which anticapsular antibody overcomes this resistance is unclear.

The capsular polysaccharide antigen of group B Neisseria meningitidis, as well as that of K1 E. coli, is a homopolymer of sialic acid. Recently, Jarvis and Vedros (21) reported that removal of sialic acid from group B meningococci correlated quantitatively with an increase in alternative pathway-mediated C3 binding to the organism. Furthermore, they reported that anticapsular antibody was required for alternative pathway-mediated serum bactericidal activity of an encapsulated strain of group B meningococcus, but not for a noncapsular variant.

Although the presence of sialic acid antigen correlates with resistance to alternative pathway-mediated complement fixation by sheep erythrocytes, K1 E. coli (18), type III, group B streptococci (19,20), and group B meningococci (21), not all organisms with surface sialic acid antigens are equally resistant to complement. Bortolussi et al., (35) have reported that virulence of K1 E. coli in newborn rats, and resistance of K1 E. coli to opsonization in vitro varies directly with the quantity of K1 capsular polysaccharide on the organism. In addition, Stevens et al., (37) have found that K1 E. coli resistance to phagocytosis not only is directly related to K1 antigen content, but strains of K1 E. coli that are relatively more sensitive to phagocytosis have a sialic acid-containing antigen that is unrelated to that found on the more resistant strains. Varki and Kornfeld (38) reported no correlation between total murine erythrycyte sialic acid content and complement sensitivity, but a good correlation between the degree of O-acetylation of murine erythrocyte sialic acid residues and their susceptibility to lysis by human complement. They proposed that the degree of O-acetylation of the sialic acid residues rather than the absolute content of sialic acid may be responsible for differences in susceptibility of murine erythrocytes to lysis by human complement. Clearly, further investigation is needed into the role of sialic acids in resistance to the alternative pathway of complement.

Haemophilus influenzae type b

Haemophilus influenzae type b is one of the most important pathogens in childhood. It is the leading cause of bacterial meningitis in children and, as such, is a major cause of acquired deafness and mental retardation. The role of complement in host defense against Haemophilus in vivo is discussed by Dr. Richard Moxon elsewhere in this publication.

It has been established that the type b capsular polysaccharide is the primary virulence factor of type b Haemophilus, and that this polysaccharide is responsible for the organism's resistance to complement-dependent serum opsonic and bactericidal activity. Specific antibody directed against the capsule or against noncapsular antigens can overcome this resistance in the presence of the classical pathway (25). However, only anticapsular antibody can overcome type b Haemophilus resistance to alternative pathway-mediated killing (24,25).

In vitro studies of the interactions between type b Haemophilus and complement proteins have yielded seemingly conflicting results. Fine et al. (22) found that type b Haemophilus did not consume complement by the alternative pathway in EGTA-treated human serum. In contrast, Quinn et al. (23) demonstrated that type b Haemophilus was able to activate and consume complement by the alternative pathway in C4-deficient guinea pig serum.

In 1982, Tarr et al. (24), who also employed C4-deficient guinea pig serum, reported the interesting finding that, in the absence of anticapsular antibody, type b Haemophilus consumed complement by the alternative pathway even though it was not killed. However, Steele et al. (25), using human agammaglobulinemic serum instead of guinea pig serum as a complement source and a different strain of type b Haemophilus, found neither killing nor complement comsumption by the alternative pathway without substantial levels of anticapsular antibody. In contrast, a recent report suggests human serum from some individuals can kill and opsonize type b Haemophilus by the alternative pathway with levels of anticapsular antibody lower than those reported by Steele (39). Several factors, such as the strain of type b Haemophilus employed, the inoculum size, the serum concentration, the use of C2- or C4-deficient versus MgEGTA-chelated serum, and the complement source (i.e., human versus guinea pig) may have contributed to the differences in the results from the different laboratories.

The type b capsular polysaccharide of Haemophilus influenzae is composed of repeating units of polyribosylribose phosphate. It has multiple hydroxyl groups, and the hydroxyl group on C-5 of the ribose moiety appears to be the one most likely to be available for covalent binding with the reactive thioester group of C3b. However, Quinn et al. (23) reported that almost no C3 was consumed by concentrations of type b capsular polysaccharide up to 500 micrograms per milliliter when C4-deficient guinea pig serum was employed as a complement source. Furthermore, Levine et al. (11), using C2-deficient human serum, found no complement consumption by this polysaccharide, either in the presence or in the absence of anticapsular antibody.

Although antibody to the polysaccharide capsule of Haemophilus influenzae type b is important in overcoming this organism's resistance to opsonization and killing by complement, particularly the alternative pathway, the means by which anticapsular antibody overcomes this resistance have not been defined. From our knowledge of mechanisms of resistance of other organisms, however, it can be speculated that anticapsular antibody could function in any of the following ways: 1) it may provide a binding site on itself for C3b, 2) it may allow C3b to bind to the capsule, perhaps by changing the orientation or conformation of the capsular polysaccharide molecule, 3) it may change the surface characteristics of the organism such that C3b can reach and bind to cell wall components such as outer membrane proteins or lipooligosaccharides, 4) it may protect C3b from regulatory proteins such as factors H and I, 5) it may provide C3b better access to other complement proteins (such as factor B for formation of $\overline{C3bBb}$, or the terminal complement components for formation of the membrane attack complex, C5b-C9), and/or 6) it may provide C3b with better access to complement receptors on phagocytes, thereby allowing more efficient phagocytosis.

Despite conflicting studies, it is clear that, in the absence of specific anticapsular antibody against type b <u>Haemophilus</u>, complement either is not activated by the alternative pathway or, if it is activated, complement proteins fail to bind, or they bind in a manner that is not effective for promoting cytolysis and opsonization.

Means of Evaluating C3 Interactions with Bacteria

Current technology and understanding of the biochemistry of complement binding permits reevaluation of the role of C3 in host defense against pathogenic organisms. First, the form and nature of C3b bonds to purified fluid-phase bacterial components can be studied in clearly defined systems using either the natural C3 convertases, $\overline{C4b2a}$ or $\overline{C3bBb}$, or serine proteases of a more generalized type such as trypsin for controlled C3 activation in the absence of the confounding influence of other serum or plasma constituents. The ability of C3b to bind to such fluid-phase antigens from organisms can be compared with its ability to bind to them when they are surface-bound on such particles such as erythrocytes or Sepharose beads, as well as on the intact bacteria themselves. Differences in C3b binding to antigens in the fluid-phase compared with the solid-phase may provide clues to factors needed for successful C3b binding to antigen, such as conformational changes to expose potential binding sites.

It must be noted, however, that results of experiments of C3b binding to intact organisms using serum as a source of antibody and/or complement may differ from those of fluid-phase experiments performed under well-defined conditions with purified reagents and bacterial antigens. For example, there can be marked differences in bacteria-complement interactions among closely related strains (13,18,24). Some investigators are using selective growth conditions (19) and transductants and recombinants of individual organisms (40,41) to study the effect of differences between closely related strains on the organism's interaction with and susceptibility to complement.

One problem in assessing complement interactions with whole organisms is that they may be altered under conditions used to evaluate C3b binding, such as treatment with detergents (12), hydroxylamine, high pH, EDTA, MgEGTA, and C3 activating enzymes other than C3 convertases (e.g., trypsin). In examining C3b binding to intact Haemophilus we have found that the organism is disrupted by the SDS and hydroxylamine treatments used to determine the nature of the organims's bonds with C3b. To circumvent this problem, many investigators have employed heat-killed (17,30) or formaldehyde- (20) or glutaraldehyde-fixed (16,32) bacteria. However, it is not clear that C3b binding characteristics of fixed and live organisms always are comparable. For example, with type b Haemophilus we have found that the washes needed to remove residual glutaraldehyde after fixation are sufficient to wash away all but tiny remnants of the capsule, thus thwarting our attempts to examine the effects of encapsulation on binding (Cates and Levine, unpublished data).

Most early studies of complement interactions with bacteria employed either human agammaglobulinemic serum or animal serum as a source of complement. The widespread availability of intravenous immune serum globulin therapy for patients with agammaglobulinemia has precluded use of this valuable source of human complement. In addition, Zollinger et al. (42) have demonstrated species differences in complement function and, thus, despite the relatively ready availability of animal serum, especially form animals deficient in specific complement components, the use of animal serum as a source of complement is not entirely satisfactory. More and more, purified human complement proteins are being used as a source of complement in vitro (9,32). The obvious advantage of this approach is that the functions of the individual complement proteins can be differentiated. However, it is possible that the activity of the complement proteins in their purified state will not accurately reflect their activity in vivo.

Finally, because of its ready availability and ease of use, some investigators are using trypsin to activate C3, instead of constructing the C3 convertases, C4b2a or C3bBb, from the less readily available individual complement components. Although, as we noted before, trypsin does cleave the same arginyl-serine bond on the C3α chain as the specific C3 convertases, it also is capable of cleaving C3 in several other locations and, thus, inactivating it if experimental conditions are not well-controlled. Also, the action of trypsin, unlike that of C4b2a and C3bBb, extends to many proteins and it is possible that it may affect the integrity of bacterial components or antiobdy in the system.

Conclusions

Activation of C3, and covalent binding of C3b to the organism's surface are critical to host defense against most pathogens. The alternative complement pathway is particularly important in protection of the nonimmune host since it usually can be activated by components of many of these organisms in the absence of specific antibody. However, many successful pathogens, particularly in the pediatric populations, are resistant to the alternative pathway.

The limitations of antibiotic therapy, immunization, and current immunotherapeutic approaches to infectious diseases necessitate further investigation of virulence factors of the causative organisms and how they can be overcome immunologically or pharmacologically. A thorough understanding of the pathogen's interactions with C3 eventually may permit the development of a new approach to immunotherapy of bacterial, protozoal, and viral diseases caused by complement-resistant organisms, namely selective activation of complement by the organisms in the absence of specific antibody.

References

1.   Tack BF (1983) The β-Cys-γ-Glu thiolester bond in human C3, C4, and α₂-macroglobulin. Springer Semin in Immunopathol 6:259-282
2.   Isenman DE, Kells DIC, Cooper NR, Muller-Eberhard HJ, Pangburn MK (1981) Nucleophilic modification of human complement protein C3: correlation of conformational changes with acquisition of C3b-like functional properties. Biochemistry 20:4458-4467
3.   Law S-KA (1983) Non-enzymic activation of the covalent binding reaction of the complement protein C3. Biochem J 211:381-389
4.   Law SK, Lichtenberg NA, Levine RP (1980) Covalent binding and hemolytic activity of complement proteins. Proc Natl Acad Sci USA 77:7194-7198
5.   Hostetter MK, Thomas ML, Rosen FS, Tack BF (1982) Binding of C3b proceeds by a transesterification reaction at the thiolester site. Nature 298:72-75
6.   Law SK, Fearon DT, Levine RP (1979) Action of the C3b-inactivator on cell-bound C3b. J Immunol 122:759-765
7.   Law SK, Lichtenberg NA, Levine RP (1979) Evidence for an ester linkage between the labile binding site of C3b and receptive surfaces. J Immunol 123:1388-1394
8.   Law S-KA, Minich TM, Levine RP (1981) Binding reaction between the third human complement protein and small molecules. Biochemistry 20:7457-7474
9.   Capel PJA, Groeneboer O, Grosveld G, Pondman KW (1978) The binding of activated C3 to polysaccharides and immunoglobulins. J Immunol 121:2566-2572
10. Law SK, Levine RP (1977) Interaction between the third complement protein and cell surface macromolecules. Proc Natl Acad Sci USA 74:2701-2705
11. Levine RP, Finn R, Gross R (1983) Interactions between C3b and cell-surface macromolecules. Annals NY Acad Sci 421:235-245
12. Weis JJ, Law SK, Levine RP, Cleary PP (1985) Resistance to phagocytosis by group A streptococci: failure of deposited complement opsonins to interact with cellular receptors. J Immunol 134:500-505
13. Hostetter MK (1986) Serotypic variations among virulent pneumococci in deposition and degradation of covalently bound C3b: implications for phagocytosis and antibody production. J Infect Dis 153:682-693
14. Pangburn MK, Muller-Eberhard HJ (1984) The alternative pathway of complement. Springer Semin Immunopathol 7:163-192
15. Law S-KA, Minich TM, Levine RP (1984) Covalent binding efficiency of the third and fourth complement proteins in relation to pH, nucleophilicity, and availability of hydroxyl groups. Biochemistry 23:3267-3272
16. Brown EJ, Joiner KA, Cole RM, Berger M (1983) Localization of complement component 3 on Streptococcus pneumoniae: anti-capsular antibody causes complement deposition on the pneumococcal capsule. Infect Immun 39:403-409
17. Brown EJ, Hosea SW, Hammer CH, Burch CG, Frank MM (1982) A quantitative analysis of the interactions of antipneumococcal antibody and complement in experimental pneumococcal bacteremia. J Clin Invest 69:85-98

18. Van Dijk WC, Verbrugh HA, Van Der Tol ME, Peters R, Verhoef J (1979) Role of Escherichia coli K capsular antigens during complement activation, C3 fixation, and opsonization. Infect Immun 25:603-609

19. Edwards MS, Nicholson-Weller A, Baker CJ, Kasper DL (1980) The role of specific antibody in alternative complement pathway-mediated opsonophagocytosis of type III, group B Streptococcus. J Exp Med 151:1275-1287

20. Edwards MS, Kasper DL, Jennings HJ, Baker CJ, Nicholson-Weller A (1982) Capsular sialic acid prevents activation of the alternative complement pathway by type III, group B streptococci. J Immunol 128:1278-1283

21. Jarvis GA, Vedros NA (1987) Sialic acid of group B Neisseria meningitidis regulates alternative complement pathway activation. Infect Immun 55:174-180

22. Fine DP, Marney SR Jr, Colley DG, DesPrez RM (1973) Haemophilus influenzae decomplementation pattern in chelated and nonchelated serum. In: Haemophilus influenzae (Sell SHW, Karzon DT, eds) Vanderbilt University Press, Nashville, pp 113-117

23. Quinn PH, Crosson FJ Jr, Winkelstein JA, Moxon ER (1977) Activation of the alternative complement pathway by Haemophilus influenzae type B. Infect Immun 16:400-402

24. Tarr PI, Hosea SW, Brown EJ, Schneerson R, Sutton A, Frank MM (1982) The requirement of specific anticapsular IgG for killing of Haemophilus influenzae by the alternative pathway of complement activation. J Immunol 128:1772-1775

25. Steele NP, Munson RS Jr, Granoff DM, Cummins JE, Levine RP (1984) Antibody-dependent alternative pathway killing of Haemophilus influenzae type b. Infect Immun 44:452-458

26. Jacks-Weis J, Kim Y, Cleary PP (1982) Restricted deposition of C3 on M+ group A streptococci: correlation with resistance to phagocytosis. J Immunol 128:1897-1902

27. Hummell DS, Berninger RW, Tomasz A, Winkelstein JA (1981) The fixation of C3b to pneumococcal cell wall polymers as a result of activation of the alternative complement pathway. J Immunol 127:1287-1289

28. Hummell DS, Swift AJ, Tomasz A. Winkelstein JA (1985) Activation of the alternative complement pathway by pneumococcal lipoteichoic acid. Infect Immun 47:384-387

29. Winkelstein JA, Bocchini JA Jr, Schiffman G (1976) The role of capsular polysaccharide in the activation of the alternative pathway by the pneumococcus. J Immunol 116:367-370

30. Brown EJ, Berger M, Joiner KA, Frank MM (1983) Classical complement pathway activation by antipneumococcal antibodies leads to covalent binding of C3b to antibody molecules. Infect Immun 42:594-598

31. Levine RP (1985) Molecular interaction between the third complement protein and bacterial cell-surface macromolecules. In: Bayer-Symposium VIII, The Pathogenesis of Bacterial Infections. Springer-Verlag, Berlin Heidelberg, pp 102-120

32. Brown EJ, Joiner KA, Gaither TA, Hammer CH, Frank MM (1983) The interaction of C3b bound to pneumococci with factor H (β1H globulin), factor I (C3b/C4b inactivator), and properdin factor B of the human complement system. J Immunol 131:409-415

33. Peterson PK, Schmeling D, Cleary PP, Wilkinson BJ, Kim Y, Quie PG (1979) Inhibition of alternative complement pathway opsonization by group A steptococcal M protein. J Infect Dis 139:575-585

34. Fearon DT (1978) Regulation by membrane sialic acid of β1H-dependent decay-dissociation of amplification C3 convertase of the alternative complement pathway. Proc Natl Acad Sci USA 75:1971-1975

35. Bortolussi R, Ferrieri P, Bjorksten B, Quie PG (1979) Capsular K1 polysaccharide of Escherichia coli: relationship to virulence in newborn rats and resistance to phagocytosis. Infect Immun 25:293-298

36. Pluschke G, Mayden J, Achtman M, Levine RP (1983) Role of the capsule and the O antigen in resistance of O18:K1 Escherichia coli to complement-mediated killing. Infect Immun 42:907-913

37. Stevens P, Chu CL, Young LS (1980) K-1 antigen content and the presence of an additional sialic acid-containing antigen among bacteremic K-1 Escherichia coli: correlation with susceptibility to opsonophagocytosis. Infect Immun 29:1055-1061

38. Varki A, Kornfeld S (1980) An autosomal dominant gene regulates the extent of 9-0-acetylation of murine erythrocyte sialic acids. J Exp Med 152:532-544

39. Musher D, Goree A, Murphy T, Chapman A, Zahradnik J, Apicella M, Baughn R (1986) Immunity to Haemophilus influenzae type b in young adults: Correlation of bactericidal and opsonizing activity with antibody to polyribosylribitol phosphate and lipooligosaccharide before and after vaccination. J Infect Dis 154:935-943

40. Zwahlen A, Rubin LG, Moxon ER (1986) Contribution of lipopolysaccharide to pathogenicity of Haemophilus influenzae: Comparative virulence of genetically-related strains in rats. Microbial Pathogenesis 1:465-473

41. Joiner KA, Grossman N, Schmetz M, Leive L (1986) C3 binds preferentially to long chain lipopolysaccharide during alternative patheay activation by Salmonella montevideo. J Immunol 136:710-715

42. Zollinger WD, Mandrell RE (1983) Importance of complement source in bactericidal activity of human antibody and murine monoclonal antibody to neningococcal group B polysaccharide. Infect Immun 40:257-264

THE MODE OF C5B-9 ATTACK ON SUSCEPTIBLE GRAM NEGATIVE BACTERIA

Peter W. Taylor

Ciba-Geigy Pharmaceuticals
Wimblehurst Road
Horsham
West Sussex
RH12 4AB
United Kingdom

Introduction

The lysis of erythrocytes by colloid osmotic deregulation upon assembly of heteropolymeric C5b-9 channels on the target membrane became the subject of intense investigation following the realization that the haemolysis assay provided a clear opportunity to unravel the complexities of the complement system (Mayer, 1984). In contrast, relatively little effort has been spent on the elucidation of the mechanism of lysis of nucleated cells and of killing of Gram-negative bacteria by complement, processes that do not follow non-cooperative or one-hit kinetics and are not due exclusively to colloid osmotic deregulation (Born and Bhakdi, 1986; Kim et al., 1987).

Gram-negative bacteria represent complex targets for complement attack, and many pathogens have evolved extremely effective mechanisms for evading the potentially lethal effects of C5-b channel formation on the cell surface (Joiner, 1985). Bacteria may be able to undertake repair of damaged sites on the cell envelope, they may display an altered phenotype in response to environmental pressure and they can reduce the number of non-lethal lesions on the cell surface by rapid division (Born and Bhakdi, 1986). Of prime significance is the presence of two distinct membranes with differing function (Lutenberg and Van Alphen, 1983) separated by a periplasmic space that may account for a significant proportion of the total cell volume (Stock et al., 1977). A large body of experimental data (reviewed by Taylor, 1983; Taylor and Kroll, 1985) provides convincing evidence that complement-mediated killing of Gram-negative bacteria is dependent upon perturbation of the cytoplasmic membrane whereas it is

NATO ASI Series, Vol. H24
Bacteria, Complement and the Phagocytic Cell
Edited by F.C. Cabello und C. Pruzzo
© Springer-Verlag Berlin Heidelberg 1988

clear from the dimensions of the lipid binding domain on C5b-9 that individual complexes interact only with single lipid bilayers. Interruption of the integrity of the outer membrane bilayer due to the intercalation of C5b-9 complexes would not cause colloid osmotic disruption although there is a small osmotic potential across the membrane due to the presence of fixed polyvalent anions in the periplasmic space (Stock et al., 1977).

Logarithmic phase cultures of rough stains of Escherichia coli and Salmonella rapidly lose viability following exposure to suitable concentrations of human serum; removal of lysozyme from serum makes little difference to the rate of viability loss (Martinez and Carroll, 1980; Taylor and Kroll, 1983) but facilitates the study of the mechanism of the complement-mediated killing process in the absence of bacteriolysis. We have studied the mechanism of killing of enterobacteria lysozyme-free serum using in the main a complement-susceptible isolate, E. coli LP1092, from a patient with a persistent infection of the urinary tract. This strain synthesizes an R-type lipopolysaccharide with a complete R2 core (Taylor, 1974) and substantial amounts of a polysaccharide K antigen containing ribose and 3-deoxy-D-manno-octulosonic acid (Taylor, 1978; Jennings et al., 1982; 1984). It is killed by 6-20% lysozyme-free human serum at a constant rate with a tenfold reduction in viability approximately every 20 min; killing is always preceded by a 120 min lag period during which there is little change in the viable count (Taylor and Kroll, 1983). Our studies placed particular emphasis on the sequence of events occurring at the cell envelope prior to the onset of viability.

The cell envelope following exposure to complement

E. coli LP1092 cells are killed predominantly by an antibody-dependent classical pathway mediated mechanism (Taylor and Kroll, 1983) that effects release of degradative enzymes such as alkaline phosphatase from the periplasmic space but leaves a cytoplasmic membrane that continues to function as a barrier to the diffusion out of the cell of cytoplasmic enzyme markers such as β-galactosidase (Kroll et al., 1983). Experiments designed to determine the locus of deposition of components of the C5b-9 complex onto the bacterial surface have been performed by exposing E. coli LP1092 cells to 20% lysozyme-free pooled human serum and separating membranes from bacteria, disrupted at various time intervals, on isopycnic sucrose density gradients (Fig. 1).

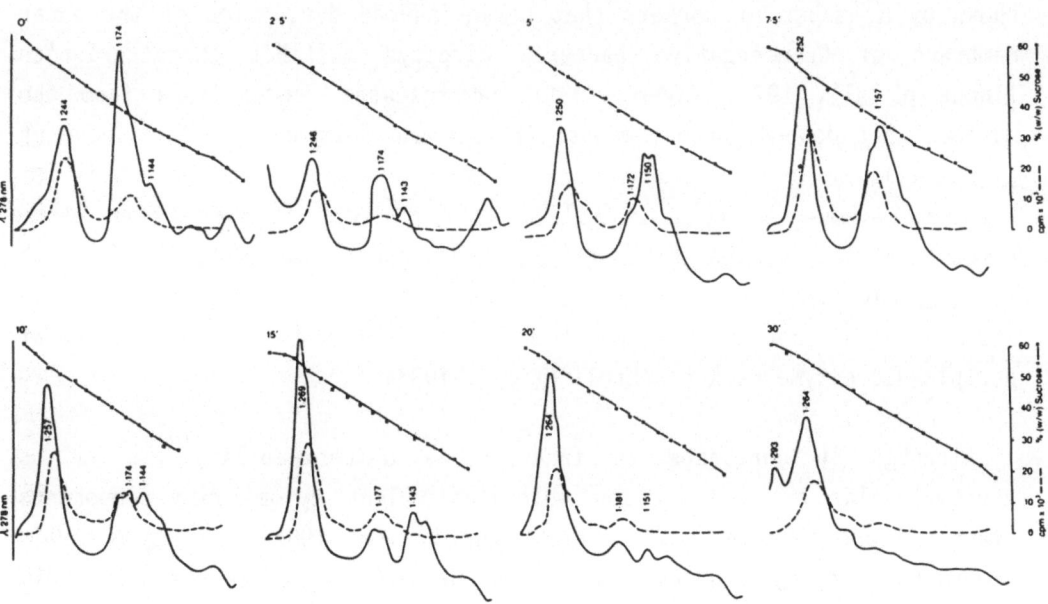

Fig. 1.   Separation of membranes from [2-³H]glycerol-grown E. coli LP1092 cells exposed to 20% lysozyme-free human serum for the indicated time periods. Membranes derived from osmotically-shocked cells were applied to 15-65% w/w sucrose density gradients and centrifuged to equilibrium; values alongside each peak represent densities in g/cm³.

Complement exposure resulted in prompt and rapid release of material from the cell envelope, as evidenced by a reduction in the amount of outer membrane (the denser of the two peaks in Fig. 1, time 0) recovered on sucrose density gradients and by the release into the reaction milieu of 30-40% of the total membrane phospholipid. In contrast, very little radiolabelled phospholipid was released from the surface of serum resistant E. coli undergoing complement-mediated attack or from E. coli 17, a mutant lacking both cytoplasmic and outer-membrane located phospholipase A, strongly suggesting that one or both of these enzymes participates in release of membrane material from susceptible strains (Taylor and Kroll, 1984). No transient reduction in outer membrane recoverability was seen with E. coli 17 (Kroll et al., 1984). It has been shown by a number of workers that serum-induced disruption of the outer membrane of Gram-negative bacteria displays multihit characteristics (Inoue et al., 1977; Mayer, 1981) and releases phospholipid from the surface in a dose-dependent manner (Wilson and Spitznagel, 1971; Inoue et al., 1977). It is clear, however, that this release is not essential for killing, because E. coli is killed by complement as rapidly as other enterobacterial strains (Kroll et al., 1984) and serum-susceptible, smooth E. coli have been shown to be efficiently killed by complement in the absence of any damage to the outer membrane sufficient to cause release of periplasmic enzyme markers (Kroll et al., 1983).

Although it continues to function as a permeability barrier to macromolecular but not to low molecular weight cytoplasmic components (Martinez and Carroll, 1980; Wright and Levine, 1981), the cytoplasmic membranes of E. coli LP1092 (Fig. 1) and other susceptible strains (Kroll et al., 1983; 1984) undergo modification during the course of the bactericidal reaction that reduces recoverability on sucrose density gradients. Determination of the distribution of markers for both membranes in fractionated gradients has established that this effect is not due to complement-induced fusion of membrane vesicles (Kroll et al., 1983) but rather to limited degradation of component phospholipids, perhaps by bacterial phospholipases, to the extent in situ they retain bilayer form and some function but that once released from the cell envelope by osmotic lysis they are subsequently unable to form vesicles (Kroll et al., 1984). It has recently been established that E. coli LP1092 cells under complement attack translocate phospholipid from the cytoplasmic to the outer membrane (Taylor and Kroll, 1985) in the absence of concomitant replacement of cytoplamsic membrane phospholipid from precursor pools (Kroll et al., 1983); this is likely to be an additional factor contributing to reduction in cytoplasmic membrane recovery.

Individual fractions from sucrose density gradients were analysed by fused rocket immunoelectrophoresis using monospecific antisera raised against C3 and against C9-specific neoantigens (Bhakdi et al, 1983). Binding of C3b to the outer membrane could be detected during the first 2.5 - 5 min of exposure to complement and the amount of C3b bound increased throughout the first 20 min of the reaction. At no stage could binding to cytoplasmic membrane fractions be detected (Kroll et al., 1983). Binding of C5b-9 to the outer membrane could be detected immediately prior to the onset of bacterial viability loss and binding to this membrane continued to increase during the killing phase of the reaction. No C5b-9 complexes could be detected using C9-neoantigen antiserum in any fractions representing cytoplasmic membrane at any time during the course of the reaction (Kroll et al., 1983). When fractions were examined by electron microscopy, C5b-9 lesions could be visualized on outer membrane vesicles obtained from E. coli LP1092 and other susceptible strains that had been incubated with lysozyme-free human serum for 10 min or more (Kroll et al., 1984). In contrast, no C5b-9 complexes were seen on cytoplasmic membrane vesicles obtained from cells exposed to complement for periods of up to 60 min. It is unlikely that complement associated loss of cytoplasmic membrane recoverability was to due to direct binding of C5b-9 complexes to this membrane, because complexes could not be found in association with partially degraded phospholipids in osmotic lysis supernatants (Taylor and Kroll, 1984). Furthermore, treatment of purified cytoplasmic membrane vesicles did not result in loss of recoverability of this material upon sucrose density gradient centrifugation (Kroll et al., 1983).

Taken together, these studies lend strong support to the hypothesis that C5b-9 complexes form exclusively on the outer membrane and effect lethal damage from this locus. With certain rough strains possessing an outer membrane-associated phospholipase A, there is substantial complement-mediated phosphilipid removal from the surface of the target cell, damage which results in the release of proteins and outer molecules from the periplasmic space but is not in itself sufficient to effect lethal killing.

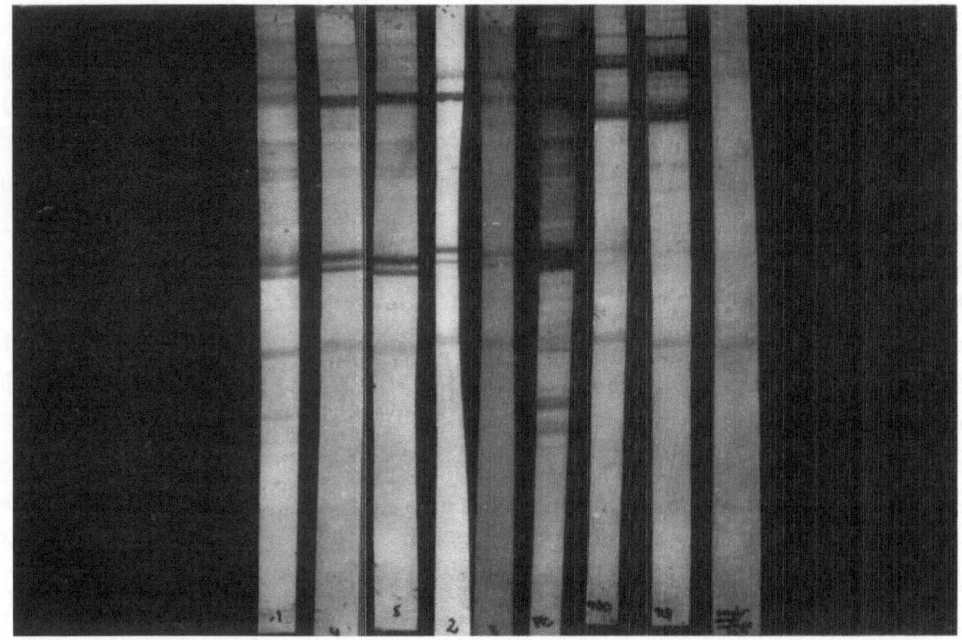

Fig. 2.   Immunoblots of SDS-PAGE OM from E. coli LP1092 cells treated for
          30 min with lysozyme-free serum. SDS-PAGE gels were loaded with
          approximately 10μg protein per lane. Strips were reacted with
          (from left to right): -anti-C9 monoclonal antibodies BK2-97-1,
          BK2-97-4, BK2-97-5, BK2-97-2, BK2-97-3, goat anti-C9 polyclonal
          (Miles Scientific Ltd., Elkhart, IN), anti-C3d (alpha chain),
          anti-C3b (both from Biotech Research Laboratories, Inc.,
          Rockville, MD) and control with no first antibody. All monoclonal
          antibodies belonged to Ig subclass IgG_1 with the exception of
          BK2-97-3 (IgG_{2_a}) and anti-C3d (IgG_{2_a}). Antibody binding to
          nitrocellulose-bound proteins was detected using peroxidase
          anti-mouse or anti-goat IgG (Miles-Yeda Ltd., Israel).

Lethal membrane damage, such as dissipation of the electrochemical
potential across the cytoplasmic membrane (Dankert and Esser, 1986), may
occur when domains at the apolar terminus of the C5b-9 cylinder make
contact with the outer surface of the cytoplasmic membrane zones of
transient adhesion between the two bilayers.

Effect of complement on bacterial metabolic parameters

It has been recognized that the cellular metabolic state can affect the degree of susceptibility of a target bacterial population to complement attack (Michael and Brown, 1959; Melching and Vas, 1971). Thus, susceptible enterobacteria in the mid-logarithmic phase of growth are significantly more serum sensitive than when in log or stationary phases (Rowley and Wardlaw, 1958). Whether growth rate per se, or some other characteristic of cells in a certain metabolic state, is responsible for interaction resulting in optimal rates of killing is not known with any certainty, although it may be significant that parameters most closely linked with cell division, such as DNA, RNA and protein synthesis, do not appear to be interrupted in the early stages of the lethal process (Melching and Vas, 1971; Martinez and Carroll, 1980; Taylor and Kroll, 1983).

Some insight into the nature of the metabolic events necessary for rapid serum killing has been gained using inhibitors and uncouplers of oxidative phosphorylation (Griffiths, 1974; Taylor and Kroll, 1983). Dinitrophenol, carbonylcyanide m-chlorophenylhydrazone and cyanide protect Gram-negative bacteria against the potentially lethal action of complement. Figure 3 shows that 5mM KCN was able to almost completely inhibit the bactericidal effect of 6.6% human serum on $E.$ coli LP1092 cells; when added to the cells 8 min into the period of serum exposure in the serum killing process, it required an input of bacterially generated energy. This is clearly substantiated by the fact that removal of low molecular weight energy sources from serum by dialysis reduces markedly the serum bactericidal properties, which can be completely restored by adding glucose or other suitable metabolisable substrates back to the serum (Taylor and Kroll, 1983). Both serum sensitive and resistant bacteria, upon exposure to human serum, convert substrates in serum to ATP during the early stages of the serum bactericidal reaction. In sensitive but not resistant strains of $E.$ coli this ATP is hydrolysed intracellularly during the active killing phase of the reaction, contributing to an as yet unidentified energy-dependent event essential for rapid killing (Kroll et al., 1983).

This requirement for bacterially-generated energy during the course of the reaction may be related to recent observations made by Dankert and Esser (1986; 1987). Formation of the C5b-8 complex on the target bacterial surface caused a transient collapse of the electrochemical potential across the cytoplasmic membrane; addition of C9 to these C5b-8-bearing cells led to irreversible loss of this potential and to cell death. A C9-derived fragment may be sufficient to cause de-energisation and is reminiscent of data obtained in studies of the mechanism of action of membrane-active colicins. Intracellular pools of ATP are depleted as the cells make a vain attempt to regenerate concentration gradients across the cytoplasmic membrane. Thus, inhibition of respiration-linked active transport systems and other metabolic changes noted in cells undergoing complement-mediated attack are a consequence of a rapid and drastic reduction of the membrane potential. Studies designed to address these key questions of complement action are underway in a number of laboratories.

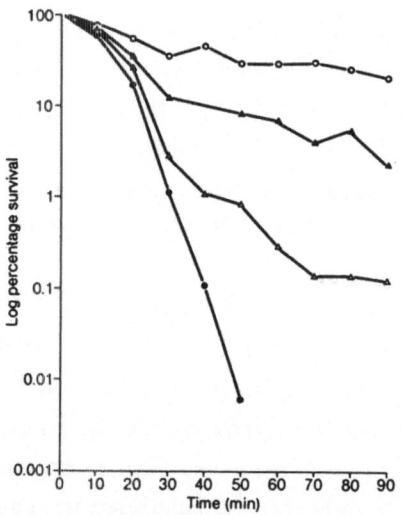

Fig. 3.   Kinetics of killing E. coli LP1092 by 6.7% human serum. Symbols: ⊕, human serum; O, serum + 5mM KCN added at the beginning of the experiment; Δ + 5mM KCN added after 18 min.

References

Bhakdi, S, Muhly, M. and Roth, M. Preparation and isolation of specific antibodies to complement components. In: Methods in Enzymology, Langone J.J. and Van Vunakis H. (eds), Vol 93. Academic, New York, pp. 409-420 (1983).

Born, J. and Bhakdi, S. Does complement kill E. coli by producing trans-mural pores? Immunology 59: 139-145 (1986).

Cavard, D., Regnier, P. and Lazdunski, C.J. A protease as a possible sensor of environmental conditions in E. coli outer membrane. Mol. Gen. Genet. 188:508-512 (1982a).

Cavard, D., Regnier, P., and Lazdunski, C.J. Specific cleavage of coli-cin A by outer membrane proteases from sensitive and insensitive strains of E. coli. FEMS Microbiol. Lett. 14:285-289 (1982b).

Dankert, J.R., and Esser, A.F. Complement-mediated killing of E. coli: dissipation of membrane potential by a C9-derived peptide. Biochemistry 25:1094-1100 (1986).

Dankert, J.R., Esser, A.F. Bacterial killing by complement. C9-mediated killing in the absence of C5b-8. Biochem. J. 244: in press (1987).

Griffiths, E. Metabolically controlled killing of Pasterurella septica by antibody and complement. Biochem. Biophys. Acta 362:598-602 (1974).

Inoue, K.,, Kinoshita, T., Okada, M., and Akiyama, Y. Release of phos-pholipids from complement-mediated lesions on the surface structures of E. coli. J. Immunol. 119:65-72 (1977).

Jennings, H.J., Rosell, K-G., Johnson, K.G. Structure of the 3-deoxy-D-manno-octulosonic acid-containing polysaccharide (K6 antigen) from E. coli LP1902. Carbohydr. Res. 105:45-56 (1982).

Jennings, H.J., Roy, R., and Williams, R.E. Chemical modification and serological properties of the 3-deoxy-α-D-manno-2-octulosonic acid-containing polysaccharide from E. coli LP1902. Carbohydr. Res. 129:243-255 (1984).

Joiner, K.A. Studies on the mechanism of bacterial resistance to comple-ment-mediated killing and on the mechanism of action of bactericidal antibody. Curr. Top. Microbiol. Immunol. 121:99-133 (1985).

Kim, S-H., Carney, D.F., Hameer, C.H., and Shin, M.L. Nucleated cell killing by complement: effects of C5b-9 channel size and extracellular Ca^{2+} on the lytic process. J. Immunol. 138:1530-1536 (1987).

Kroll, H-P., Bhakdi, S., and Taylor, P.W. Membrane changes induced by exposure of E. coli to human serum. Infect. Immun. 42:1055-1066 (1983).

Kroll, H-P., Voigt, W-H., Taylor, P.W. Stable insertion of C5b-9 comple-ment complexes into outer membrane of serum-treated, susceptible E. coli cells as a prerequisite for killing. Zentralbl Bakteriol Mikrobiol Hyg A 258:316-326 (1984).

Lugtenberg, B., and van Alphen, L. Molecular architecture and function-ing of the outer membrane of E. coli and other Gram-negative bacteria. Biochim. Biophys. Acta 737:51-115 (1983).

Martinez, R.J., and Carroll, S.F. Sequential metabolic expressions of the lethal process in human serum-treated E. coli: role of lysozyme. Infect. Immun. 28:735-745 (1980).

Mayer, M.M. Membrane damage by complement. Johns Hopkins Med J 148:243-258 (1981).

Mayer, M.M. Complement: Historical perspectives and some current issues. Complement 1:2-26 (1984).

Melching, L., and Vas, S.I. Effects of serum components on Gram-negative bacteria during bactericidal reactions. Infect. Immun. 3:107-115 (1971).

Michael, J.G., and Braun. Modification of bactericidal effects of human sera. Proc. Sco. Exp. Biol. Med. 102:486-490 (1959).

Rowley, D., and Wardlae, A.C. Lysis of Gram-negative bacteria by serum. J. Gen. Microbiol. 28:529-533 (1958).

Stock, J.B., Rauch, B., and Roseman, S. Periplasmic space in _Salmonella typhimurium_ and _E. coli_. J. Biol. Chem. 252:7850-7861 (1977).

Taylor, P.W. An unusual acidic polysaccharide produced by a rough strain of _E. coli_. Biochem. Biophys. Res. Commun. 61:148-154 (1974).

Taylor, P.W. Biosynthetic and structural studies on a 3-deoxy-D-manno-octulosonic acid-containing acidic polysaccharide produced by a urinary strain of _E. coli_. Proc. IX Intern. Symp. Carbohydr. Chem. pp. 411-412 (1978).

Taylor, P.W. Bactericidial and bacteriolytic activity of serum against Gram-negative bacteria. Microbiol. Rev. 47:46-83 (1983).

Taylor, P.W., and Kroll, H-P. Killing of an encapsulated strain of _E. coli_ by human serum. Infect. Immun. 39:122-131 (1983).

Taylor, P.W., and Kroll, H-P. Interaction of human complement proteins with serum-sensitive and serum-resistant strains of _E. coli_. Molec. Immunol. 21:609-623 (1983).

Taylor, P.W., and Kroll, H-P. Effect of lethal doses of complement on the functional integrity of target enterobacteria. Curr. Top. Microbiol. Immunol. 121:135-158 (1986).

Wilson, L.A., Spitznagel, J.K. Characteristics of complement-dependent release of phospholipid from _E. coli_. Infect. Immun. 4:23-28 (1971).

SALMONELLA AND COMPLEMENT: THE CRITICAL INFLUENCE OF O-POLYSACCHARIDE WITHIN LPS

K. Joiner[1] V. Jiménez-Lucho[2], N. Grossman[2], J. Foulds[2], M. Frank[2] and
L. Leive[2]

[1]Laboratory of Clinical Investigation
National Institute of Allergy and Infectious Diseases
National Institutes of Health
Bethesda, Maryland 20892

[2]Laboratory of Structural Biology
National Institute of Diabetes, Digestive and Kidney Disease
National Institutes of Health
Bethesda, Maryland 20892

This paper is dedicated to the memory of Loretta Leive

Virulent enteric gram negative bacteria generally produce lipopolysaccharide molecules bearing O-antigen side chains. This scenario applies despite the fact that nearly all of the in vitro and in vivo toxic effects of LPS are ascribed to the lipid A portion of the molecule. Thus, there is an apparent survival advantage to the organism in producing LPS with O-polysaccharide side chains. Although the presence of O-polysaccharide confers some advantage to the cell in fending off harmful exogenous hydrophobic molecules, the overall fluidity and permeability of the outer membrane is not altered substantially merely as a consequence of O-antigen addition (1). Certainly, one survival advantage to cells bearing O-antigen resides in their capacity to evade the host humoral immune system. The presence within virulent organisms of O-antigen serotypes not commonly encountered insures that the native host will not have developed specific antibody to the organism at the time of invasion. A second factor, and the subject of this paper, is that O-antigen within LPS is the major determinant influencing the interaction of the complement system with the gram negative cell wall. In fact, it could be argued that this is the most clearly demonstrated consequence of O-antigen within LPS. The theme of this manuscript is that both the carbohydrate composition of the O-antigen, which dictates the extent of opsonic C3 fragment deposition by the alternative complement pathway, and the size and distribution of

NATO ASI Series, Vol. H24
Bacteria, Complement and the Phagocytic Cell
Edited by F.C. Cabello und C. Pruzzo
© Springer-Verlag Berlin Heidelberg 1988

O-antigen, which determines the susceptibility to direct killing by the alternative complement pathway, are major virulence determinants for LPS within enteric organisms.

Another paper in this symposium by Dr. Helen Mäkelä describes ongoing work from her laboratory studying three isolates of Salmonella which vary only in the O-antigen composition of their LPS. By transformation at the rfb locus, Valtonen and Mäkelä replaced the original O-4,12 serotype O-antigen of Salmonella typhimurium with either the O-9,12 serotype of S. enteritidis or the O-6,7 serotype of S. montevideo, leaving all other outer and inner membrane constituents identical (2,3,4). When these three isogenic strains were tested for virulence in the mouse after intraperitoneal challenge, there was a substantial difference in virulence, with LD_{50} of $5x10^7$ for O-6,7, $2x10^6$ for O-9,12 and $3x10^5$ for O-4,12, respectively.

These Salmonella isolates have been the subject of extensive study to define the mechanism for their differences in virulence. Liang-Takasaki, Mäkelä and Leive (5) compared the rate and extent of phagocytosis of the strains by thioglycolate elicited mouse peritoneal macrophages as well as by the macrophage-like cell line J774 (6). The rate of phagocytosis showed an inverse relationship with the virulence of the bacteria: O-6,7 which is the least virulent, was ingested the fastest; O-4,12, which is the most virulent, was ingested the slowest, and O-9,12, which is intermediate in virulence, was ingested at an intermediate rate. Interestingly, when the uptake rate at different concentrations of bacteria was plotted according to Lineweaver-Burk kinetics, the Vmax for ingestion did not differ between organisms whereas the Km for half maximal uptake varied proportionally to the virulence of the organism, suggesting that the differences in uptake were due to differences in affinity of attachment to the macrophages, with equal rates of ingestion once attachment occurred. Saxén, Riema and Mäkelä (7) have recently demonstrated that this difference in rate of phagocytosis operates for peritoneal but not for hepatic macrophages, thus explaining why the organisms have a similar LD_{50} when injected intravenously, since organisms are cleared by hepatic and splenic macrophages in this circumstance. Mäkelä's group has therefore presented a coherent model explaining the difference in virulence of the organisms when injected intraperitoneally. Greater killing of O-6,7 than of O-9,12 than of O-4,12 by peritoneal macrophages after intraperitoneal injection dictates the number of organisms entering the bloodstream and becoming trapped in the reticuloendothelial macrophages of the liver and spleen, where all three isolates multiply at an equal rate.

Experiments from Loretta Leive's laboratory have focused on the mechanisms by which the O-antigen differences influence the rate of phagocytosis. Liang-Takasaki, Grossman and Leive (8) demonstrated that phagocytosis was complement dependent and antibody independent and was mediated through the alternative pathway, since the differential uptake occurred in serum deficient in the classical pathway component C4 as well as in normal serum. The least virulent organism, O-6,7, consumed the most C3 from serum; the most virulent organism, O-4,12, consumed the least C3, and O-9,12 was intermediate in both respects. These differences in extent of complement consumption were exactly paralleled by differences in extent of C3 deposition on the bacterial surface, as assessed by binding of purified, radiolabelled C3. Grossman and Leive (9) then formally proved that only the LPS differences between the organisms were responsible for the differences in complement activation and deposition, and not other unknown differences among the strains. LPS was purified from the three organisms and used to coat erythrocytes, which were then tested for the capacity to consume C3 from serum. In the absence of any other bacterial constituent, the LPS conferred upon the erythrocytes exactly the same relative ability to consume serum C3 and to bind radiolabelled C3 as occurred with whole bacteria. Furthermore, this differential activation was independent of the length of the O polysaccharide. Salmonella species are known to contain LPS molecules with O polysaccharide chains varying in length from 0 to more than 40 repeating units. When different fractions of purified LPS were used to coat erythrocytes, each of them with different O polysaccharide side chain lengths, all lengths of O polysaccharide from 2 up to more than 40 O polysaccharide repeating units consumed C3 equally well. Thus, differences in complement activation and deposition on the bacteria are determined solely by differences in the primary carbohydrate structure of the O polysaccharide and not by differences in its length or density.

A more detailed analysis was then initiated to identify the form of C3 on the surface of the three strains. Our goal was to understand the mechanism of differential complement activation as a function of carbohydrate composition. In the initial experiments (10), the form of C3 bound to O-6,7, O-9,12 and O-4,12 after incubation in serum, was tested by two separate methods: a) analysis by SDS-PAGE of [125I]C3 chain structure on the bacterial surface after incubation of organisms in serum bearing [125I]C3; b) the rate and extent of binding of conglutinin, which attaches only to iC3b, was examined on bacteria incubated in serum. By both approaches, we demonstrated that the relative ratio of the hemolytically active C3b fragment to the hemolytically inactive iC3b fragment was identical on all three strains at all time points measured, with approximately 35% of bound C3 present as C3b. The mechanisms controlling this C3 fragmentation were studied further, as described below.

The process of C3 deposition on any surface involves the initial binding of C3b or a "C3b-like" molecule followed either by 1) formation of a C3 convertase which amplifies the effect by depositing further C3 or 2) cleavage of C3b to iC3b by factors H and I, a process which blocks further C3 deposition (Figure 1).

Thus, it is possible to partition the analysis of C3 deposition into three related but separate steps: 1) initial deposition of C3b, which depends upon the efficiency of binding of nascent C3b to expose OH or NH_2 groups on the bacterial surface, 2) amplification of C3 deposition, which hinges on the efficiency with which factor B binds to C3b, leading ultimately to formation of the stabilized C3 convertase PC3bBb, and 3) degradation of C3b, which depends on the affinity of the interaction of factor H with bound C3b, leading to I-dependent cleavage of C3b to iC3b and a block in further C3 deposition.

Victor Jiménez-Lucho systemically analyzed and compared each of these steps for the O-6,7, O-9,12 and O-4,12 strains (11). The initial deposition step was examined by comparing binding of C3b to the 3 strains during fluid phase, trypsin-mediated C3 cleavage, using increasing inputs of C3. As anticipated, the amount of C3 deposited on each strain was proportional to the amount of C3 added to the reaction mixture.

Surprisingly, however, the initial deposition of C3b was 3-4 times greater at each C3 input on 0-6,7 and 0-4,12, than on 0-9,12.

C3 DEPOSITION AND CLEAVAGE
Analysis of Sequential Steps

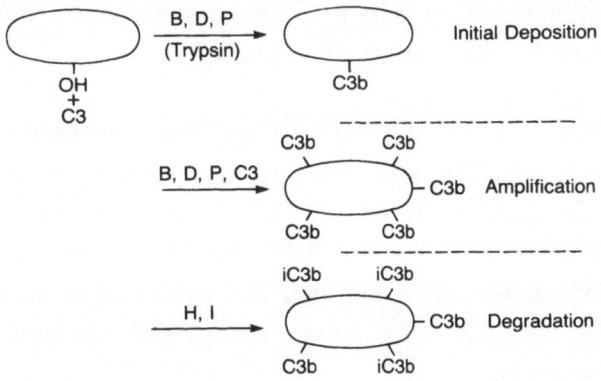

Fig. 1.   Amplification of C3 Deposition: Analysis of Sequential Steps. The process of C3 deposition and cleavage on any surface can be divided operationally into three separate steps, each of which is amenable to experimental testing. Initial deposition of C3b via a fluid phase C3 convertase (which can also be mimicked by trypsin cleavage of C3) depends on the availability of exposed hydroxyl groups of LPS sugars to form a covalent ester linkage. Amplification of the initial C3b deposition occurs as a consequence of C3 convertase formation by factors B,D,P and C3. The most critical step in this reaction is the interaction of B with C3b. The final step is one of C3b degradation, which is most influenced by the affinity of H for bound C3b. Although these processes are all occurring simultaneously and although the amplification and degradation processes compete with one another, they can be analyzed as separate steps.

Jiménez-Lucho next compared the amplification step of C3 deposition for the three bacterial strains. Cells bearing C3b initially deposited with trypsin were incubated with sequential cycles of B, D, and P, followed by

C3. When the extent of amplification of C3b deposition over the previous step was analyzed, a consistent pattern emerged. Amplification was 2 to 3.4 times higher for cells bearing 0-6,7 or 0-9,12 LPS than for cells bearing 0-4,12 LPS. In contrast, the amplification on cells with 0-9,12 LPS was as efficient as that on cells with 0-6,7 LPS.

These differences in extent of C3 deposition during amplification were mirrored in the analysis of [125]I B binding to C3b on the three strains. Binding curves showed that only 25% of C3b on cells with 0-4,12 LPS bound C3b with an affinity of $3-5 \times 10^6 M^{-1}$, in comparison to 90% of C3b on 0-6,7 or 0-9,12 binding B with this high affinity.

These results, which suggested that the intermediate level of C3 binding on 0-9,12 results from low initial deposition rather than from an alteration of the amplification process, were confirmed by studies showing that total binding of C3b on 0-9,12 equalled that on 0-6,7 after 3-4 cycles of amplification. In contrast, the differences in C3b deposition on cells carrying 0-4,12 LPS were maintained throughout the entire amplification process.

Finally, Jiménez-Lucho examined the degradation step of C3 binding by testing binding of factor H to the Salmonella strains bearing C3b. Binding affinities ($K_a = 7 \times 10^7 M^{-1}$) and maximum percentage of C3b recognized by H (40-60%) were similar on all three strains, indicating that the degradation step was similar, confirming serum experiments showing an equivalent fraction of C3b/iC3b on all three strains.

A model illustrating our current understanding of how complement interacts with the three Salmonella strains is shown in Figure 2. Although the complement receptors which are involved in the process of phagocytosis have not been identified in this system, we presume that either CR1 (the C3b receptor) or CR3 (the iC3b receptor) or both are involved in the interiorization process, an issue which we are currently evaluating. Of particular importance and interest, especially in the context of the work from Mäkelä's laboratory, will be a comparison of the effects of different receptors in influencing the fate of Salmonella after phagocytosis by different cell populations.

145

A second major way in which LPS is involved in evasion of host defense is in preventing direct complement mediated killing of the bacterial cell. Serum resistance, or the capacity to evade direct killing by the cytolytic C5b-9 terminal attack complex of complement, is a clear virulence determinant for gram negative organisms. LPS is the microbial virulence determinant which was first associated with serum resistance, and remains the constituent most clearly related to this trait. Smooth organisms bearing 0-antigen are generally serum resistant where rough strains lacking 0-antigen are invariably serum sensitive (12,13). However, some isolates containing 0-antigen may be killed in normal human serum (14). In these circumstances, serum sensitivity generally, but not invariably, correlates with the presence of lesser amounts of 0-antigen than are present in the strains capable of evading complement lysis. Overall, it is reasonable to conclude that serum resistance is a function not only of the presence of 0-antigen but also of the amount of 0-antigen. Nonetheless, it has been difficult until recently to relate differences in LPS profile with the mechanism of serum resistance.

CARBOHYDRATE COMPOSITION OF O-Ag in LPS CONTROLS EXTENT OF OPSONIC C3 FRAGMENT DEPOSITION

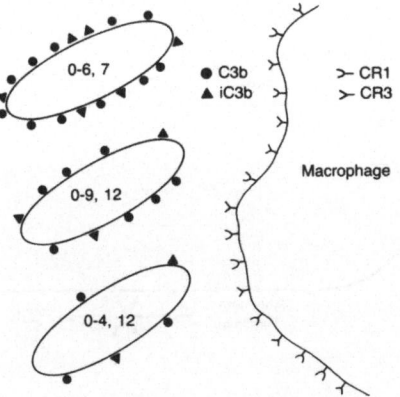

Fig. 2.   Model for C3 deposition on <u>Salmonella</u> strains 0-6,7, 0-9,12 and 0-4,12. The putative receptors involved in the phagocytosis process are shown but have not yet been examined in this system.

In order to address some of these issues, we undertook a series of experiments to investigate the mechanism of 0-antigen dependent serum resistance in gram negative enteric organisms. In the initial studies (15,16), we compared complement activation and binding on an isogenic pair of organisms consisting of a smooth, serum sensitive <u>Salmonella minnesota</u> strain (S218) and the deep rough, serum sensitive Re 595 mutant of this parent strain. Neither of these organisms bears a polysaccharide capsule nor outer membrane proteins known to be associated with serum resistance and thus, affords a model system in which to study the mechanism by which 0-antigen confers serum resistance.

Our initial studies examined the kinetics and extent of complement consumption and binding by the isogenic paris varying in serum sensitivity. Experiments were designed to distinguish between three conceptual mechanisms for serum resistance; 1) inefficient complement activation, 2) a block in the complement cascade after initial activation, or 3) a failure of the formed C5b-9 complex to insert effectively into the outer membrane (Figure 3).

MECHANISMS OF SERUM RESISTANCE IN GRAM NEGATIVE BACTERIA

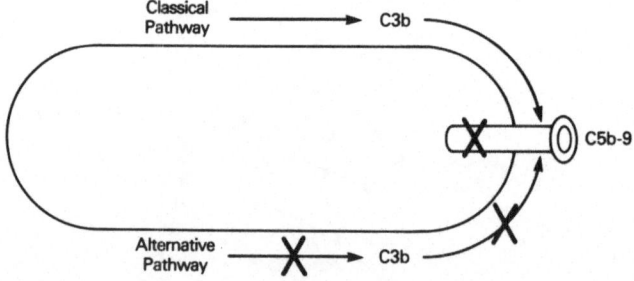

Fig. 3. Potential Mechanism for Serum Resistance in Gram Negative Bacteria. Organisms may be resistant to indirect complement mediated lysis based on inefficient complement activation with failure of C3b generation, due to a block in C5b-9 formation after formation of C3b, or due to a failure of the formed C5b-9 complex to effectively insert into and damage critical sites on the bacterial outer or inner membrane.

Serum resistant and serum sensitive Salmonella minnesota isolates activated complement to an equivalent extent, since they consumed similar amounts of C3 during incubation in serum. Furthermore, C3 binding on the serum resistant strain exceeded that on the serum sensitive isolate. These results excluded the first possibility (i.e., inefficient activation) mentioned above. The second possibility was ruled out by the surprising observation that utilization of terminal complement components was much more extensive for the resistant than for the sensitive strain. Direct binding experiments with purified, radiolabelled terminal complement components indicated that the C5b-9 complex was rapidly deposited on both organisms. Whereas the complex was stably bound on the sensitive strain, C5b-9 was spontaneously and rapidly released from the resistant organism. Thus, it appeared that the third possibility above was the most likely explanation for serum resistance in the Salmonella minnesota strains we were studying.

Further experiments using different strains of Salmonella and E. coli, and different approaches to this question have further clarified the nature of complement interaction with the surface resistant organisms. Results for C5b-9 binding similar to those described above for S. minnesota were obtained with serum resistant, unencapsulated strains of E. coli 0111 (17) as well as with the 0-6,7, 0-9,12 and 0-4,12 strains used in the phagocytosis studies, all of which are serum resistant. The mechanism of C5b-9 release from the cell surface was ascribed to the failure of C5b-9 to insert into hydrophobic domains of the outer membrane, in comparison to its effective insertion into hydrophobic domains of the rough, serum sensitive strains. It is known that as C5b-9 forms, with progressive addition of C7, C8 and multiple C9 molecules per complex, the forming complex binds more detergent and phospholipid as it acquires hydrophobic binding sites (18). In the absence of an insertion site for these hydrophobic domains, the complex is released.

Alterations in the LPS composition and phenotype of the outer membrane alter serum sensitivity. The unstable binding of C5b-9 to resistant cells is rendered stable by pre-treatment of cells with EDTA prior to serum incubation. This procedure was shown by Loretta Leive to remove a large

subset of LPS from the outer membrane (19) and it is also known to render
resistant cells sensitive to the serum bactericidal reaction. We also
studied the interaction of complement with serum resistant mutants of E.
coli 0111 which were derived by serial passage of a serum sensitive strain
in the presence of bactericidal IgG and normal human serum (20). When
outer membranes from these mutants were examined for interaction with
C5b-9 (21), we found that the unstable association of C5b-9 with the outer
membrane was a function of the amount of LPS/cell, the average length of
the 0-polysaccharide side chain and the percentage of LPS with 0-antigen
polysaccharide covering the lipid A core oligosaccharide moeity. Although
we were thus able to clearly show that 0-antigen rendered C5b-9 binding
unstable, neither the exact mechanism for this observation nor the precise
0-antigen configuration necessary to mediate this effect were clear.

In order to more precisely define the role of 0-antigen polysaccharide
in preventing C5b-9 insertion, we sought to identify the exact site of
complement activation on the bacterial surface. We hoped to distinguish
between the possibility that smooth LPS might provide a steric barrier to
complement attack, as suggested nearly 20 years ago, and the possibility
that the reduced fluidity of the smooth gram negative outer membrane might
preclude C5b-9 insertion even in the absence of a steric barrier. We
initially chose to identify the site of C3 deposition, since a) C3 is an
obligatory component of the C5 convertase and, hence, determines the site
of C5 cleavage and C5b-9 deposition and b) C3 binds covalently to the
activating surface, affording a powerful tool for immunochemical
localization studies. We first showed that C3 bound almost exclusively to
LPS during complement activation by whole cells in serum (22). More
importantly for studies of serum resistance, experiments indicated that C3
bound only to the small subset of LPS molecules (approximately 2-3% of the
total) bearing the longest 0 polysaccharide side chains (23). Using a
strain of Salmonella in which the LPS coverage could be manipulated by
regulating the supply of mannose in the growth media, we found that C3
bound to molecules with shorter 0 polysaccharide side chains as the extent
of coverage was decreased, concomitant with an increase in serum
sensitivity.

These experiments led to the concept that the C3 and hence C5b-9 deposition on smooth, serum resistant <u>Salmonella</u> strains was sterically constrained from interacting with hydrophobic domains of the outer membrane. We next undertook experiments to analyze the LPS topography and configuration required to mediate serum resistance (24). For these experiments, we used a mutant strain of <u>Salmonella</u> <u>montevideo</u>. This mutant lacks the enzyme phosphomannoseisomerase and, hence, requires added mannose to make its O-antigen, which contains predominantly mannose. The strategy of the experiments was to vary the supply of mannose during growth, in order to vary the degree and pattern of O-antigen substitution.

In the first series of experiments, cells were grown continuously with concentrations of mannose varying downward from 5.0 mM, which we showed was optimal for O-antigen synthesis, to 2.5, 1.0, 0.5, 0.25 and 0 mM mannose (condition A). Cells grown under these varying mannose concentrations were tested for susceptibility to killing during incubation in 5% normal human serum. As expected, cells grown in the presence of 5.0 mM mannose were totally resistant, whereas cells grown without mannose were sensitive. Of note, cells grown at 2.5 mM mannose were as resistant to killing as were cells grown at 5.0 mM mannose, but there was a dramatic increase in serum susceptibility following growth in 1.0 mM mannose. The LPS profile of cells grown under each condition was assessed by SDS-PAGE. From quantitative densitometry of LPS autoradiograms we calculated: i) the average number of O-antigen units per LPS molecule, ii) the percentage of molecules with greater than 14 O-antigen units, iii) the percentage of LPS molecules bearing O-antigen, and iv) the median number of O-antigen units/molecule in the peak of long chain LPS containing >14 O-antigen units. As expected, growth with increasing concentrations of mannose led to an increase in the average number of O-antigen units/molecule, in the percentage of long chain LPS with >14 O-antigen units, and in the percent of LPS molecules bearing O-antigen. However, in contrast to the sudden change seen in serum sensitivity as mannose input varied, these parameters of LPS coverage all changed more gradually. Surprisingly, the median number of O-antigen units/molecule in the peak of long chain LPS did not change with alteration in mannose input, indicating a preferential synthesis of long chain LPS by cells restricted in mannose.

We next compared two additional methods for limiting mannose supply during growth: in condition B) cells were grown without mannose for at least 15 generations, then optimal mannose was added and cells were maintained in logarithmic growth until harvested; in condition C) cells were grown with 5.0 mM mannose for at least 15 generations, then removed from mannose containing media by rapid filtration, suspended in medium lacking mannose and growth continued. Our rationale for comparing these two conditions with the cells grown continuously with varying limited mannose (condition A) was that any particular condition chosen might alter the surface topography of LPS molecules bearing long O-antigen side chains in different ways. Newly synthesized LPS molecules are inserted into the outer membrane at discrete points on the cell surface, and diffuse laterally with continued incubation at 37°C. We would predict in our experiments that cells grown under condition A (continuous, limited mannose) would have a relatively uniform distribution of LPS over the surface, whereas cells grown under conditions B or C (introduction or removal of mannose) may have a heterogeneous distribution of LPS on the cell surface. Therefore, when comparing different conditions giving comparable values for O-antigen units/LPS molecule, for percent of LPS molecules bearing long O-antigen side chains, and for percent coverage of lipid A core with O-antigen, the surface topography of the LPS molecules presumably differs, according to the growth conditions used to arrive at the indicated values.

Growth of cells under condition B leads to a sudden shift from a serum sensitive to a serum resistant phenotype after 1 complete doubling time with mannose (Figure 4). When the LPS profile of cells grown under this condition was tested by densitometric scanning of SDS-PAGE gels, all parameters of LPS coverage changed gradually in comparison to the sudden alterations in serum sensitivity.

During growth of bacteria under condition C, loss of serum resistance did not begin until 1 doubling time and was not complete until 2 doubling times in the absence of mannose. Again, under this circumstance, parameters of LPS coverage changed gradually whereas serum sensitivity changed more precipitously.

LPS PROFILE AND KILLING
vs. DOUBLING TIME WITH MANNOSE

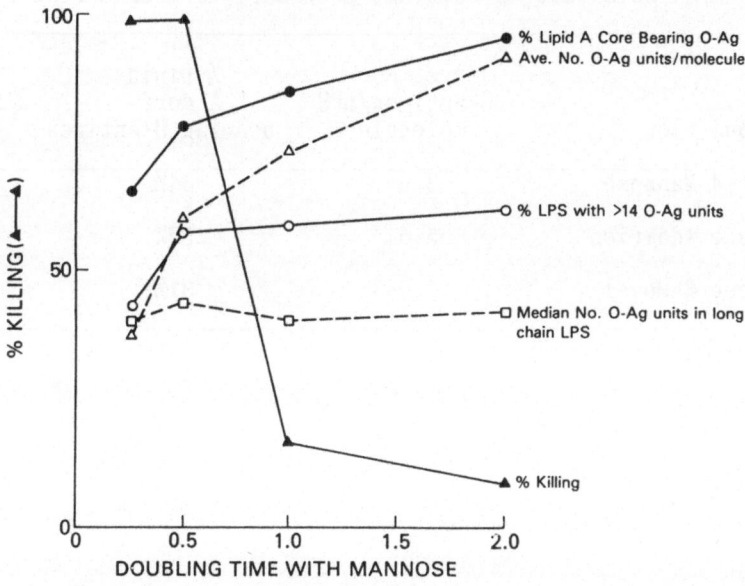

Fig. 4.   LPS Profile and Killing versus Doubling Time with Mannose. LPS
from bacteria grown under condition B as described in the text
was analyzed by quantitative densitometry of dried gels. The
parameters of O-antigen coverage are shown as a function of the
doubling time in mannose and are compared with the extent of
serum killing when the organisms are incubated in 5% PNHS.

    During growth of bacteria under condition C, loss of serum resistance
did not begin until 1 doubling time and was not complete until 2 doubling
times in the absence of mannose. Again, under this circumstance,
parameters of LPS coverage changed gradually whereas serum sensitivity
changed more precipitously.

Table 1

Parameters of O-Ag coverage necessary for serum resistance

Growth Condition	Ave. No O-antigen/LPS molecule	% Lipid A core bearing O-antigen	% LPS with 14 O-antigen (long chain)
(A) Limited Mannose	3.9	34%	20%
(B) Mannose Addition	5.6	52%	20%
(C) Mannose Removal	4.0	31%	20%

BINDING OF C3 AND C5b-9 TO GNB

ROUGH SERUM SENSITIVE

SMOOTH SERUM RESISTANT

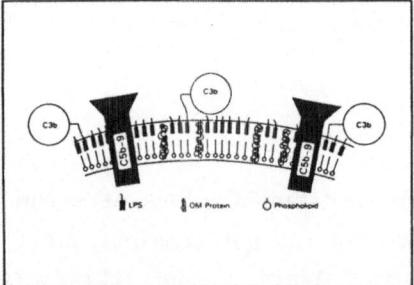 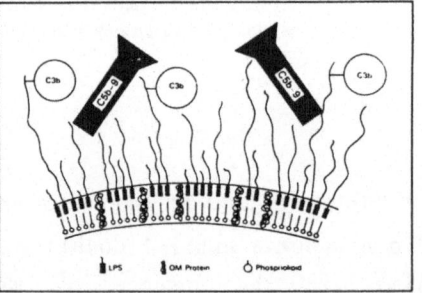

Fig. 5.   Model of Serum Resistance in Gram Negative Enteric Bacteria. Illustrated in the model is the concept that long O polysaccharide side chains of LPS sterically hinder access of C3 and, hence, of C5b-9 to hydrophobic domains in the outer membrane. Thus, the C5b-9 complex is shed and is not bactericidal.

Finally, we wished to directly relate quantitative changes in LPS profile with the serum susceptibility of cells grown under varying mannose inputs. As expected, all parameters associated with addition of more or longer O-antigen to lipid A led to an increase in percent survival. We determined the LPS parameters associated with greater than 10% survival in serum (Table 1).

In comparison to other parameters, the percent coverage with long chain LPS correlates best with the relative survival during incubation in serum. We postulate, based on this and previous work, that a critical density (>20%) of long O-antigen side chains is required to protect the cells from serum killing and that the O-antigen protection is a result of steric hindrance to C5b-9 access to hydrophobic domains of the outer membrane (Figure 5).

In conclusion, O-antigen polysaccharide within LPS is the major determinant of the extent, form and site of complement deposition on enteric gram negative bacteria. In turn, the extent, form and site of complement deposition dictate the opsonic and bactericidal effects of the bound complement molecules, and, thus, play a major role in influencing the outcome of the interaction between the organism and the host.

References

1. Nikaido, H. Nonspecific transport through the outer membrane. In Bacterial Outer Membranes. M. Inouye (ed). Wiley and Sons, New York (1979).
2. Valtonen, V.V. Mouse virulence of Salmonella strains: the effect of different smooth-type O-side chains. J Gen Microbiol 64:255-268 (1970).
3. Valtonen, M.V., Plosila, M., Valtonen, V.V., Mäkelä, P.H. Effect of the quality of the lipopolysaccharide on mouse virulence of Salmonella enteritidis. Infect Immun 12:828-832 (1975).
4. Valtonen, M.V. Role of phagocytosis in mouse virulence of Salmonella typhimurium recombinants with O antigens 6,7 or 4,12. Infect Immun 18:574-582 (1977)
5. Liang-Takasaki, C-J., Mäkelä, P.H., Leive, L. Phagocytosis of bacteria by macrophages: changing the carbohydrate of lipopolysaccharide alters interaction with complement and macrophages. J Immunol 128:1229-1235 (1982).

6. Ling-Takasaki, C-J., Saxén, H., Mäkelä, P.H. and Leive, L. Complement activation by polysaccharide of lipopolysaccharide: an important virulence determinant of Salmonellae. Infect Immun 41:563-569 (1983).

7. Saxén, H., Reima, I. and Mäkelä, P.H. Alternative complement pathway activation by Salmonella O polysaccharide as a virulence determinant in the mouse. Microb Pathogen 2:15-28 (1987).

8. Liang-Takasaki, C-J., Grossman, N. and Leive, L. Salmonella activate complement differentially via the alternative pathway depending on the structure of their lipopolysaccharide O-antigen. J Immunol 130:1867-1870 (1983).

9. Grossman, N. and Leive, L. Complement activation via the alternative pathway by purified Salmonella lipopolysaccharide is affected by its structure but not its O-antigen length. J Immunol 132:376-385 (1984).

10. Grossman, N., Joiner, K.A., Frank, M.M. and Leive, L. C3b binding, but not its breakdown, is affected by the structure of the O-antigen polysaccharide in lipopolysaccharide from Salmonellae. J Immunol 136:2208-2215 (1986).

11. Jiménez-Lucho, V.E., Joiner, K.A., Foulds, J., Frank, M.M. and Leive, L. C3b generation is affected by the structure of the O-antigen polysaccharide in lipopolysaccharide from Salmonellae. (Submitted for publication).

12. Rowley, D. Sensitivity of rough gram-negative bacteria to the bactericidal action of serum. J Bacteriol 95:1647-1650 (1968).

13. Muschel, L.H., Larsen, L.J. The sensitivity of smooth and rough gram-negative bacteria to the immune bactericidal reaction. Proc Soc Exp Biol Med 133:345-348 (1970).

14. Taylor, P.W. Sensitivity of some smooth strains of Escherichia coli to the bactericidal action of normal human serum. J Clin Pathol 27:626-629 (1974).

15. Joiner, K.A., Hammer, C.H., Brown, E.J., Cole, R.J., Frank, M.M. Studies on the mechanism of bacterial resistance to complement-mediated killing. I. Terminal complement components are deposited and released from Salmonella minnesota S218 without causing bacterial death. J Exp Med 155:797-808 (1982).

16. Joiner, K.A., Hammer, C.H., Brown, E.J. and Frank, M.M. Studies on the mechanism of bacterial resistance to complement-mediated killing. II. C8 and C9 release C5b67 from the surface of Salmonella minnesota S218 because the terminal complex does not insert into the bacterial outer membrane. J Exp Med 155:809-819 (1982).

17. Joiner, K.A., Goldman, R.C., Hammer, C.H., Leive, L. and Frank, M.M. Studies on the mechanism of bacterial resistance to complement-mediated killing. VI. IgG increases the bactericidal efficiency of C5b-9 for E. coli O111B4 by acting at a step before C5 cleavage. J Immunol 131:2570-2575 (1983).

18. Podack, E.R., Biesecker, G., Muller-Eberhard, H.J. Membrane attack complex of complement: Generation of high-affinity phospholipid binding sites by fusion of five hydrophilic plasma proteins. Proc Natl Acad Sci 76(2):897-901 (1979).

19. Leive, L. Release of lipopolysaccharide by EDTA treatment of E. coli. Biochem Biophys Res Commun 21:290-296 (1965).

20. Goldman, R.C., Joiner, K.A. and Leive, L. Serum-resistant mutants of _Echerichia coli_ 0111 contain increased lipopolysaccharide, lack an 0 antigen-containing capsule, and cover more of their lipid A core with 0 antigen J. Bacteriol. 159:877-882 (1984).

21. Joiner, K.A., Schmetz, M.A., Goldman, R.C., Leive, L. and Frank, M.M. Mechanism of bacterial resistance to complement-mediated killing: inserted C5b-9 correlates with killing for _Escherichia coli_ 0111B4 varying in 0-antigen capsule and 0 polysaccharide coverage of lipid A core oligosaccharide. Infect Immun 45:113-117 (1984).

22. Joiner, K.A., Goldman, R., Schmetz, M., Berger, M., Hammer, C.H., Frank, M.M. and Leive, L. A quantitative analysis of C3 binding to 0-antigen capsule, lipopolysaccharide, and outer membrane protein of _E. coli_ 0111B4. J Immunol. 132:369-375 (1984).

23. Joiner, K.A., Grossman, M., Schmetz, M. and Leive, L. C3 binds preferentially to long chain lipopolysaccharide during alternative pathway activation by _Salmonella montevideo_. J Immunol 136:710-715 (1986).

24. Grossman, N., Schmetz, M.A., Foulds, J., Klima, E.N., Jiménez, V., Leive, L. and Joiner, K.A. Lipopolysaccharide size and distribution determines serum resistance in _Salmonella_. J Bacteriol 169:856-863 (1987).

ABILITY TO ACTIVATE THE ALTERNATIVE COMPLEMENT PATHWAY AS A VIRULENCE DETERMINANT IN SALMONELLAE

P. Helena Mäkelä, Marianne Hovi, Harri Saxen, Matti Valtonen and Ville Valtonen

National Public Health Institute
Mannerheimintie 166
SF-00280 Helsinki
Finland

This work was started many years ago as an attempt to chart the mechanisms of pathogenesis in a well established experimental infection, salmonellosis of the mouse. As a first approach we asked whether relatively small differences in the quality of the cell surface O antigen of the bacteria would affect the outcome of the infection, and if so, by which mechanisms (Valtonen, 1970). This approach had the advantage that both the chemical structure and the genetic determination of the O antigenic part of the lipopolysaccharide (LPS) molecule were known, and it was possible to construct nearly isogenic pairs of strains for comparison. These studies showed that differences were indeed present, and in the course of working out their mechanism of action we have learned a lot of the events that take place in the mouse during the infection. Since we recently could also wind up the story (Saxen et al., 1987) it is now possible to describe the complete sequence of events from a defined molecule on the surface of the bacteria to the outcome of the infection.

The bacterial aspect

We constructed a number of nearly isogenic pairs of strains that differed in known ways in the O-antigenic polysaccharide chains of the LPS. The constructions were made by conjugation or transduction. In the set of strains that form the basis of this paper the rfb region - a cluster of more than ten genes that determine the synthesis of the repeating oligosaccharide units of O antigen (Mäkelä and Stocker, 1984) - of one Salmonella strain was exchanged for the rfb region derived from

NATO ASI Series, Vol. H24
Bacteria, Complement and the Phagocytic Cell
Edited by F. C. Cabello und C. Pruzzo
© Springer-Verlag Berlin Heidelberg 1988

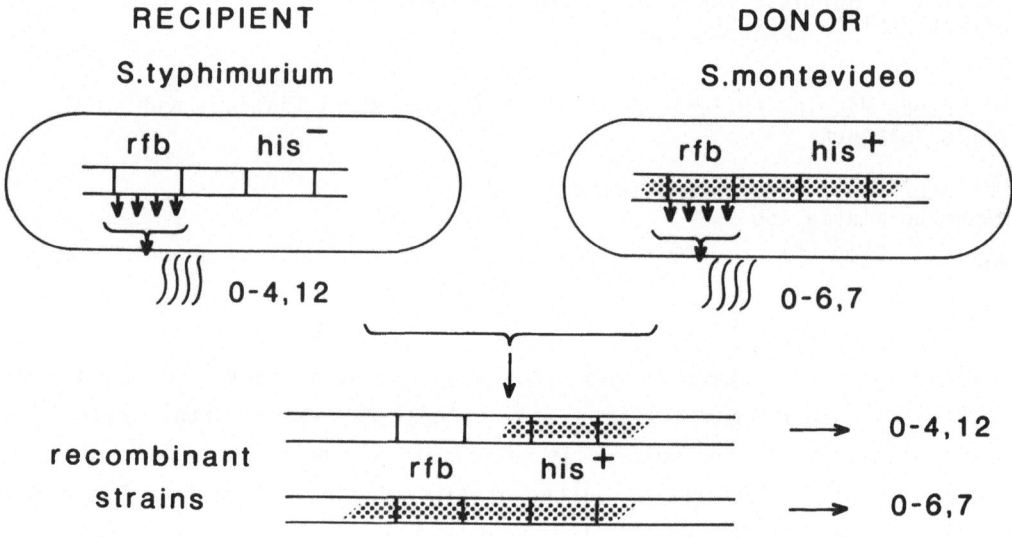

Fig. 1. Derivation of the sister strains studied (Valtonen, 1970). The parent strains were Salmonella typhimurium (0-4,5,12), and S. montevideo (0-6,7) or S. enteritidis (0-9,12) [not shown in the figure]. The chromosomal area involved in the exchanges is shown for each strain (white for S. typhimurium, shaded for S. montevideo). The more than ten genes in the rfb cluster determine the synthesis and structure of the 0 polysaccharide of the LPS on the cell surface. The nearly isogenic sister strains were constructed by selecting for the replacement of the mutant his⁻ gene cluster of the recipient (in this case S. typhimurium) strain by the his⁺ of the donor. Among the recombinants, some had retained the rfb, and thus the 0 antigen, of the recipient, whereas others had received also the closely linked rfb cluster of the donor, and therefore synthesized donor-type 0 antigen. In some recombinants a cross-over inside the rfb cluster had resulted in rfb gene products that could not collaborate to synthesize an 0 polysaccharide (Nikaido et al., 1966), resulting in rough (R) recombinants (not shown, and not discussed in this paper).

another strain with a different 0 antigen (Fig. 1). The exchanges were made so that the nearby situated his genes of the sister strains were the same, whereas the strains could differ in genes flanking the rfb region.

The size of DNA in a P22 transducing particle (at most 40 genes), and the size of _rfb_ (over 10 genes) and _his_ put a definite upper limit to this difference. Furthermore, since the virulence differences to be described shortly always corresponded to the 0-antigenic type of the transductants or exconjugants it seemed most probable that they were actually caused by the differences in the 0 antigen. We considered this likely enough to go ahead and look for the mechanism by which the relatively small differences in the polysaccharide structure could influence virulence. The results of this effort do indeed show that the quality of the 0 antigen is a determinant of mouse virulence of these bacteria.

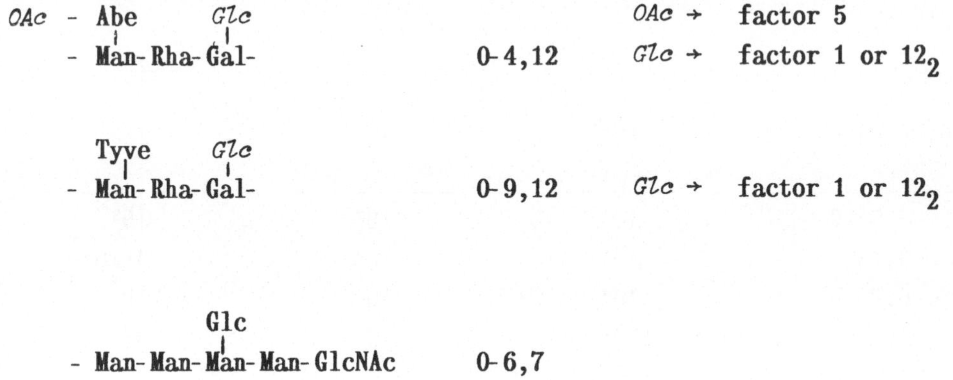

$$
\begin{array}{lll}
OAc \ - \ Abe \quad Glc & & OAc \ \rightarrow \quad \text{factor 5} \\
\quad \ - \ \text{Man-Rha-Gal-} & \text{O-4,12} & Glc \ \rightarrow \quad \text{factor 1 or } 12_2 \\
\\
\quad \ \ \ \ Tyve \quad Glc & & \\
\quad \ - \ \text{Man-Rha-Gal-} & \text{O-9,12} & Glc \ \rightarrow \quad \text{factor 1 or } 12_2 \\
\\
\\
\quad \ \ \ \ \ \ \ \ \ Glc & & \\
\quad \ - \ \text{Man-Man-Man-Man-GlcNAc} & \text{O-6,7} &
\end{array}
$$

Fig. 2. The chemical structure of the three types of 0-antigenic units. In the LPS they are polymerized and linked to the LPS core - see Fig. 3. Abbreviations: Abe, abequose; Gal; galactose; Glc, glucose; GlcNAc, N-acetylglucosamine; Man, mannose; OAc, 0-acetyl; Rha, rhamnose; Tyve, tyvelose.

These _rfb_-region exchanges produced a series of nearly isogenic strains with the 0 antigens 4,5,12; 9,12 or 6,7 (Fig. 2). The numbers denote antigenic factors, i.e. epitopes of the 0 polysaccharide recognized by monospecific absorbed antisera; in most cases the chemical structure corresponding to the epitope is now known (Luderitz _et al._, 1971). These strains formed a series of increasing virulence (6,7 < 9,12 < 4,5,12), as seen by decreasing doses (LD_{50}) needed to kill 50% of the mice after intraperitoneal inoculation (Table 1). As a point of reference, a sister rough (R) strain devoid of O polysaccharide because of mutation in a <u>rfb</u> gene is also shown: it is the least virulent. It is noteworthy that the virulence difference between the strains was dependent on their O antigen irrespective of the natural O antigen of the recipient bacteria: compare the 4,5,12 and 9,12 derivatives of <u>S</u>. <u>typhimurium</u> (originally O-4,5,12) and <u>S</u>. <u>enteritidis</u> (originally O-9,12). Both parent strains are mouse virulent and cause natural infections in the mouse (Collins, 1970), yet in both cases the O-4,5,12 derivative was more virulent than its O-9,12 sister (Valtonen <u>et al.</u>, 1975).

Table 1[1]

	LD_{50} of derivatives of lines		
LPS character	LT2	IH2	IH4
S; O-4,5,12	10^4	10^3	10^5
S; O-9,12	10^5	10^4	4×10^5
S; O-6,7	10^6	10^5	ND
R	10^7	ND	ND

[1]Virulence of nearly isogenic derivatives of naturally O-4,5,12 <u>S</u>. <u>typhimurium</u> and naturally O-9,12 <u>S</u>. <u>enteritidis</u>, measured as LD_{50} in 10 days after ip injection of graded doses of bacteria into groups of 10 mice each (Valtonen, 1970; Valtonen <u>et al.</u>, 1975). ND, not determined.

The chemical structures of the three types of O antigenic units are shown in Fig. 2. In the LPS these units are linked to a common core oligosaccharide, and their numbers vary from zero to 100 or more (Fig. 3). The difference between O-4,5,12 and O-9,12 is very small, since the backbone is the same and the branch sugar in both cases is a dideoxyhexose (about the O-acetyl in the O-4,5,12 structure, see below). The O-6,7 structure differs more from the above, yet the difference in virulence between successive members of the series (4,5,12 > 9,12 > 6,7) is about ten-fold as shown in Table 1.

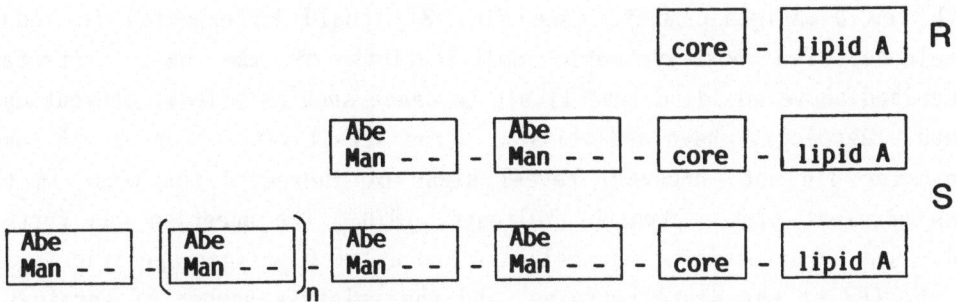

Fig. 3. LPS molecules of different types present in 0-4,12 bacteria. R
forms have only lipid A and core, but no 0 antigen units. In S
form molecules varying numbers of 0 antigen units (whose
structures are shown in Fig. 2) are polymerized to the 0
polysaccharide linked to the core.

However, all differences in the structure of the 0 polysaccharide do
not affect virulence. We studied several variations of the 0-4,5,12
structure (Valtonen and Mäkelä, 1971), in which the rfb-determined basic
structure is modified after its synthesis by products of modification
genes outside rfb (Mäkelä and Stocker, 1984). Thus the 0-acetyl, which
corresponds to factor 5, could be removed (by introducing a mutant allele
of the corresponding oafA gene) without a change in virulence. Likewise, a
glucose branch could be introduced to position C4 (factor 12_2) or position
C6 (factor 1) of the galactose residue without a change in virulence. Even
the linkage between the 0-antigenic units could be altered (from α 1-2 to
α 1-6, creating factor 27) without an effect on virulence.

Possible mechanisms of action

Why then did some features of the 0 polysaccharide affect virulence,
others not? A possibility that seemed most likely a priori was that the
rfb exchanges also affected the "smooth" character of the bacteria. Since
R forms are practically avirulent (Table 1), a relative increase of R form
LPS molecules (those without any 0-antigenic material), or of molecules

with few O-antigenic units (see Fig. 3), could be expected to reduce virulence. The post-synthetic modifications of the basic structure described above would be less likely to cause such an effect. Conventional tests - serology, phage sensitivity, serum sensitivity - for smooth-rough character did not, however, reveal signs of increased roughness in the less virulent sister strains (Valtonen, 1970). The question was further addressed by determining the relative amount of O antigen-specific sugars in the LPS of the sister strains, and the relative number of terminal O units as compared to internal units (see Fig. 3) - neither method supported the suggested difference in smooth character (Nurminen et al., 1971). A final proof was provided when methods to separate and visualize LPS-molecules with different numbers of O units were developed (Palva and Mäkelä, 1980), and the length distribution of LPS chains was found identical in the sister strains that differed nearly 100-fold in virulence (Grossman and Leive, 1984). Thus the decisive factor did not seem to be any quantitative aspect of LPS but rather the quality of its polysaccharide chains.

The presence or absence of a cell surface glycolipid, ECA or the enterobacterial common antigen (Mäkelä and Mayer, 1976) also affects bacterial virulence in mouse salmonellosis (Valtonen et al., 1976). In some rfb recombinants between O-6,7 and O-4,5,12 bacteria the ECA is also affected, and this could contribute to their differences of virulence. However, recombinants specifically constructed to ensure the same fully expressed ECA but differing in their O antigen still displayed the above-described O antigen-dependent differences of virulence (Valtonen et al., 1976).

If the differences in virulence were indeed due to qualitative differences in O antigen structure, the most likely mechanism mediating this seemed to be one involving the immune system, since this system is specifically designed to recognize structural features of this sort. Would mice have "natural" antibodies to O-6,7 but not to O-4,5,12 (with O-9,12 being intermediate)? Or would they respond quicker to O-6,7 than to the other antigens? No evidence to support either of these hypotheses could be found, no natural antibodies could be detected in the mice (within the limitations of the assay methods), and the same O antigen-dependent difference was seen in immunosuppressed mice inspite of a great reduction in the LD$_{50}$ of all the strains (Valtonen et al., 1971; Valtonen and Hayry, 1978). Although such negative evidence could not be conclusive, other mechanisms seemed more likely, and were therefore looked for as described below.

Overall course of salmonella infection in the mouse

Salmonellosis - both typhoid fever in man and mouse typhoid in the mouse - is a complex disease with several clinically separable phases (Fig. 4). The initial infection takes place by mouth, after which the bacteria traverse the intestinal epithelium, most probably at the specialized lymphoid organs, the Peyer's patches (Hohmann et al., 1978). Soon after, some bacteria are found in the local lymph nodes and very transiently in blood, but most of them end up in the liver and spleen, where they multiply for several days. All this time the infection is without symptoms, which appear only after the bacteria in the liver and spleen have multiplied to a high level (appr. 10^8 bacteria/liver in the mouse (Collins, 1969)). Then the infection becomes fulminant, with fever, bacteremia and toxic symptoms from the central nervous system and elsewhere. At this stage the mouse soon succumbs to the infection; in many recovery is slow and uncertain. This complex picture itself suggests that different host defense mechanisms are operative in the different phases, and the bacterium will use different means - recognizable as virulence factors - to avoid them. Our first question then was to find out in which phase of the infection the quality of O antigen is important.

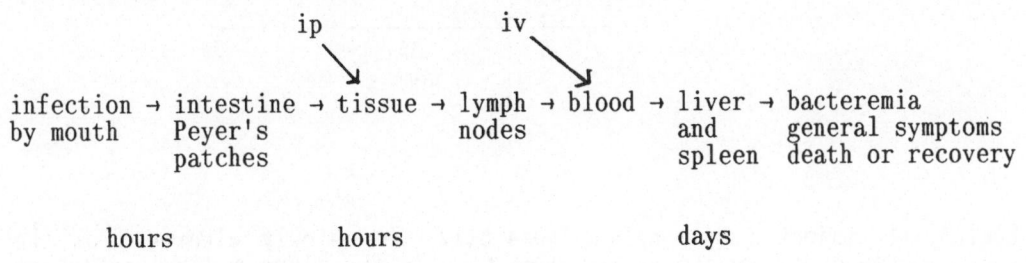

Fig. 4. Course of mouse typhoid.

The infection in the mouse can be very nicely monitored experimentally by following bacterial multiplication (or destruction) in serial samples from the main sites of infection. It is also possible to focus on a certain phase of the infection by injecting the bacteria intraperitoneally (ip) or intravenously (iv) and thus bypassing previous phases (Fig. 4). The iv-injected bacteria are slowly cleared from the blood and appear at the same time in the liver and spleen (Fig. 5). Some hours after injection the majority of the bacteria are found in these organs (Collins, 1969), where their numbers then increase for several days (Fig. 6). This multiplication of the bacteria is believed to take place in the local macrophages, called Kupffer cells in the liver. This view is supported by the fact that at this stage the infection is unaffected by antibiotics such as streptomycin that do not enter animal cells, and also by circulating antibodies (Blanden et al., 1966). For the ip-injected bacteria it takes longer to reach the liver and spleen, but 24 hrs after injection they are found in these organs in approximately similar numbers

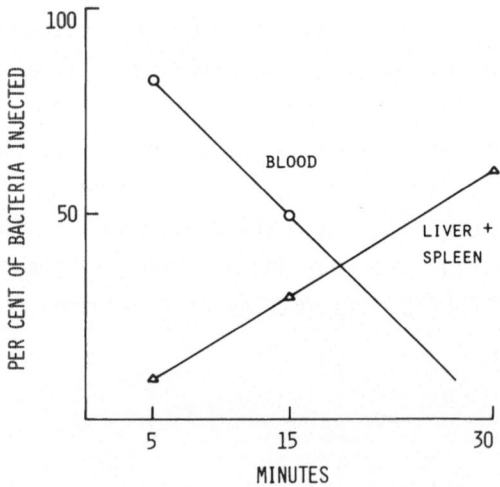

Fig. 5. Iv-injected Salmonellae (0-4,5,12) are slowly cleared from the blood and appear at the same time in the liver and spleen (after Saxen, 1984).

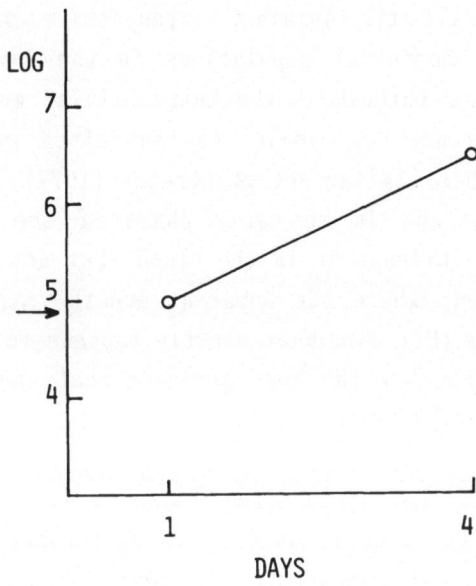

Fig. 6. The numbers of viable bacteria increase in the liver (and spleen) for several days (after Saxen, 1984). The figure shows 0-4,5,12 bacteria injected iv or ip (arrow = inoculum size), but a similar figure is obtained for their 0-6,7 sister strains injected iv (Saxen et al., 1984).

as iv-injected bacteria (with the same dose). The numbers of bacteria in the peritoneal cavity stay level through the next days while the bacteria in the liver and spleen multiply (Fig. 6) like they do after iv injection.

The fate of the 0 antigen variants in the mouse

When the series of bacteria described above were tested by the ip and iv injection routes, the 0 antigen-dependent difference in virulence (Table 1) was seen after ip but not after iv injection (Saxen et al., 1984). Although the clearance rate from the blood after iv injection was faster for the 0-6,7 bacteria than for the 0-4,12 bacteria (Valtonen, 1977), this difference was not reflected in the subsequent fate of the bacteria. Instead, all the strains, irrespective of their 0 antigen, established similar bacterial populations in the liver and spleen, and grew at the same rate throughout the intracellular growth phase. In oral infection the difference was similar to that after ip injection as shown by Lyman et al. with a similar set of strains (1977). Thus, it seemed that 0 antigen was important in an early phase of the infection when the bacteria were in the tissues or in the blood, but not yet in the cells of the liver and spleen, where the apparent growth rate of all the sister strains was the same (Fig. 6). What exactly happens to the bacteria at the tissue stage, and what are the host defenses that the 0-4,5,12 can avoid better than the 0-6,7 bacteria?

When the fate of the ip-injected bacteria was followed in samples withdrawn from the peritoneal cavity, the difference between the sister strains was very marked (Saxen et al., 1987). Only 15 per cent of the inoculum of 0-6,7 bacteria could be recovered as colony forming units (CFU) half an hour after ip injection, whereas the corresponding figure for 0-4,5,12 was 54%. The data suggested rapid killing of the 0-6,7 bacteria in the peritoneal cavity, but other explanations were also possible. Thus, the bacteria might have adhered to the walls of the peritoneal cavity and therefore escaped being withdrawn in the peritoneal washing. Another possibility considered was that they had been removed from the peritoneal cavity altogether. Both these possibilities were, however, disproven when the bacteria were radiolabeled by growth with ^{35}S-methionine before ip injection. Whereas the CFU behaved as before, with a large difference between the 0-4,5,12 and the 0-6,7 bacteria, almost all the label of both 0-6,7 and 0-4,5,12 bacteria was recovered in the peritoneal wash fluid (Fig. 7).

Rapid intraperitoneal killing of the 0-6,7 bacteria as compared to 0-4,5,12 thus seems to be the decisive difference between the two strains. It resulted in much lower numbers of the 0-6,7 bacteria surviving in the peritoneal cavity, seeping out to the blood, and establishing themselves in the liver and spleen. The intracellular growth rate of the bacteria in these organs was not, however, affected by the 0 antigen (Fig. 8) as already seen after iv injection.

How were the ip-injected 0-6,7 bacteria killed? Was it after phagocytosis or in the fluid phase? Centrifugation of the peritoneal wash fluid showed that the radiolabel of 0-6,7 bacteria was pelleted together with the peritoneal cells, whereas the 0-4,5,12 bacteria (and the associated label) remained in the supernatant, unassociated to cells (Saxen et al., 1987). To identify the cells to which the 0-6,7 bacteria had been associated, and the mode of the association, the pellet from the above experiment was fixed, thin sectioned, and examined by electron microscopy (Saxen et al., 1987). In the pellet from the 0-6,7 injected mice the bacteria were seen associated with morphologically typical macrophages (and not with polymorphonuclear cells also present in the pellet).

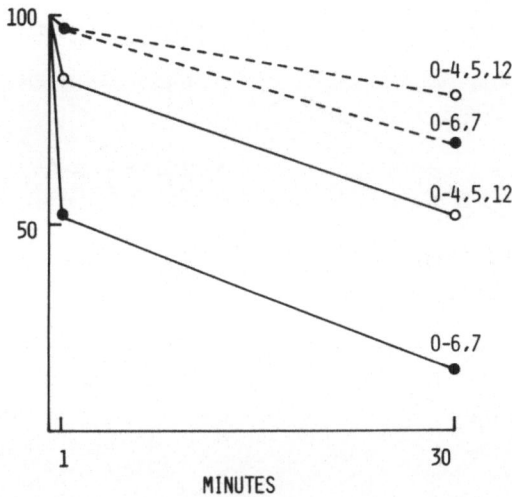

Fig. 7. The fate of the sister bacteria in the peritoneal cavity (after Saxen et al., 1987). Peritoneal washings were obtained 1 and 30 min. after ip injection of the ^{35}S-methionine-labeled 0-4,5,12 (open symbols) or 0-6,7 (filled symbols) bacteria and analyzed for radioactivity (broken lines) and viable bacteria (solid lines).

Most of them were in vacuoles inside the macrophage (Fig. 9, C-E). In samples taken very quickly, one minute after injection, bacteria could also be seen on the surface of the macrophages (Fig. 9B) or entangled in its projections. In later samples, e.g. 30 min. after injection, the bacteria in the macrophages already appeared dead, in the process of disintegration (Fig. 9E). Thus, the morphological findings were consistent with the previous findings of rapid association of the 0-6,7 bacteria with peritoneal cells, and their subsequent killing, and showed that the killing took place in the phagocytic vesicles of the peritoneal macrophages. As expected, extremely few 0-4,5,12 bacteria were seen in association with peritoneal cells.

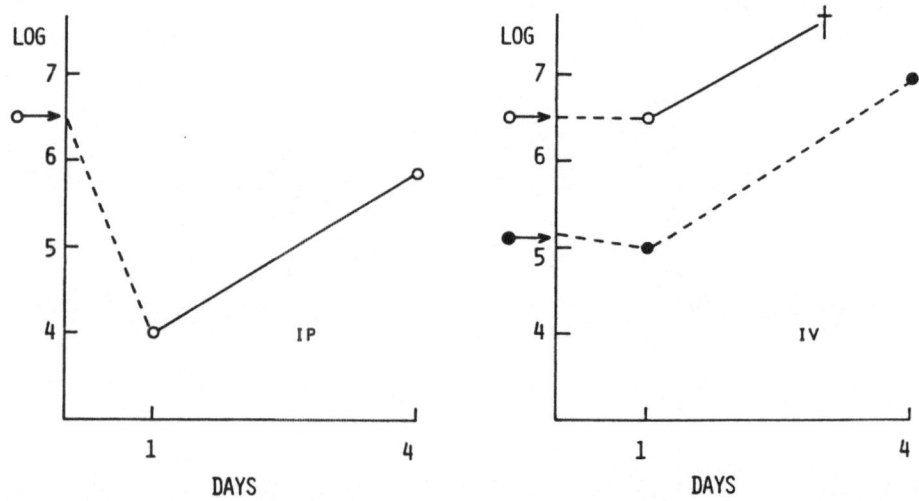

Fig. 8. The numbers of viable (0-6,7) bacteria in the liver after ip (left panel) and iv (right panel) injection (after Saxen et al., 1984). The arrows to the left indicate the inoculum size. The larger inoculum given iv killed the animals before day 4 (+).

When looking at the electron micrographs we were struck by a homogenous relatively electron-dense layer around the bacteria that separated them from the membrane of the phagocytic vesicle or from the macrophage surface. A similar layer was not seen on the broth-grown bacteria before injection (Fig. 9A). What was this layer, and was it perhaps connected with the phagocytosis of these bacteria? And was it also connected with the quality of their O polysaccharide? Before describing our experiments to answer these questions, let us return for awhile to another series of experiments that had provided a clue to this problem.

Role of complement

Loretta Leive and her coworkers studied phagocytosis of the same series, 0-6,7; 0-9,12; and 0-4,5,12 bacteria. They showed, that these bacteria were taken up into macrophages (both a cultured macrophage cell line, and freshly isolated peritoneal macrophages of the mouse were used) in a manner corresponding to their virulence properties - the less virulent taken up more rapidly (Liang-Takasaki et al., 1982). They then showed that the uptake was dependent on the presence of complement - in its absence all the bacteria were taken up only very slowly - and required only the alternative complement pathway but not antibodies (Liang-Takasaki et al., 1983). This strongly suggested that the different O polysaccharides would differ in their ability to activate the alternative pathway of complement.

This was soon proven by elegant experiments in which the LPS of these bacteria was isolated and coated onto red blood cells, which then became complement activators (Grossman and Leive, 1984). The rate of activation was dependent on the quality (0-6,7; 0-9,12; or 0-4,5,12) of the O polysaccharide, but not on its length in the LPS molecules. These experiments and the subsequent work on the mode of complement activation and the resultant deposition of C3b and C3bi on the bacteria are described in the paper by Keith Joiner in this volume. The important conclusion for the purposes of the present paper is that 0-6,7 (and 0-9,12) bacteria activate complement at a faster rate than do 0-4,5,12 bacteria, and this results in more C3b and C3bi deposited on their surface in a given time.

We therefore wanted to test whether the layer that we had seen on the 0-6,7 bacteria taken from the peritoneal cavity was in fact a deposit of C3b (and/or C3bi). We first approached this by asking whether or not we could see a similar deposit on the bacteria incubated in vitro with a

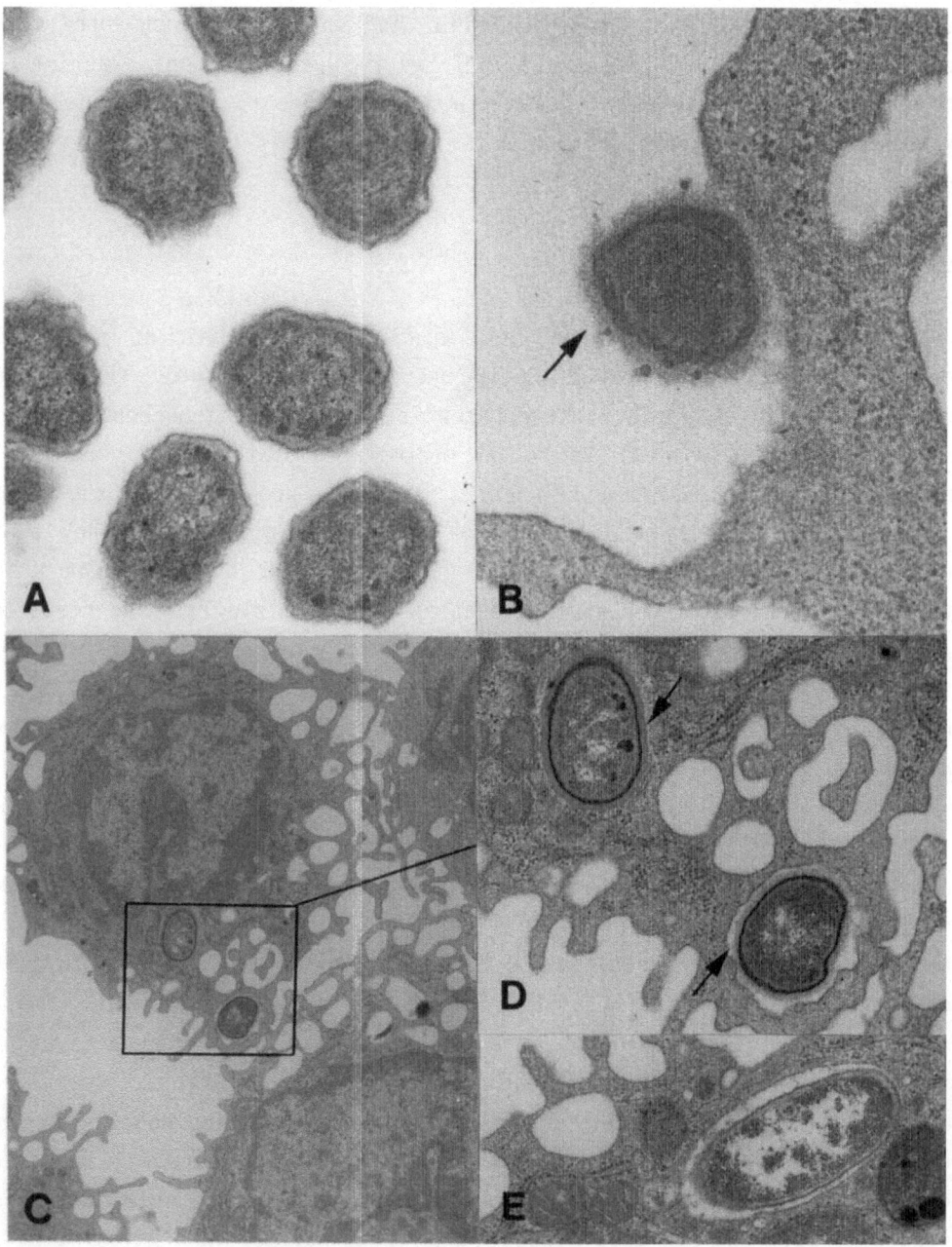

Fig. 9. Transmission electron micrograph of the O-6,7 bacteria before ip
        injection (A), in the process of being phagocytized (B), and
        inside a macrophage retrieved from the peritoneal cavity of a
        mouse 1 min (C and D) or 30 min (E) after ip inoculation (from
        Saxen et al., 1987), with permission). Note the electron dense
        layer around the bacteria, separating them from the phagosomal
        vesicle membrane.

complement source (Saxen et al., 1987). The 0-6,7 bacteria were incubated
for 20 minutes in fresh mouse serum (we wanted to use mouse serum, since
the infection experiments were also done in the mouse, although most of
the complement studies cited above were done with other complement
sources, bovine, guinea pig, and human) and prepared for electron
microscopy as above. Indeed, all the bacteria were now coated with a layer
of relatively electron-dense deposit that looked exactly like the one seen
in the peritoneal samples (Fig. 10). Bacteria incubated in mouse serum
heated to 56°C for 30 min to destroy the complement system were completely
bald, like those grown in broth only (Fig. 9A). We also showed that C3 was
present on the surface of these bacteria by their staining with
peroxidase-conjugated goat anti-mouse C3 (Saxen et al., 1987).

If the complement coat on the 0-6,7 bacteria was the reason for their
phagocytosis and killing in the peritoneal cavity, the more virulent
0-4,5,12 should not show a similar coating. However, when the 0-4,5,12
bacteria were incubated for 20 min. in fresh mouse serum, a large part of
them, up to 80%, were seen with a similar coat, and only 20% were without
such a deposit (Saxen et al., 1987). After this unexpected result we
realized that incubation in mouse serum was not a fair experiment - the
concentration of complement in the tissues, including the peritoneal
fluid, is believed to be much less than in the blood.

When we then incubated the 0-6,7 and the 0-4,5,12 bacteria for
different periods of time in lower serum concentrations we found that they
indeed behaved differently - in 15% serum almost no 0-4,5,12 but almost
all 0-6,7 bacteria acquired the complement coat in 15 minutes (Saxen et
al., 1984). This difference is consistent with the previously shown
differential rate of complement activation by these bacteria
(Liang-Takasaki et al., 1984), and can eminently explain the different
fate of the bacteria in the peritoneal cavity.

We could also perform the opposite experiment, in which 0-4,5,12
bacteria were rendered susceptible to intraperitoneal killing. We
incubated the 0-4,5,12 bacteria for 20 minutes in fresh mouse serum
(during which time most of them acquire a coat of complement), and then
injected them ip in the mouse, and found that they were killed rapidly

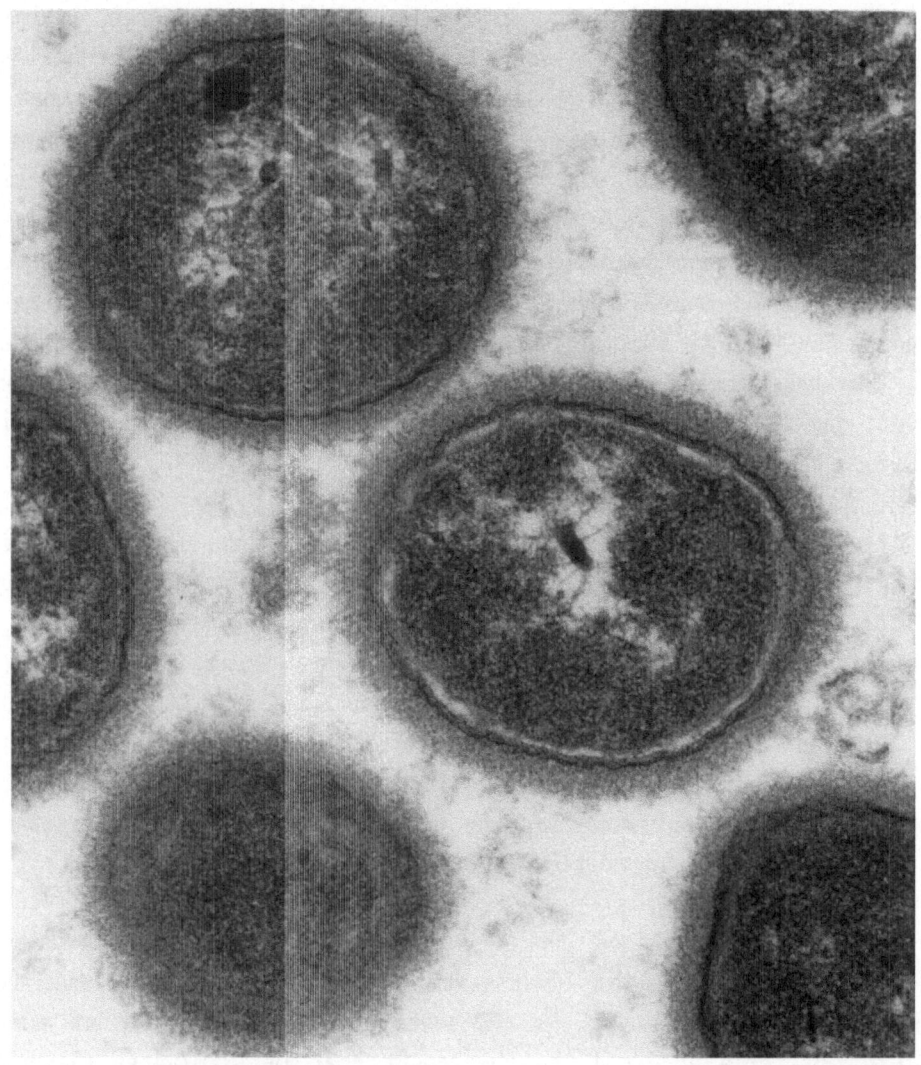

Fig. 10. Transmission electron micrograph of 0-6,7 bacteria incubated for 20 min. in fresh mouse serum (Saxen et al., 1987). Compare the electron-dense layer around the bacteria with that seen in Fig. 9 B-D, and with its absence in Fig. 9A.

just like 0-6,7 bacteria (Saxen et al., 1987). The same was seen when the 0-4,5,12 bacteria were opsonized with antibody: again, they were killed rapidly in the peritoneal cavity, and the numbers of the bacteria reaching the liver and spleen were reduced by a factor of 100 (Saxen, 1984).

The kinetics of infection with 0-6,7 bacteria, that are rapidly opsonized by activating the alternative pathway of complement, and with 0-4,5,12 bacteria preopsonized in vitro or injected in the mouse with antibody capable of opsonizing them in vivo, were found to be very similar: increase of LD_{50} by a factor of 100 or so by the ip route, but full virulence by the iv route (Saxen et al., 1984).

Conclusion

From the above it is clear that the deposition of C3b is a key event in deciding the fate of Salmonella in the peritoneal cavity of the mouse. It is furthermore clear that the quality of the 0-antigenic polysaccharide is decisive in promoting this event through the alternative complement pathway. This is in fact not surprising, knowing that many polysaccharides, of which the mannan of yeast cells ("Zymosan") is the best known example, are activators of this pathway. What is surprising is the specificity of this reaction, especially the difference between the chemically very similar 0-9,12 and 0-4,12 structures. This difference could be shown by all tests, including the molecular level studies of Leive, Joiner and coworkers (see the paper of Keith Joiner in this volume). However, as pointed out above, all changes of the polysaccharide structure did not affect the interaction with complement (Valtonen and Makela, 1971). It now seems possible to go one more step ahead and learn exactly which aspects of the polysaccharide are important in this interaction by studying the interactions of a series of synthetic oligosaccharides with the relevant purified components of the alternative complement pathway.

Smooth Salmonellae are intrinsically resistant to the lytic action of complement, and also the sister strains used in these studies have been shown to be resistant to killing in fresh mouse serum. Thus phagocytosis is the primary mechanism by which complement deposition could affect the course of the Salmonella infection studied. This has now been shown to be the case, and the resident peritoneal macrophage implicated as the main effector cell. Both C3b and C3bi are detectable on the bacterial surface after reaction with complement, and the different O polysaccharide structures studied did not have a major influence on their relative proportions (see Joiner, in this volume). Thus, we may assume that macrophage receptors to both of these (CR1 and CR3, respectively) could mediate the binding and internalization of these bacteria, but their relative contributions have not been studied.

Again, there were two surprising findings. Firstly, the peritoneal macrophages phagocytosed the complement-opsonized bacteria without the apparent need of another surface receptor being engaged. The involvement of the Fc-receptor is pretty much excluded by the studies of ip-injected O-4,5,12 bacteria that were not phagocytosed as such, but were made readily phagocytisable by preincubation in fresh (but not heat inactivated) mouse serum, or by minute amounts of IgM antibodies. Much higher concentrations of IgG were needed for the same effect (Saxen, 1984). The involvement of a mannose receptor is a possibility, in a manner shown to work in the phagocytosis of similarly complement-opsonized Leishmaniae (Blackwell et al., 1985); in a preliminary experiment, yeast mannan injected ip seemed indeed to impair phagocytosis of the O-6,7 bacteria (Saxen et al., 1987).

The second unexpected finding was that the opsonized Salmonellae were, in fact, killed in the peritoneal macrophages. Since Salmonellae are facultative intracellular parasites that both survive and multiply in the macrophages of the liver and spleen, one expected that opsonization would not impair their survival. And in fact it did not have an effect on survival when the bacteria were injected iv and thus, were primarily taken up by the macrophages in these sites. Thus the peritoneal macrophages differ from the macrophages of the liver and spleen in at least this way. Since we do not know what allows the Salmonellae to survive in the liver macrophages it is not possible to speculate further on the mechanism of this difference. The same difference is, however, seen in the relation of these cells towards another facultatively intracellular bacterium, Listeria monocytogenes, which survives and grows very well after iv injection in the mouse but is killed ip like the O-6,7 Salmonellae. Furthermore, a similar complement deposit is seen on the Listeriae incubated in fresh mouse (Saxen et al., manuscript in preparation).

Acknowledgements

Our sincere thanks are due to Kaija Helisjoki for her patient help with this as well as previous manuscripts on which this paper is based, and to Liisa Phyala, M.Sci., and her crew who raised the numbers of mice used over the years.

References

Blackwell JM, Ezekowitz RAB, Roberts MB, Channon JY, Sim RS, Gordon S (1985) Macrophage, complement and lectin-like receptors bind Leishmania in the absence of serum. J Exp Med 162:324-331.

Blanden RV, MacKaness GB, Collins FM (1986) Mechanisms of acquired resistance in mouse typhoid. J Exp Med 124:585-600.

Collins FM (1969) Effect of immune mouse serum on the growth of Salmonella enteritidis in non-vaccinated mice challenged by various routes. J Bacteriol 97:667-675.

Collins FM (1970) Immunity to enteric infection in mice. Infect Immun 1: 243.

Grossman N, Leive L (1984) Complement activation via the alternative pathway by purified Salmonella lipopolysaccharide is affected by its structure but not its O-antigen length. J. Immunol 132:376-385.

Hohmann AW, Schmidt G, Rowley D (1978) Intestinal colonization and virulence of Salmonella in mice. Infect Immun 22:763-770.

Liang-Takasaki C-J, Grossman N, Leive L (1983) Salmonellae activate complement differentially via the alternative pathway depending on the structure of their lipopolysaccharide O-antigen. J Immunol 130: 1867-1870.

Liang-Takasaki C-J, Mäkelä PH, Leive L (1983) Phagocytosis of bacteria by macrophages: Changing the carbohydrate of lipopolysaccharide alters interaction with complement and macrophages. J Immunol 128:1229-1235.

Luderitz O, Westphal O, Staub AM, Nikaido H (1971) Isolation, chemical and immunological characterization of bacterial lipopolysaccharides. In: Weinbaum G, Kadis S, Aijl SA (eds) Microbial Toxins. Vol IV: Bacterial Endotoxins, pp 145-233. Academic Press New York.

Lyman MB, Stocker BAD, Roantree RJ (1977) Comparison of the virulence of 0:9,12 and 0:4,5,12 Salmonella typhimurium his$^+$ transductants for mice. Infect Immun 15:491-499.

Mäkelä PH, Mayer H (1976) Enterobacterial common antigen. Bact Rev 40: 591-632.

Mäkelä PH, Stocker BAD (1984) Genetics of lipopolysaccharide. In: Rietschel ET (ed) Handbook of Endotoxin. Vol. 1: Chemistry of Endotoxin, pp 59-137. Elsevier Science Publishers BV Amsterdam.

Nikaido H, Nikaido K, Mäkelä PH (1966) Genetic determination of enzymes synthesizing O-specific sugars of Salmonella lipopolysaccharides. J Bacteriol 91:1126-1135.

Nurminen M, Hellerqvist CG, Valtonen VV, Mäkelä PH, The smooth lipopoly-saccharide character of 1,4 (5),12 and 1,9,12 transductants formed as hybrids between groups B and D <u>Salmonella</u>. Eur J Biochem 22:500-505 (1971).

Palva ET, Mäkelä PH (1984) Lipopolysaccharide heterogeneity in <u>Salmonella typhimurium</u> analyzed by sodium dodecyl sulfate/polyacrylamide gel electrophoresis. Eur J. Biochem 107:137-143.

Saxen H (1984) Mechanisms of the protective action of anti-<u>Salmonella</u> IgM in experimental mouse salmonellosis. J Gen Microbiol 130:2277-2283.

Saxen H, Hovi M, Mäkelä PH (1984) Lipopolysaccharide and mouse virulence of <u>Salmonella</u> O antigen is important after intraperitoneal but not intravenous challenge. FEMS Microbiol Lett 24:63-66.

Saxen H, Reima I, Mäkelä PH (1987) Alternative complement pathway activation by <u>Salmonella</u> O polysaccharide as a virulence determinant in the mouse. Microbial Pathogenesis 2:15-28.

Valtonen MV (1977) Role of phagocytosis in mouse virulence of <u>Salmonella typhimurium</u> recombinants with O-antigen 6,7 or 4,12. Infect Immun 18:574-578.

Valtonen MV, Hayry P (1978) O antigen as a virulence factor in mouse typhoid - the effect of B cell suppression. Infect Immun 19:26-28.

Valtonen MV, Plosila M, Valtonen VV, Mäkelä PH (1975) Effect of the quality of the lipopolysaccharide on mouse virulence of <u>Salmonella enteritidis</u>. Infect Immun 12:828-835.

Valtonen VV (1970) Mouse virulence of <u>Salmonella</u> strains: the effect of different smooth-type O-side chains. J Gen Microbiol 64:255-268.

Valtonen VV, Mäkelä PH (1971) The effect of lipopolysaccharide modifications - antigenic factors $1,5,12_2$ and 27 - on the virulence of <u>Salmonella</u> strains for mice. J Gen Microbiol 69:107-115.

Valtonen VV, Aird J, Valtonen M, Makela O, Mäkelä PH (1971) Mouse virulence of <u>Salmonella</u>: antigen-dependent differences are demonstrable also after immunsuppression. Acta Path Microbiol Scand 79B:715-718.

Valtonen MV, Larinkari UM, Plosila M, Valtonen VV, Mäkelä PH (1976) Effect of enterobacterial common antigen on mouse virulence of <u>Salmonella typhimurium</u>. Infect Immun 13:1601-1605.

INTERACTION OF HAEMOPHILUS INFLUENZAE WITH COMPLEMENT

E.R. Moxon[1] and J.A. Winkelstein[2]

[1] Infectious Diseases Unit
Department of Paediatrics
John Radcliffe Hospital
Headington, Oxford OX3 9DU
United Kingdom

[2] Division of Pediatric Immunology
The Johns Hopkins Hospital
600 N. Wolfe Street
Baltimore, Maryland 21205

Interaction of complement and H. influenzae in vitro

H. influenzae, whether capsulated or not, is able to activate both the classical and alternative pathways (Anderson et al, 1972; Quinn et al, 1977) resulting in the potential to generate bactericidal and opsonising activities. Most studies have investigated these activities in relationship to type b strains because of the importance of this serotype as a cause of life-threatening invasive infections in childhood. Anderson et al (1972) showed that either purified type b capsular polysaccharide (PRP) or non-capsulated H. influenzae organisms could absorb factors from human serum which were necessary for complement dependent bactericidal activity. Subsequent studies by Steele et al used affinity purified IgG antibodies and agammaglobulinaemic human serum from which factor D and properdin had been selectively inactivated to demonstrate that H. influenzae type b could be killed through classical pathway activities mediated by antibodies to PRP. Similarly, bactericidal activity could also be demonstrated using an IgG fraction from human sera that had been pre-adsorbed with PRP to remove anticapsular antibodies. Taken together, these studies suggested that antibodies to either the PRP capsule or to cell envelope constituents are able to activate the classical pathway and generate bactericidal activity.

H. influenzae b is also able to activate the alternative pathway, although the specific component(s) of the cell wall involved is not known.

NATO ASI Series, Vol. H24
Bacteria, Complement and the Phagocytic Cell
Edited by F.C. Cabello und C. Pruzzo
© Springer-Verlag Berlin Heidelberg 1988

Evidently, PRP is not responsible since unencapsulated organisms, lacking detectable PRP, activate the alternative pathway whereas purified PRP does not (Quinn et al, 1977). Tarr et al (1984) showed that the bactericidal activity resulting from activation of the alternative pathway was antibody dependent. C4-deficient guinea-pig serum alone did not kill H. influenzae b, although C3 was consumed. However, after the addition of a human serum IgG fraction obtained from an individual previously immunised with type b capsule, bactericidal activity could be demonstrated. These observations indicated a requirement for type b specific antibodies for alternative pathway killing. Since prior adsorption of the IgG fraction to remove anticapsular antibodies decreased the bactericidal activity, and because a variant strain lacking capsule was not killed, it was concluded that antibodies to PRP were an absolute requirement for alternative pathway killing. In subsequent studies, Steele et al (1984) used affinity-purified anticapsular IgG antibodies and agammaglobulinaemic serum (both derived from humans) and were able to demonstrate that anticapsular antibodies were required to kill H. influenzae b by the alternative pathway. Using C2-deficient human serum or agammaglobulinaemic-Mg EGTA serum, they were unable to detect complement consumption via the alternative pathway, whereas they and others (Quinn et al, 1977) observed consumption of complement using C4-D guinea-pig serum (GPS). These findings emphasise the importance of species differences in alternative pathway activities. In summary, there is a requirement for antibody to PRP in the generation of bactericidal activity through the alternative pathway, but antibodies to either capsular PRP or cell-wall antigens may be lethal to H. influenzae b through the classical pathway.

Whether or not complement mediated bactericidal activity plays a biologically relevant role in the clearance of H. influenzae b in human infections is not certain, but capsular and non-capsular antibodies are also able to confer protection by opsonisation (Anderson et al, 1972). At least three distinct mechanisms exist. Antibody of the IgG class may act as a primary opsonic ligand, independent of its ability to activate complement. Antibody may also promote opsonisation by classical pathway activation and deposition of C3b. Finally, C3b may be fixed on the bacterial cell surface by activation of the alternative pathway. Johnston et al (1973a) compared opsonisation of H. influenzae b in normal serum,

serum from a patient with genetically determined C2 deficiency and serum
from a patient with renal disease and acquired C3 deficiency. The fact
that C3 dependent opsonisation can can occur in serum from patients with
genetically determined total deficiency of C2 indicates that at least some
of the opsonic activity may be mediated via the alternative pathway.
Purified anticapsular IgG is capable of opsonising H. influenzae b;
however, its full potential can only be demonstrated in the presence of an
intact complement system (Johnston et al, 1972). Antibodies to cell wall
antigens are also able to initiate complement dependent opsonisation
(Anderson et al, 1972).

Susceptibility of encapsulated H. influenzae strains, representative of
the five other serotypes, to complement mediated killing was investigated
by Sutton et al (1982) using serum from a colostrum deprived calf which
contained no detectable gammaglobulin when analysed by cellulose acetate
electrophoresis. Type d and type e strains were rapidly killed in 20%
colostrum deprived calf serum. When the serum concentration was raised to
90%, type c and type f strains were also killed. Type a organisms were
more resistant to killing by this serum, but only type b strains showed no
net reduction in bacterial counts even when undiluted serum was used. It
was proposed, on the basis of these observations, that the relatively
greater resistance of type b strains to complement mediated killing
correlated with the greater virulence displayed by type b strains.

The interactions of complement and non-typable H. influenzae have
received sparse attention. Activation of both alternative and classical
pathways provides the potential for both bactericidal and opsonic
activities but the striking heterogeneity of non-typable clinical isolates
has precluded general conclusions based on data derived from studies on
arbitrarily selected and uncharacterised strains. Musher et al (1983)
demonstrated bactericidal and opsonic activity in pooled adult human serum
although this activity varied substantially depending on the strain under
study. Adsorbtion of the serum pool by any one non-typable strain resulted
in removal of some, all or none of the bactericidal activity thus
emphasising the heterogeneity of antigens which are targets for the
relevant serum antibodies. In general, bactericidal activity correlated
with opsonic activity, but this was not invariably so. For example,

patients with pneumonia caused by non-typable H. influenzae whose sera were found to possess ample bactericidal activity, but relatively modest serum opsonic activity, acquired significantly greater opsonic activity over the ensuing three weeks (Musher et al, 1983). It must be concluded that the roles of the alternative and classical pathway, and indeed the role of other serum factors, in host defense against non-typable H. influenzae remains poorly understood.

Role of Complement in Experimental H. influenzae Infections:

The interaction of the host and an invading micro-organism is complex and dynamic. The host possesses a number of different defence mechanisms, each acting in a somewhat different manner, often at a different site of infection and at different times during the infection. In addition, a number of different immunological factors accomplish the same biological end result, such as opsonisation. Accordingly, although in vitro studies can give a great deal of insight into the potential of the complement system to participate in the host defence against H. influenzae, studies in experimental animals and observations in patients are vital to establish the biological significance of the complement system and provide insight into the mechanisms by which it might exert its protective effects. To examine these issues, several experiments have made use of a model of H. influenzae infection in young rats, a biologically relevant model of human disease.

The protective role of complement in experimental type b infections of infant rats was examined following intranasal challenge of animals rendered C3 deficient by injection of cobra venom factor (CVF). (Crosson et al, 1976). CVF is cobra C3b and as such is insusceptible to the action of mammalian C3b inactivator (Factor I). As a result, when purified CVF is injected into an experimental animal, it leads to the uncontrolled formation of the alternative pahtway C3 convertase, C3bBb, and results in the pharmacological depletion of C3. Since C3 is required for the activation of C5-9, C3 depletion renders the animal functionally depleted of all activities mediated by C3-9. Compared to control animals (injected with saline), complement deficient rats developed a significantly higher incidence and magnitude of bacteraemia after intranasal challenge. Deaths occurred in the complement depleted rats but not in control rats. Thus, one effect of complement is apparently to limit the magnitude of bacteraemia to a sub-lethal level. This protective activity could then allow time for other immune mechanisms, such as specific antibody synthesis, to begin and thus result in survival. The less efficient clearance of organisms in complement depleted rats also resulted in a higher incidence of meningitis, although the magnitude of bacteraemia at which the complement depleted rats developed meningitis was not changed and the number of bacteria in the CSF and CVF treated rats was similar to controls. Therefore, neither protection against CNS bacterial invasion nor control of bacterial multiplication within the CNS was a complement mediated effect in this model. The lowered CSF leukocyte counts in the CVF treated rats could reflect lowered chemotactic activity due to C5 depletion as well as decreased C3 dependent C5 activation. Such complement related chemotactic activity may be important in the early stages of the inflammatory response and, hence, may be involved in the early entry of leukocytes into the CSF. Alternatively, the lowered CSF leukocyte counts may reflect the depressed peripheral leukocyte response caused by overwhelming bactereamia. However, serial measurements of blood leukocytes in both CVF and saline injected rats did not support this latter hypothesis. The potential for complement to play an important role in the non-immune host (lacking antibodies to the b capsule), was therefore demonstrated.

Similar experiments performed on different serotypes of encapsulated H. influenzae were performed using clinical isolates (Corrall et al, 1982) or genetically related transformants (Zwahlen et al, 1983a). For example, when type b and type c clinical isolates and genetically related transformants expressing either the b or c capsule were compared, C3 depletion resulted in relatively greater virulence enhancement of the type c strains so that they resembled the type b strains. These findings suggested that type b strains have a phenotype that is more resistant to complement-mediated activities than when the capsule is, for example, of the type c phenotype.

The potential role of complement on intravascular and intrapulmonary clearance of non-typable H. influenzae has also been studied in experimentally infected rats and mice respectively. Zwahlen et al (1983b) showed that complement deficiency is important in host defense against non-typable H. influenzae and that complement deficiency provided a model for demonstrating differences in virulence potential that were attributable to determinants other than capsule. Thus, depletion of C3 significantly enhanced early intravascular bacterial survival, so that bacteraemia developed in CVF treated rats but not those injected with saline. Indeed, a type b variant lacking capsule was significantly more virulent than two non-typable strains, suggesting that the virulence phenotype of type b strains is determined, in part, by its cell wall antigens.

Toews et al (1985) studied the role of complement in pulmonary clearance using congenic C-5 deficient (C-5D) and normal mice following endobronchial challenge with a non-typable H. influenzae strain. Pulmonary clearance was significantly impaired in the C-5D mice and the authors concluded that the more efficient clearance of H. influenzae occurred through the prompt recruitment of PMNs and that C-5D mice lacked the capacity to generate sufficient amounts of intra-alveolar chemotoxins (e.g. C5a).

Biological Implications:

There is firm evidence which indicates an important role for complement in human host defenses against H. influenzae. Non-typable and capsulated H. influenzae can activate both the alternative and classical pathways in vitro to generate bactericidal and opsonic activities. The potential biologic relevance of these activities is supported by experimental infection of animals; complement deficiency can be shown to affect intravascular and intra-alveolar clearance of H. influenzae. Finally, "experiments of nature" in the form of inherited or acquired deficiencies of complement provide direct support for a biologically relevant role for complement in human host defense against infection. Thus, among the more than forty reported patients with C2 deficiency, two young girls presented

with H. influenzae meningitis (Thong et al, 1980). Two patients with C3b
inactivator have been described (Alper et al, 1970; Eng et al, 1980) who
contracted H. influenzae b infections and at least one patient with
homozygous C3 deficiency had recurrent infection which included H.
influenzae (Ballow et al, 1975). Amalgamating the evidence, it seems
reasonable to suggest that the major contribution of complement relates to
its opsonic, rather than bactericidal, activity as far as H. influenzae
infections are concerned. The increased susceptibility of patients with
deficiencies of early complement components (C2,C3) contrasts with the
fact that terminal complement deficiencies, which are known to predispose
to recurrent neisserial infections (Alper et al, 1970), do not increase
susceptibility to H. influenzae. Thus, an inability to assemble and insert
the membrane attack complex does not seem to invite unusual susceptibility
to either encapsulated or non-typable H. influenzae infections, whereas
the lack of opsonically active C3 or either of the two pathways necessary
for its activation, does. This hypothesis receives further support from
experimental infection of rats in which H. influenzae organisms labelled
with ^{32}P were inoculated intravenously (Weller et al, 1978). Clearance
curves were obtained in which the decline in ^{32}P paralleled that of viable
bacteria. The loss of label from blood could have arisen through either
phagocytic uptake of lysis of H. influenzae and subsequent uptake of
"free" label by the reticulo-endothelial system. However, uptake of "free"
^{32}P by spleen or liver was found to be less than 5% of total, whereas
bacteriolysis liberated approximately 70% of the label when the bacteria
were subjected to bacteriolysis in vitro. Thus, the ^{32}P found in the liver
and spleen of rats (which accounted for 80-90% of that expected) probably
represented phagocytosis of organisms.

How does the role of complement fit into the pathogenesis of systemic
H. influenzae b infections in young children? There is a lack of
information concerning its participation in the events which result in
colonisation, or in the translocation of organisms from the respiratory
tract to the blood. There is substantial evidence to suggest that when H.
influenzae type b reaches the blood, intravascular clearance is less
efficient in the absence of complement components. In the non-immune host,
lacking antibodies to PRP, the type b capsule might act as a shield
limiting the extent to which cell envelope antigens are able to activate

the alternative pathway or interact with antibodies to cell envelope antigens. Alternatively or in addition, PRP may favor the binding or β1H, relative to B, thus increasing the rate of C3b degradation or alter the site at which C3b is deposited on the bacterical cell surface.

Additional factors need to be considered. First, there is considerable heterogeneity among the polypeptides that comprise the major outer membrane proteins and the LPS molecules of the outer leaflet. Second, antigenic variation among LPS molecules has been demonstrated in which the loss or gain of surface exposed (and therefore antibody accessible) epitopes occurs at high frequency (circa 1%) (Kimura and Hansen, 1986). This switching seems to be of the classic on-off-on variety and occurs presumably at the transcriptional level. This antigenic variation correlates with changes in the susceptibility of these strains to complement-mediated bactericidal activity and virulence expression. Phenotypic variations in H. influenzae have also been demonstrated and the susceptibility of any given strain may depend on environmental factors. For example, organisms obtained from bacteraemic animals or patients are relatively resistant to complement dependent lysis when compared to the same organisms grown in broth (Shaw et al, 1976). Conversely, organisms grown in broth can be made relatively resistant to the bactericidal action of complement by incubating them for brief periods of time in normal serum or a lower molecular weight dialysate of serum. This acquired resistance of H. influenzae to antibody-dependent serum bactericidal activity appears to be the result, at least in part, of the acquisition of resistance to anti-LPS antibodies (Anderson et al, 1980) rather than to the action of complement per se.

An understanding of the complex mechanisms involved in the regulation of the bacterial cell surface would appear to be essential for a clear understanding of the interactions of H. influenzae with complement. The use of molecular techniques for engineering genetically defined strains affords a powerful tool to assist in these analyses. The feasibility to this approach has been demonstrated in recent work from our laboratories (Moxon et al, 1984; Ely et al, 1986).

Acknowledgements

We wish to thank Dr. Dan Musher for his critical review of the manuscript and Mrs. Sheila Hayes for the manuscript preparation.

References

Alper, C., Abramson, N., and Johnston, R.B. Increased susceptibility to infection associated with abnormalities of complement-mediated functions and of the third component of complement (C3). N. Engl. J. Med. 282:349-354 (1970).

Anderson, P., Johnston, R.B. Jr., and Smith, D.H. Human serum activities against Haemophilus influenzae type b. J. Clin. Invest. 51:31-38 (1972).

Anderson, P., Flesher, A., Shaw, S., Harding, L.A., and Smith, D.H. Phenotypic and genetic variation in the susceptibility of Haemophilus influenzae type b to antibodies to somatic antigens. J. Clin. Invest. 65:885-891 (1980)

Ballow, M. Shira, J.E., Harden, L., Yang, S.Y., and Day, N.K. Complete absence of the third component of complement in man. J. Clin. Invest. 56:703-710 (1975)

Corrall, C.J., Winkelstein, J.A., Moxon, E.R. Participation of complement in host defense against encapsulated Haemophilus influenzae types a, c, and d. Infect. Immun. 35:759-763 (1982).

Crosson, F.J. Jr., Winkelstein, J.A., and Moxon, E.R. Participation of complement in the non-immune host defense against Haemophilus influenzae type b septicemia and meningitis. Infect. Immun. 14:882-887 (1976).

Ely, S., Tippett, J., and Kroll, J.S., and Moxon, E.R. Mutations affecting expression and maintenance of genes encoding the serotype b capsule of Haemophilus influenzae. J. Bact. 167:44-48 (1986).

Eng, R.H.K., Seligman, S.J., Arnout, M.A., and Alper, C.R. Variable expression of homozygous C3b inactivator deficiency. Clin. Res. 26:394A (1978).

Johnston, R.B. Jr., Anderson, P., and Newman, S.L. Opsonization and phagocytosis of Haemophilus influenzae type b. In: Sell, S.H.W., Karzon, D.T. (eds.) Haemophhilus influenzae. Nashville: Vanderbilt University Press, pp. 99-112 (1973a).

Johnston, R.B. Jr., Anderson, P., Rosen, F.S., and Smith, D.H. Characterization of human antibody to polyribose-phosphate, the capsular antigen of Haemophilus influenzae type b. Clin. Immunol. Immunopath. 1:234-240 (1973b).

Kimura, A. and Hansen, E.J. Antigenic and phenotypic variations of Haemophilus influenzae type b lipopolysaccharide and their relationship to virulence. Infect. Immun. 51:69-79 (1986).

Moxon, E.R., Deich, R.A. and Connelly, C.J. Cloning of chromosomal DNA from Haemophilus influenzae: Its use for studying the expression of type b capsule and virulence. J. Clin. Invest. 73:298-306 (1984).

Musher, D.M., Hague-Park, M., Baughn, R.E., Wallace, R.J., and Cowley, B. Opsonizing and bactericidal effects of normal human serum on non-typable Haemophilus influenzae. Infect. Immun. 39:297-304 (1983).

Musher, D.M., Kubitschek, K.R., Crennan, J., and Baughn, R.E.   Pneumonia
    and acute febrile tracheobronchitis due to Haemophious influenzae. Ann.
    Intern. Med. 99:444-450 (1983).
Quinn, P.H., Crosson, F.J., Winkelstein, J.A., and Moxon, R.W.   Activat-
    ion of the alternative complement pathway by Haemophilus influenzae,
    type b. Infect. Immun. 16:400-402 (1977).
Shaw, S., Smith, A.L., Anderson, P., and Smith, D.H.   The paradox of
    Haemophilus influenzae, type b, bacteremia in the presence of serum
    bactericidal activity. J. Clin. Invest. 58:1019-1029 (1976).
Steele, N.P., Munson, R.S. Jr., Granoff, D.M., and Cummins, J.E., and
    Levine, R.P.   Antibody-dependent alternative pathway killing of
    Haemophilus influenzae type b. Infect. Imun. 44:452-458 (1984).
Sutton, A.R., Schneerson, R., Kendall-Morris, S., and Robbins, J.B.   Dif-
    ferential complement resistance mediates virulence of Haemophilus
    influenzae type b. Infect. Immun. 35:95-104 (1982).
Tarr, P.I., Hosea, S.W., Brown, E.J., Schneerson, R., Sutton, A., Frank,
    M.M.   The requirement of specific anticapsular IgG for killing of
    Haemophilus influenzae by an alternative pathway of complement
    activation. J. Immunol. 120:1772-1775 (1982).
Thong, Y.H., Simpson, D.A., and Muller-Eberhard, H.J.   Homozygous defi-
    ciency of the second component of complement presenting with recurrent
    bacterial meningitis. Arch. Dis. Child. 55:471-473 (1980).
Toews, G.B., Vial, W.C. and Hansen, E.J.   Role of the C5 and recruited
    neutrophils in early clearance of non-typable Haemophilus influenzae
    from murine lungs. Infect. Immun. 50:207-212 (1985).
Weller, P.F., Smith, A.L., Smith, D.H., and Anderson, P.   Role of immunity
    in the clearance of bacteremia due to Haemophilus influenzae. J.
    Infect. Dis. 138:427-436 (1978).
Zwahlen, A., Winkelstein, J.A. and Moxon, E.R.   Surface determinants of
    Haemophilus influenzae pathogenicity: Comparative virulence of capsular
    transformants in normal and complement-depleted rats. J. Infect. Dis.
    148:385-394 (1983a).
Zwahlen, A., Winkelstein, J.A., and Moxon, E.R.   Participation of comple-
    ment in host defense against capsule-deficient Haemophilus influenzae.
    Infect. Immun. 42:708-715 (1983b).

INTERACTION OF PSEUDOMONAS AERUGINOSA WITH COMPLEMENT

Neal L. Schiller

Division of Biomedical Sciences
University of California, Riverside
Riverside, CA 92521

Introduction

Pseudomonas aeruginosa is a ubiquitous gram-negative organism which is
not ordinarily pathogenic for individuals with normal host defenses.
However, this organism is a major cause of morbidity and mortality for
patients with serious burn or wound injuries, neoplastic disease,
granulocytopenia, immunological deficiencies or cystic fibrosis (CF), and
for patients receiving intensive immunosuppressive chemotherapy (Bodey et
al. 1983). Once established, this opportunistic pathogen can produce a
variety of extracellular virulence factors, including toxins, hemolysins,
aggressins, adherence factors, and a slime exopolysaccharide (prevalent
amongst isolates from CF patients), any of which can contribute to the
disease manifestations observed and/or to the resistance of this organism
to host clearance mechanisms (Lory and Tai, 1985). Furthermore, infections
with P. aeruginosa remain extremely difficult to control and/or eradicate
with currently available antibiotics. As a result, a more comprehensive
understanding of the interaction between this organism and host defense
mechanisms is needed to guide efforts aimed at developing alternative
prophylactic or immunotherapeutic treatment regimens.

The importance of complement in host defense against P. aeruginosa
infection has been demonstrated in a number of studies. Optimal
phagocytosis of P. aeruginosa by human polymorphonuclear leukocytes and/or
alveolar macrophages requires the participation of complement (Young and
Armstrong, 1972; Bjornson and Michael, 1973; Peterson et al, 1978; Murphey
et al 1979; Hammer et al, 1981; Nguyen et al, 1982). These in vitro
observations are supported by research using C5 deficient mice (Larsen et
al, 1982) or mice rendered hypocomplementemic using cobra venom factor

NATO ASI Series, Vol. H24
Bacteria, Complement and the Phagocytic Cell
Edited by F. C. Cabello und C. Pruzzo
© Springer-Verlag Berlin Heidelberg 1988

(Gross et al, 1978; Heidbrink et al, 1982). In either case, pulmonary clearance mechanisms for P. aeruginosa were totally ineffective in complement deficient mice in contrast to results seen in complement sufficient control mice, demonstrating the crucial role of complement as an opsonin and/or chemotaxin in these animal models. Further support for the role of C3 as an important opsonin for phagocytosis of this gram negative microorganism has been provided by recent studies which have correlated the resistance of some P. aeruginosa isolates to polymorphonuclear leukocyte phagocytosis to the inadequate binding of C3 onto these strains (Baltimore and Shedd, 1983; Engels et al, 1985 a,b). Complement, activated via either the classical or alternative pathway, has also been found to be directly bactericidal for P. aeruginosa strains isolated from the sputum of CF patients (discussed in more detail below). Taken together, these reports document the central role which complement plays in host defense against P. aeruginosa infection.

The focus of this report will be to: 1) characterize the mechanism of complement-mediated killing of a mucoid serum-sensitive strain; 2) define how serum-resistant strains avoid the complement-mediated bactericidal activity of serum; and 3) examine the apparent inconsistency between complement activation and C3 binding by certain strain of P. aeruginosa.

Mechanism of complement-mediated bactericidal activity

Whereas most strains P. aeruginosa isolated from blood, wounds, urine or burns are serum resistant (Young and Armstrong, 1972; Reyes et al, 1979); Schiller and Hatch, 1983), strains isolated from the sputum of patients with CF are usually serum sensitive and deficient in lipopolysaccharide (LPS) 0 side chains (Muschel et al, 1969; Hoiby and Olling, 1977; Thomassen and Demko, 1981; Meschulam et al, 1982; Schiller and Hatch, 1983; Hancock et al, 1983). Since this serum susceptibility clearly distinguishes CF strains from non-CF strains, and since serum susceptibility is related to the composition of the bacterial surface of gram-negative organisms, it was of interest to compare the nature of serum susceptibility and resistance in these strains.

There have been a number of studies which have examined the nature of serum sensitivity in P. aeruginosa. Several investigators have presented evidence suggesting the importance of the classical pathway activated by antibody in the killing of some P. aeruginosa strains (Thomassen and Demko, 1981; Schiller et al, 1984a). In contrast, some reports have stated that certain strains are killed by complement activated via the alternative pathway in the apparent absence of antibodies (Offredo-Hemmer et al, 1983; Pier and Ames, 1984), while others have suggested a role for either or both pathways (Meshulam et al, 1982). Most likely, these differences represent strain variations in serum susceptibility amongst P. aeruginosa isolates.

Previous studies in this laboratory using P. aeruginosa strain 144M have suggested that the bactericidal activity of pooled normal human serum (PNHS) for this strain is due to an antibody-depedent activation of complement via classical pathway (Borowski and Schiller 1983; Schiller et al 1984a). These observations are supported by the following experiment. Strain 144M was incubated at 37°C in the presence of purified ^{125}I-labeled human C3 and various serum preparations, and at various time intervals bacterial survival and C3 binding was determined as previously described (Schiller and Joiner, 1986). As shown in Table 1, 10% PNHS led to a rapid deposition of C3 onto the surface of 144M (3.55×10^4 molecules of C3 per colony forming unit (CFU) in 60 min), and produced a 2.67 $\log_{10}$ kill of strain 144M. Whereas most of the bactericidal activity of PNHS was blocked by the addition of 2mM Mg^{2+} and 10 mM EGTA (MgEGTA), which inhibits the activation of the classical pathway, heating PNHS at 50°C for 20 min, which blocks the alternative pathway, had no appreciable effect on either serum bactericidal killing of C3 deposition. These studies support the important role of the classical pathway for serum bactericidal activity for this strain. Furthermore, pre-absorption of PNHS with 144M at 0°C (MabsS) removed essentially all bactericidal activity from PNHS. This loss of bactericidal activity is believed to be due to removal of specific bactericidal antibodies since CH_{50} levels of MabsS were $\geq 90\%$ of that in PNHS. Furthermore, the activity lost by preabsorption with 144M could be restored by the addition of affinity purified anti-144M LPS antibodies (αLPS Ab) isolated from PNHS.

Table 1. Measurement of bactericidal activity for and C3 binding on strain 144M in the presence of various serum preparations

		molecules C3/CFU	
Serum source[2]	$\log_{10}$ kill[2]	15 min	60 min
10% PNHS	2.67	1.22×10^4	3.55×10^4
10% PNHS (50°C, 20 min)	2.43	1.01×10^4	2.81×10^4
10% PNHS (MgEGTA)	0.15	0	0.26×10^4
10% MabsS	0.18	0.23×10^4	1.12×10^4
10% MabsS + 10% αLPS Ab	2.09	1.15×10^4	3.16×10^4
10% MabsS + 10% HIS	2.22	n.d	n.d

[1] See text for description of these items.
[2] Data represent the difference between $\log_{10}$ CFU/ml in PNHS- heated at 56°C for 30 min (HIS) and $\log_{10}$ CFU/ml in test serum after 60 min incubation at 37°C.

Table 2. Comparison of complement activation by serS and serR P.aeruginosa strains

	% Complement Consumed[1]				
serS strains	OD = 0.1[2]	OD = 0.2[2]	serR strains	OD = 0.1	OD = 0.2
144M	18.2%	27.0%	144M- SR	67.7%	100%
WcM	16.1%	30.6%	WcM- SR	47.4%	100%
ByM	26.6%	31.4%	ByM- SR	64.0%	100%

[1] Complement consumption was determined by comparing the no. of CH_{50} units remaining in PNHS after 60 min incubation at 37°C with bacteria to that in PNHS incubated under similar conditions without bacteria. % of complement consumed = [1- (CH_{50} units in test serum/CH_{50} units in control serum)] x 100%.
[2] Bacterial concentrations measured by optical density at 550 nm. An OD_{550} nm of $0.1 \simeq 2 \times 10^8$ bacteria/ml.

Similar bactericidal activity was observed with the addition of 10% HIS to MabsS, suggesting that αLPS Ab are the principal bactericidal antibodies in PNHS. Note that the addition of αLPS Ab to MabsS caused a more rapid deposition of C3 onto 144M than MabsS alone, indicating that the role of bactericidal antibodies in PNHS directed to the rough type LPS on 144M is to facilitate the activation and deposition of C3 onto the surface of this strain, which in turn leads to the generation of the lethal terminal membrane attack complex, C5b-9.

Mechanism of serum resistance

To examine the nature of serum resistance of P. aeruginosa strains, a series of serum-resistant (serR) derivatives were isolated from serum-sensitive (serS) parental strains initially obtained from the sputum of CF patients. This was accomplished by passing serS strains through increasing concentrations of PNHS until strains totally resistant to serum killing were isolated. Six pairs of serS and serR strains were prepared in this manner (Schiller et al, 1984b). The only difference noted between the serS and serR strains was that whereas the parental serS strains were deficient in the expression of LPS O side chains (as expected for isolates from CF patients), the serR derivatives contained LPS bearing long O side chains as evidenced by their migration characteristics in SDS-polyacrylamide gels (Schiller et al, 1984b). It should be noted that resistance to complement-mediated killing has also been observed in certain Escherichia coli and Salmonella spp. strains possessing LPS with long O side chains whereas isogenic mutants with LPS deficient in O side chains were serS (Nelson and Roantree, 1967; Rowley, 1968; Muschel and Larsen, 1970).

The first experiment was designed to determine whether these serR derivatives were resistant to complement-mediated killing due to an inability to activate complement. However, as shown in Table 2, the serR derivatives actually activated or consumed more complement than equivalent numbers of serS strains. Since this assay only measured total complement activity as measured by lysis of antibody-coated sheep red blood cells (CH_{50}), additional experiments were done to examine activation of individual complement components C3, C5, and C9 by 144M (serS) and 144M-SR (serR). While strain 144M activated 54.9% and 27.4% of the available C3 and C5 respectively, equivalent numbers of the serR strain 144M-SR activated 88.7% of the C3 and 96.4% of the C5 during the same 60 min incubation period (Schiller and Joiner, 1986). Furthermore, as shown in Figure 1, the serR strain also activated C9 much more rapidly and completely than its serS counterpart. However, this activation did not result in equivalent levels of C9 binding. In this experiment each strain was incubated at 37°C in 10% PNHS containing ^{125}I-labeled human C9 and the kinetics of C9 uptake and C9 consumption by the 2 strains was compared. There was an initial deposition of C9 on the surface of 144M-SR which peaked at 15 min, followed by a gradual decline over the next 105 min (Figure 1). In contrast, a lag in C9 binding occurred with 144M (consistent with a lag in C9 activation), but deposition increased rapidly

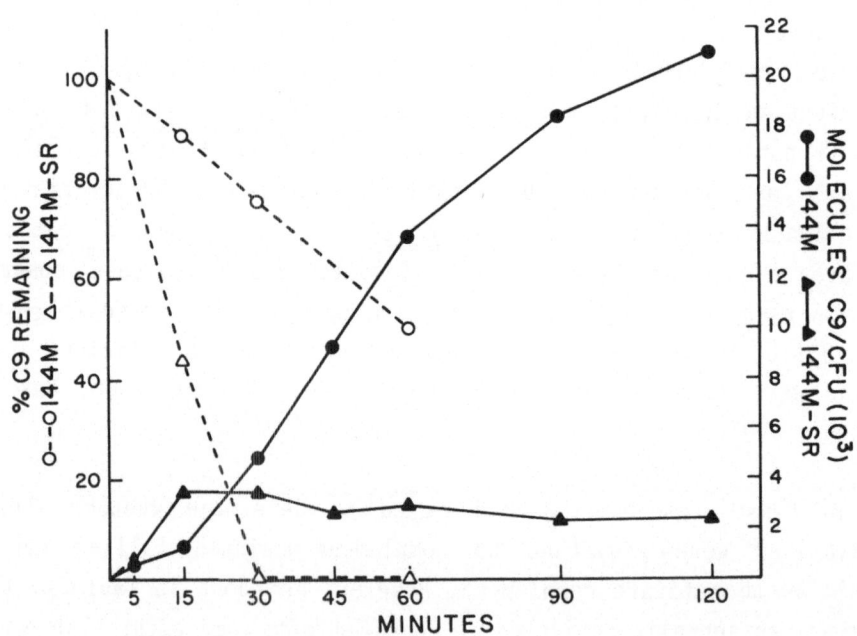

Fig. 1. Consumption and uptake of C9 by P. aeruginosa 144M and 144M-SR. All values for C9 consumption were expressed relative to a control tube containing serum without bacteria incubated under similar conditions. Values for C9 binding were corrected for non-specific binding, using control tubes with bacteria incubated with 10% HIS.

after 15 min and continued to increase over the 120-min incubation period. At 60 min there was 4.6 times more C9 on 144M (13,700 molecules of C9/CFU) than on 144M-SR (2,990 molecules of C9/CFU).

Table 3. Comparison of complement activation and C3 binding activity by serS and serR P. aeruginosa strains

strain	serS strains % Complement consumed	molecules C3/CFU	strain	serR strains % Complement consumed	molecules C3/CFU
144M	32.1%[1]	4.79×10^4	144M-SR	100%	2.74×10^4
P1M	40.3%	1.05×10^4	Mc208	100%	0.90×10^4
P7b	35.2%	3.43×10^4	Mc209	58.2%	0.28×10^4
P10	52.8%	5.65×10^4	Mc210	60.7%	0.08×10^4
P11	38.0%	n.d.			

[1] Complement consumption was determined by comparing the no. of CH_{50} units remaining in PNHS after 60 min incubation at $37^{0}C$ with ~ 4 x 10^8 bacteria/ml of serum to that in PNHS incubated under similar conditions without bacteria. % complement consumed = [1-(CH_{50} units in test serum/CH_{50} units in control serum)] x 100%.

When the bacterial pellets with C9 attached were treated with 0.1% trypsin for 30 min at $37^{0}C$, 93.4% of the C9 on 144M-SR was released, indicating that the C9 on the surface of 144M-SR was not inserted into the outer membrane of these bacteria. In contrast, more than 5,700 molecules of C9 remained on the surface of 144M after similar trypsin treatment (Schiller and Joiner, 1986). These experiments suggest that the serum resistance of P. aeruginosa strain 144M-SR does not represent a failure to activate complement effectively, but instead reflects a failure of the assembled terminal complement complex C5b-9 to insert stably into the outer membrane of this strain. This mechanism of serum resistance is similar to that previously described for other strains bearing long LPS O side chains, suggesting that this is one common strategy employed among these gram-negative bacteria for avoiding the effects of this important host defense mechanism (Goldman et al., 1984; Joiner et al., 1982a, 1982b, 1984; Taylor and Kroll, 1984).

Table 4. Summary of results observed after 60 min incubation of each
strain with 10% PNHS at 37°C

Strain	Time	³H-C4 bound/CFU	%C3 consumed	C3 bound/ CFU	%C5 consumed
144M	5	48.1	--	5,200	--
	15	97.9	26.4	25,787	5.1
	30	112.5	37.8	37,390	20.1
	60	141.8	54.9	51,980	27.4
144M-SR	5	3.5	--	2,596	--
	15	47.7	61.0	17,612	33.8
	30	76.1	76.7	22,843	76.8
	60	55.6	88.7	26,841	96.4

Discrepancy between complement activation and C3 binding

During the course of the investigations described above, the apparent
contrast between complement activating ability and the binding of C3 onto
the surface of 144M and 144M-SR was noted (Schiller and Joiner, 1986). To
determine if this observation was unique to these 2 strains, we compared
the complement activating ability of 4 other serS CF patient sputum
strains (P1M, P7b, P10 and P11) and 3 serR strains isolated from patients
with P. aeruginosa bacteremia (Mc208, Mc209 and Mc210). As shown in Table
3, when comparing equivalent numbers of serS and serR strains, the serR
strains generally activated complement better than the serS strains.
However, this activating ability did not translate directly into effective
C3 binding. This was especially noticeable with the bacteremic strains
Mc208, Mc209 and Mc210.

A closer examination of the relationship between C3 activation and
binding was done using strains 144M and 144M-SR. Earlier experiments had
demonstrated that MgEGTA treatment of PNHS completely blocked C3
consumption and C3 binding by both 144M and 144M-SR, indicating the

importance of the classical pathway for activation of C3. In this study, each strain was incubated at 37°C in 10% PNHS containing either ^{3}H-C4, ^{125}I-C3 or ^{125}I-C9, and samples were removed at the indicated times and examined for C4, C3 or C9 binding and C3, C5 or C9 consumption as previously described (Schiller and Joiner, 1986). As shown in Table 4, 144M had 112.5 molecules of ^{3}H-C4 bound/CFU after 30 min of incubation, with 37.8% of the available C3 consumed and 37,390 molecules of C3 attached/CFU. In contrast, with only 76.1 molecules of ^{3}H-C4 bound/CFU, 76.7% of the available C3 had been activated by 144M-SR; however, only 22,843 molecules of C3 were detected on the surface of 144M-SR.

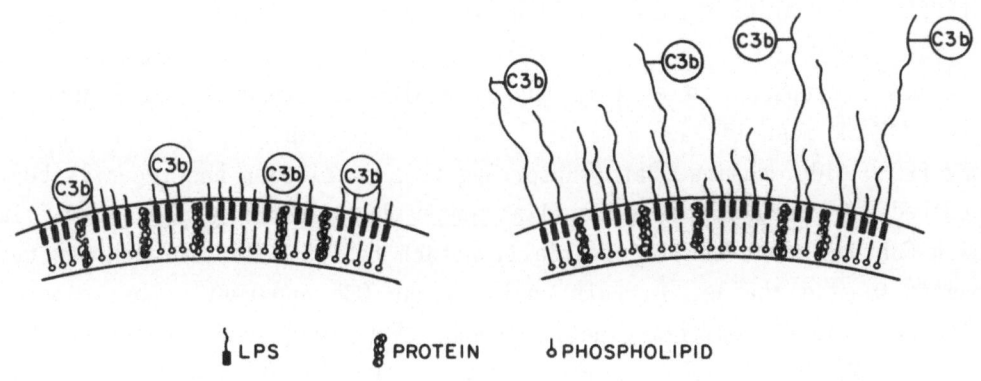

Fig. 2. Model of the interaction of C3b with the surface of strain 144M, a serS strain with LPS deficient in O side chain (shown on the left) and strain 144M-SR, a serR strain with LPS bearing long O side chains (shown on the right).

A lack of correlation between C3 consumption and C3 fixation by P. aeruginosa strains varying in LPS composition has also been observed by other investigators (Engels et al, 1985a,b). These authors postulated that C3 might be released from the surface of bacteria by the solubilizing effect of complement on preformed IgG-C3b complexes, which have been reported to be more effective in alternative pathway consumption of C3 than fluid phase C3b (Fries et al, 1984). An alternative explanation for this discrepancy might be that C3 had been enzymatically digested by bacterial proteases. P. aeruginosa elastase has been reported to be able to inactivate several complement components, including C3 (Schultz and Miller, 1974). However, neither 144M nor 144M-SR bacterial pellets nor culture supernatants were able to directly inactivate or consume either C3 or C5 over the 60 min incubation period at 37°C. Therefore, it would appear that there is a marked discrepancy between the C3 activating ability and the efficiency of C3 binding by these 2 strains. Furthermore, since the C3b attached to the cell surface serves to bind C5 for cleavage by C2a, one might expect more C5 consumption by 144M which has a greater number of C3 molecules on its surface than 144M-SR. That this is not the case implies that not only do these 2 strains differ in C3 binding ability, but the effectiveness of the bound C3 as a participant in C5 convertase activity also varies significantly between these 2 strains.

A model proposed to explain these results is presented in Figure 2. With 144M-SR most of the activated C3 might be expected to bind to the long LPS O side chains. Data supporting this prediction has recently been reported for strains of Salmonella montevideo (Joiner et al, 1986) in which C3b was found to preferentially attach to LPS molecules bearing the longest O side chains. In this position the C3b apparently participates effectively in C5 convertase activity and, therefore, leads to further C5, C6, C7, C8 and C9 activation. However, the C5b-9 complex would be generated far from the outer membrane surface and, hence, be sterically hindered from insertion into the outer membrane by the long LPS O side chains on this strain. Support for this hypothesis comes from preliminary studies using EDTA treatment of 144M-SR, a procedure known to release LPS (Leive, 1965), which dramatically increased C9 attachment to the surface of 144M-SR. In contrast, C3b on the surface of 144M is presumably covalently attached to either the rough type LPS or a surface-exposed outer membrane protein. Preliminary studies suggest that the C3b on 144M-SR is attached to a high moleculer weight component, as revealed by its migration on SDS-polyacrylamide gels (consistent with the hypothesis that it is attached to LPS molecules bearing long O side chains), whereas the C3b on 144M is covalently attached to a small molecular weight component. Although the identity of either C3b acceptor site remains to be defined, the data would suggest that the association of the C3b with its acceptor on 144M somehow inhibits its effectiveness in C5 convertase activity, thus generating less C5a (a potent chemotaxin). This could prove to be a distinct advantage for strains isolated from CF patients, since fewer neutrophils and macrophages would be recruited to the site of bacterial infection. This would be consistent with the chronic nature of pulmonary infections with <u>Pseudomonas</u> <u>aeruginosa</u> in CF patients. Current research efforts are focused on identifying the C3 acceptor site on these 2 strains, and characterizing the molecular form of C3 bound to this acceptor site.

Summary

Studies were described which demonstrate that <u>P</u>. <u>aeruginosa</u> strain 144M, which contained LPS deficient in 0 side chains, is killed by complement activated by the classical pathway initiated by antibody to the rough type LPS of this strain. These antibodies are present in normal human serum, which is consistent with the lack of bacteremic spread of these serS organisms in CF patients. In contrast, strains with LPS bearing long 0 side chains are resistant to complement-mediated killing. This resistance does not represent an inability to activate complement effectively, but rather reflects the inhibition of the insertion of the C5b-9 complex into the outer membrane, presumably due to steric hindrance by long LPS 0 side chains. This type of serum resistance appears to be a common mechanism present amongst several genera of gram-negative bacteria. Finally, strains from CF patients (serS) differ dramatically from those of non-CF patients (serR) in their activation and binding of C3. CF strains generally activate less complement per CFU than non-CF strains, thereby producing less C5a, a potent chemotaxin. These strains are therefore less inflammatory, consistent with their tendency to produce chronic infections in CF patient lungs. In addition, the CF and non-CF strains bind C3 differently, perhaps due to the covalent attachment of C3 to different acceptor sites on the cell surface. The full clinical significance of these differences remains to be determined.

Acknowledgements

The author gratefully acknowledges the assistance and generosity of Keith A. Joiner, M.D. who contributed materials and expertise needed for this project, much of which was done while the author was on sabbatical in Dr. Joiner's laboratory at the National Institutes of Health. This study was supported in part by grants from the National Institutes of Health, the Cystic Fibrosis Foundation, and the Academic Senate of the University of California, Riverside.

References

Baltimore, R.S. and Shedd, D.G. The role of complement in the opsonization of mucoid and non-mucoid strains of Pseudomnas aeruginosa. Pediat. Res. 17:952-958 (1983).

Bjornson, A.B. and Michael, J.G. Factors in normal human serum that promote bacterial phagocytosis. J Infect Dis 130:S119-131 (1973).

Bodey, G.P., Bolivar, R., Fainstein, V. and Jadeja, L. Infections caused by Pseudomonas aeruginosa. Rev Infect Dis 5:279-313 (1983).

Borowski, R.S., Schiller, N.L. Examination of the bactericidal and opsonic activity of normal human serum for a mucoid and non-mucoid strain of Pseudomonas aeruginosa. Curr Microbiol 9:25-30 (1983).

Engels, W., Endert, J., Kamps, M.A.F., and van Boven, C.P.A. Role of lipopolysaccharide in opsonization and phagocytosis of Pseudomonas aeruginosa. Infect Immun 49:182-189 (1985a).

Engels, W., Endert, J., and van Boven, C.P.A. A quantitative method for assessing the third complement factor (C3) attached to the surface of opsonized Pseudomonas aeruginosa: interrelationship between C3 fixation, phagocytosis, and complement consumption. J Immunol Methods 81:43-53 (1985b).

Goldman, R.C., Joiner, K., and Leive, L. Serum-resistance mutants of Escherichia coli contain increased lipopolysaccharide, lack an O antigen capsule, and cover more of their lipid A core with O antigen. J Bacteriol 159:877-882 (1984).

Gross, G.N., Rehm, S.R., and Pierce, A.K. The effect of complement depletion on lung clearance of bacteria. J Clin Invest 62:373-378 (1978).

Hammer, M.C., Baltch, A.L., Sutphen, N.T., Smith, R.P., and Conroy, J.V. Pseudomonas aeruginoa: quantitation of maximum phagocytic and bactericidal capabilities of normal human granulocytes. J Lab Clin Med 98:938-948 (1981).

Hancock, R.E.W., Mutharia, L.M., Chan, L., Darveau, R.P., Speert, D.P., and Pier, G.B. Pseudomonas aeruginosa isolates from patients with cystic fibrosis: a class of serum-sensitive, nontypable strains deficient in lipopolysaccharide O side chains. Infect Immun 42:170-177 (1983).

Heidbrink, P.J., Toews, G.B., Gross, G.N., and Pierce, A.K. Mechanisms of complement-mediated clearance of bacteria from the murine lung. Am Rev Respir Dis 125:517-520 (1982).

Hoiby, N., and Olling, S. Pseudomonas aeruginosa infection in cystic fibrosis. Bactericidal effect of serum from normal individuals and patients with cystic fibrosis on Pseudomonas aeruginosa strains from patients with cystic fibrosis and other diseases. Acta Path Micrbiol Scant Sect C 85:107-114 (1977).

Joiner, K.A., Hammer, C.H., Brown, E.J., Cole, R.J., and Frank, M.M. Studies on the mechanism of bacterial resistance to complement-mediated killing. I. Terminal complement components are deposited and released form Salmonella minnesota S218 without causing bacterial death. J Exp Med 155:797-808 (1982a).

Joiner, K.A., Hammer, C.H., Brown, E.J., and Frank, M.A. Studies on the mechanism of bacterial resistance to complement-mediated killing. II. C8 and C9 release C5b67 from the surface of Salmonella minnesota S218 because the terminal complex does not insert into bacterial outer membrane. J Exp Med 155:809-815 (1982b).

Joiner, K.A., Schmetz, M.A., Goldman, R.C., Leive, L., and Frank, M.M. Mechanism of bacterial resistance to complement-mediated killing: inserted C5b-9 correlates with killing for Escherichia coli 0111B4 varying in 0-antigen capsules and 0-polysaccharide coverage of lipid A core oligosaccharide. Infect Immun 45:113-117 (1984).

Joiner, K.A., Grossman, N., Schmetz, M., and Leive, L. C3 binds preferentially to long-chain lipopolysaccharide during alternative pathway activation by Salmonella montevideo. J Immunol 136:710-715 (1986).

Larsen, G.L., Mitchell, B.C., Harper, T.B., and Henson, P.M. The pulmonary response of C5 sufficient and deficient mice to Pseudomonas aeruginosa. Am Rev Respir Dis 126:306-311 (1982).

Leive, L. Release of lipopolysaccharide by EDTA treatment of Escherichia coli. Biochem Biophys Res Commun 21:290-296 (1965).

Lory, S., and Tai, P.C. Biochemical and genetic aspects of Pseudomonas aeruginosa virulence. Curr Top Microbiol Immunol 118:53-69 (1985).

Meshulam, T., Verbrugh, H., and Verhoeff, J. Serum induced lysis of Pseudomonas aeruginosa. Eur J Clin Microbiol 1:1-6 (1982).

Murphey, S.A., Root, R.K., and Schreiber, A.D. The role of antibody and complement in phagocytosis by rabbit alveolar macrophages. J Infect Dis 140:896-903 (1979).

Muschel, L.H., Ahl, L.A., and Fisher, M.W. Sensitivity of Pseudomonas aeruginosa to normal serum and polymyxin. J Bacteriol 98:453-457 (1969).

Muschel, L.H., and Larsen, L.J. The sensitivity of smooth and rough gram-negative bacteria to the immune bactericidal reaction. Proc Soc Exp Biol Med 133:345-348 (1970).

Nelson, B.W., and Roantree, R.J. Analyses of lipopolysaccharides extracted from penicillin-resistant, serum-sensitive Salmonella mutants. J Gen Microbiol 48:179-188 (1967).

Nguyen, B-YT, Peterson, P.K., Verbrugh, H.A., Quie, P.G., and Hoidal, J.R. Differences in phagocytosis and killing by alveolar macrophages from humans, rabbits, rats and hamsters. Infect Immun 36:504-509 (1982).

Offredo-Hemmer, C., Berche, P., and Veron, M. A complement-sensitive mutant of Pseudomonas aeruginosa. Ann Microbiol (Paris) 134A:281-294 (1983).

Peterson, P.K., Kim, Y., Schmeling, D., Lindemann, M., Verhoef, J., and Quie, P.G. Complement-mediated phagocytosis of Pseudomonas aeruginosa. J Lab Clin Med 92:883-894 (1978).

Pier, G.B., and Ames, P. Mediation of the killing of rough, mucoid isolates of Pseudomonas aeruginosa from patients with cystic fibrosis by the alternative pathway of complement. J Infect Dis 150:223-228 (1984).

Reyes, M.P., El-Khatib, M.R., Brown, W.J., Smith, F., and Lerner, A.M. Synergy between carbenicillin and an aminoglycoside (gentamicin or tobramycin) against Pseudomonas aeruginosa isolates from patients with endocarditis and sensitivity of isolates to normal human serum. J Infect Dis 140:192-202 (1979).

Rowley, D. Sensitivity of rough gram-negative gram-negative bacteria to the bactericidal action of serum. J Bacteriol 95:1647-1650 (1968).

Schiller, N.L., and Hatch, R.A. The serum sensitivity, colonial morphology, serogroup specificity, and outer membrane protein profile of Pseudomonas aerginosa strains isoalted from several clinical sites. Diagn Microbiol Infect Dic 1:145-157 (1983).

Schiller, N.L., Alazard, M.J., and Borowski, R.S. Serum sensitivity of a Pseudomonas aeruginosa mucoid strain. Infect Immun 45:748-755 (1984a).

Schiller, N.L., Hackley, D.R., and Morrison, A. Isolation and characterization of serum-resistant strains of Pseudomonas aeruginosa derived from serum-sensitive parenteral strains. Curr Microbiol 10:185-190 (1984-b).

Schiller, N.L., Joiner, K.A. Interaction of complement with serum-sensitive and serum-resistant strains of Pseudomonas aeruginosa. Infect Immun 54:689-694 (1986).

Schiller, D.R., and Miller, K.D. Elastase of Pseudomonas aeruginosa: inactivation of complement components and complement-dervied chemotactic factors. Infect Immun 10:128-135 (1974).

Taylor, P.W., and Kroll, H-P. Interaction of human complement proteins with serum-sensitive and serum-resistant strains of Escherichia coli. Mol Immunol 21:609-620 (1984).

Thomassen, M.J., and Demko, C.A. Serum bactericidal effect on Pseudomonas aeruginosa isolates from cystic fibrosis patients. Infect Immun 33:512-518 (1981).

Young, L.S., and Armstrong, D. Human immunity to Pseudomonas aeruginosa: I. In vitro interaction of bacteria, polymorphonuclear leukocytes, and serum factors. J Infect Dis 126:257-275 (1972).

COMPLEMENT-MEDIATED OPSONIZATION OF GROUP A STREPTOCOCCI INHIBITED BY THE BINDING OF FIBRINOGEN TO SURFACE M PROTEIN FIBRILLAE

Ellen Whitnack, Thomas P. Poirier, and Edwin H. Beachey

Veterans Administration Medical Center and
University of Tennessee
1030 Jefferson Avenue
Memphis, Tennessee 38104
USA

The M protein radiating from the surface of group A streptococci is the principal virulence factor of these organisms. Streptococci lacking M protein are readily opsonized by complement through the alternate pathway (1) and as a result are rapidly ingested and killed by blood phagocytes. Surface M protein renders the organisms resistant to opsonization, ingestion and killing by phagocytic cells (2). Two hypotheses have been advanced to account for the antiphagocytic properties of M protein: it may be directly toxic to phagocytic cells, or it may prevent opsonization by complement. A mechanism for direct toxicity has been proposed by Manjula and Fischetti (3), who noted structural similarities between type 6 M protein and alpha-tropomyosin and postulated that M protein might disable the phagocyte by interfering with its contractile proteins. To date, however, no evidence of toxicity of M protein for phagocytes has been obtained. Hot acid extracts of group A streptococci do, indeed, exert cytotoxic effects on human polymorphonuclear leukocytes, but these appear to be mediated by the formation of immune complexes involving streptococcal antigens other than M protein (4). Furthermore, M-positive streptococci do not attach to phagocytes in whole blood (5), nor do they inhibit the phagocytosis of M-negative organisms mixed with them (6).

The second hypothesis, that M protein interferes with opsonization, has been studied in some detail. Avirulent group A streptococci, which lack M protein, are readily opsonized by complement activated via the alternate pathway (1). Organisms possessing M protein activate complement less efficiently (7) and bind less of it (i.e., C3) on their surface (8). This difference, however, may not be sufficient to explain the resistance to

NATO ASI Series, Vol. H24
Bacteria, Complement and the Phagocytic Cell
Edited by F.C. Cabello und C. Pruzzo
© Springer-Verlag Berlin Heidelberg 1988

phagocytosis of M-protein-bearing organisms. By diluting the serum used to treat an M-negative strain, Jacks-Weis et al. (8) were able to reduce the amount of C3 bound to those organisms to equal the amount bound by the M-positive parent strain, despite which the M-negative strain remained fully susceptible to phagocytosis. Therefore, there must be some interference with phagocytosis other than simple reduction in the amount of complement deposited. These investigators found that C3 was deposited in an uneven, patchy pattern on the surface of the M-positive organisms. They proposed that this unevenness interfered with the "zipper" mechanism of enclosure of the bacterial cell in the phagosome. We have noticed that type 24 streptococci, when suspended in fresh serum, deposit C3 on their surface quite evenly, but are still at least somewhat resistant to phagocytosis compared with M-negative organisms (9). A simple explanation would be that complement bound on the cell wall---perhaps to the peptidoglycan component (10)---is poorly accessible to the phagocyte's complement receptors because of interference by the projecting M protein fibrillae; but so far little information has been obtained about the site of complement deposition on intact streptococcal cells (11).

Recent studies in our laboratory suggest that the puzzling ability of group A streptococci to remain resistant--or at least partially resistant-- to phagocytosis when opsonized by serum may not be relevant to the situation in the infected host. We have discovered that the binding of human fibrinogen to streptococcal M protein, an interaction reported some years ago (12,13), contributes to the anti-opsonic properties of the M protein molecule by totally preventing the deposition of complement. In this paper, we report some of the fibrinogen binding properties of streptococci M protein and the biological consequences of this interaction. We show that such binding renders the organisms resistant to phagocytosis in nonimmune human blood. Moreover, M protein cloned and expressed on the surface of Streptococcus sanguis, a species that normally does not express this molecule, binds fibrinogen which renders the organisms resistant to opsonization and phagocytosis. These experiments provide the most definitive evidence yet obtained of the antiphagocytic function of M protein-fibrinogen complexes.

The effect of fibrinogen on phagocytosis of group A streptococci

When a small inoculum of virulent, mouse- and blood-passed streptococci is rotated in fresh nonimmune human blood, a 2^6-2^8 fold increase in the number of colony-forming units may be obtained, as in cell-free media, indicating that little or no phagocytosis has occurred. Similarly, when large inocula are used so that phagocytosis may be assessed by light-microscopic scoring of the leukocytes, fewer than 5% of the neutrophils will be found with attached or ingested cocci. When serum and washed blood cells are substituted for whole blood in these two assays, only one or two generations of growth are obtained, and up to 95% of the neutrophils (depending on inoculum size) may be associated with bacteria (9). If plasma or fibrinogen is added back to the serum, bacterial uptake is inhibited in a dose-dependent manner. Inhibition may likewise be obtained if the streptococcal cells are pretreated with plasma or fibrinogen and washed before incubation with serum and blood cells. These results have been obtained with each of several other M serotypes tested including types 5,6,19, and 24 (9,14). The loss of resistance to phagocytosis that occurs in a fibronogen-free medium is incomplete, in that even greater phagocytosis may be obtained if the organisms are pretreated with trypsin to remove the surface fibrillae (9). These observations indicate that the binding of fibrinogen accounts in part for the resistance to phagocytosis of group A streptococci in whole blood.

The receptor for fibrinogen on the streptococcal cell surface

Analysis of binding isotherms (15) suggests that fibrinogen binds to a single population of high affinity receptor sites with a dissociation constant of 5 nM. Since extracted M proteins precipitate fibrinogen (12,13), one could reasonably expect that M protein would be the fibrinogen receptor. However, M protein is not the only streptococcal structure that interacts with fibrinogen; T protein also bind (16,17), as does lipoteichoic acid (unpublished observation). Evidence that M protein serves as the receptor for fibrinogen has been obtained in several experiments (9,15,18). On electron micrographs, fibrinogen appears to bind to the M protein-bearing fibrillae, markedly increasing their density. Treatment of the cells with trypsin removes the fibrillae and abolishes binding of fibrinogen, and the cells so treated are taken up by neutrophils equally well in serum and in plasma. Treatment with dilute pepsin at suboptimal pH, which leaves the proximal portion of the fibrillae and the surface lipoteichoic acid intact, reduces binding by 90% and likewise abolishes the inhibition of phagocytosis by fibrinogen. Antisera to pepsin-extracted M protein (pep M, the distal half of the M protein molecule) competitively inhibits binding, whereas anti-LTA and anti-T do not. Finally, pep M proteins inhibit binding by 50 to 80 percent, depending on M type, and partially inhibit binding to heterologous M types (15 and unpublished observations).

Recently, we have obtained more definitive evidence for the antiopsonic properties of cell-surface M protein-fibrinogen complexes through molecular cloning experiments (19,20). Cloned type 5 streptococcal DNA, including the M protein gene (21), was inserted into a shuttle plasmid and used to transform Streptococcus sanguis, a non-group A species that does not normally have surface fibrillae or M protein. The transformants expressed surface fibrillae to a variable extent; only about one-third had detectable M protein as determined by ELISA-inhibition and opsonization-inhibition tests. That third, however, bound fluorescein-labeled fribrinogen and could therefore be separated from the rest by fluorescence-activated cell sorting. The sorted cells resisted phagocytosis and could be opsonized by anti-M antibody.

How does bound fibrinogen inhibit phagocytosis?

Since opsonization by complement is required for phagocytosis of group A streptococci, we investigated the possibility that bound fibrinogen impedes the deposition of complement on the cell surface (9). This appears to be the case; organisms incubated in fresh serum fluoresce brightly when treated with fluorescein-labeled anti-human C3, whereas after incubation in plasma no fluorescence is obtained. Cells pretreated with trypsin fluoresce brightly after incubation in either serum or plasma. We do not believe that fibrinogen binds directly to complement-binding structures, since there is no evidence that M protein activates complement (22). Rather, we suspect that the bound fibrinogen inhibits complement activation and/or deposition sterically, by virtue of its large size and its location on and amongst the fibrillae. However, the binding site(s) of C3 has yet to be determined with precision (11).

Degradation products of fibrinogen and fibrin

Because both clotting and fibrinolysis may be part of the inflammatory process, we were concerned to see if the antiopsonic properties of fibrinogen extend to plasmic degradation products of fibrinogen and fibrin (23). Runehagen et al. (24) had already shown that group A streptococci bind fibrinogen D fragments. We confirmed this observation, and extended it by showing that D fragments interact with purified pep M24. Digested fibrinogen inhibited the binding of fibrinogen to streptococcal cells, blocked C3 deposition, and blocked uptake of the bacteria by neutrophils in a concentration-dependent manner. The binding affinity of digested fibrinogen was ~ 30-fold lower than that of intact fibrinogen, and the digest was correspondingly less potent in inhibiting uptake by neutrophils, but nonetheless inhibited phagocytosis completely at physiologic concentrations. Fibrin degradation products bound somewhat better than fibrinogen degradation products, perhaps because they contain fragment D in dimeric form. Despite this, they were less effective in inhibiting complement deposition and phagocytosis, an observation that we have yet to explain.

Streptococci and fibrin

In the course of our studies of fibrinogen degradation products, we have found that pep M24 binds to all plasmic fragments of fibrinogen that react with antiserum to fragment D, including X, Y, D_1, D_2, D_3, and several smaller fragments (23). This finding implies that pep M binds to a site different from any of the previously described functional sites on the D domain, all of which are lost when fragment D_1 is converted to D_3; in particular, the sites involved in fibrin polymerization are absent in D_3(25). Furthermore, pep M does not bind to fragment E, the site of thrombin action. Therefore, fibrinogen bound to M protein should remain susceptible to thrombin cleavage and perhaps to fibrin polymerization. Accordingly, we used a radioimmunoassay to measure the release of fibrinopeptide A (FPA) from fibrinogen-pep M precipitates treated with thrombin. No change in the appearance of the precipitate occurred, but FPA was released in the expected 2:1 molar ratio. Next, ^{3}H-labeled fibrinogen was mixed with a large excess of pep M so that soluble complexes rather than precipitates would be formed (15) and then treated with thrombin. The counts could be pelleted, implying that polymerization had occurred (in these experiments the concentration of fibrinogen was too low for a visible clot). Similarly, thrombin treatment released FPA from fibrinogen-coated streptococci. These experiments were done in the presence of soybean trypsin inhibitor, and no ^{3}H counts were released, showing that the release of FPA was not due to the action of contaminating proteases. To confirm that fibrinopeptide B was also released by the action of thrombin, fibrinogen-coated, thrombin-treated streptococci were incubated with a monoclonal antibody raised against a synthetic copy of the amino terminal heptapeptide of the fibrin beta chain (26; gift of Dr. Gary Matsueda, Harvard University). Organisms so treated bound antibody as shown by immunofluorescence microscopy, whereas control organisms not treated with thrombin did not. Thus, fibrinogen bound to the streptococcal cell surface could be converted to fibrin by the action of thrombin.

Because phagocytic cells have fibrin receptors (27,28) that promote internalization of fibrin (29), we next wanted to see if fibrin-coated streptococci would interact with neutrophils and macrophages. When fibrin-coated streptococci were incubated with washed blood cells and either fresh or heat-inactivated serum, no uptake of bacteria by neutrophils occurred, showing that fibrin inhibits complement deposition just as fibrinogen does, and is not itself opsonic. However, when the fibrinogen-coated organisms were incubated with rabbit peritoneal macrophages, there was a 30% increase in the number of cells with attached bacteria compared with fibrinogen- or albumin-treated controls. We have yet to determine if the fibrin actually promotes the internalization and killing of the streptococci, but if it does, the process could be a mechanism for non-immune clearance of streptococci in tissues.

What do M proteins of different serotypes have in common that allows them to bind fibrinogen?

When the precipitation of fibrinogen with streptococcal extracts was first discovered (12,13), most but not all of the M-types tested reacted with fibrinogen. An extract of type 24, for example, failed to precipitate; but in our studies, a more highly purified type 24 M protein preparation readily precipitated fibrinogen (15), suggesting that the failure to observe the reaction in the earlier study may have been due to inhibitors in the rather crude M protein preparations employed. The M protein-fibrinogen interaction thus appears to be common to most, and perhaps all, M-types. In our initial studies we attempted to see if the fibrinogen-binding site of different serotypes of M protein could be identified immunologically. Indeed, we identified several opsonic cross-reactions between types 5,6 and 19 that were partially or completely blocked by fibrinogen (14). Binding of fibrinogen may thus account for the well-known fact that tests for opsonic antibody performed using whole blood are far more type-specific than various in vitro antibody assays such as ELISA and immunoprecipitation. However, none of the cross-reactive sera that we have studied reacts with type 24 M protein, so that we cannot claim to have identified an antibody against a common fibrinogen-binding epitope.

What structural feature of the various serotypes of M protein is responsible for the interaction with fibrinogen? The current model of M protein is of a noncovalent coiled-coil dimer projecting from the cell surface as a fibril, with carboxy terminal anchor sequences at the cell wall and uncoiled amino terminal "floppy ends" at the tip (30,31). The coiled-coil, fibrillar conformation is achieved through a seven-residue periodicity in the placement of hydrophobic amino acid residues rather than a particular amino acid sequence, although there is, in fact, considerable homology in the sequences of the M proteins so far studied, particularly as the carboxy end is approached (32,33). Although we have not identified the fibrinogen-binding region of M protein, we suspect that this region will prove to be a fairly extended portion of the fibril somewhere near the middle of its length. This supposition is based on several pieces of evidence (9,14,15). Fibrinogen-coated streptococci can be opsonized by homologous anti-M antibody, implying that some portion of the M protein molecule remains exposed. We would expect this portion to be at the amino terminus, that is, at or near the tips of the fibrillae, where the antibody would be accessible to interaction with phagocytes. Second, pep M proteins--the amino terminal halves of M proteins--bind fibrinogen and partially inhibit its binding to streptococcal cells, implying that the binding region is at least partly composed of this moiety. However, the "stubble" left on the streptococcal cell after pepsin extraction also binds fibrinogen weakly, which could mean that it contributes to the binding region of the intact molecule. Third, small peptide fragments of pep M of $\leq$3,500 kDa, prepared either by enzymatic cleavage or by synthesis, do not interact with fibrinogen, implying that the binding region does not consist of a short amino acid sequence; in fact, we have not identified any fibrinogen-binding peptide smaller than 20 kDa, although the search has not been exhaustive. If, indeed, the fibrinogen-binding region of M protein is a fairly large portion of the molecule, definition of this region may best be approached by manipulation of the cloned M protein gene.

References

1. Petersen PK, Schmeling D, Cleary PP, Wilksinson BJ, Kim Y, Quie P (1979) Inhibition of alternative complement pathway opsonization by group A streptococcal M protein. J Infect Dis 139:575-585.
2. Lancefield R (1962) Current knowledge of type-specific M antigens of group A streptococci. J Immunol 89:307-313.
3. Manjula BN, Fischetti VA (1980) Tropomyosin-like seven residue periodicity in three immunologically distinct streptococcal M proteins and its implications for the antiphagocytic property of the molecule. J Exp Med 151:695-708.
4. Beachey EH, Stollerman GH (1973) Mediation of cytotoxic effects of streptococcal M protein by nontype-specific antibody in human sera. J Clin Invest 52:2563-2570.
5. Beachey EH, Cunningham M (1973) Type-specific inhibition of preopsonization versus immunoprecipitation by streptococcal M proteins. Infect Immun 8:19-24.
6. Manjula BN, Schmidt ML, Fischetti VA (1985) Unimpaired function of human phagocytes in the presence of phagocytosis-resistant group A streptococci. Infect Immun 50:610-613.
7. Bisno AL (1979) Alternate complement pathway activation by group A streptococci: role of M protein. Infect. Immun. 26:1172-1176.
8. Jacks-Weiss J, Kim Y, Cleary PP (1982) Restricted deposition of C3 on M+ group A streptococci: correlation with resistance to phagocytosis. J Immunol 128:1897-1902.

9. Whitnack E, Beachey EH (1982) Antiopsonic activity of fibrinogen bound to M protein on the surface of group A streptococci. J. Clin. Invest. 69:1042-1045.
10. Greenblatt J, Boackle RJ, Schwab JH (1978) Activation of the alternate complement pathway by peptidoglycan from streptococcal cell wall. Infect Immun 19:296-303.
11. Weis JJ, Law SK, Levine RP, Cleary PP (1985) Resistance to phagocytosis by group A streptococci: failure of deposited complement opsonins to interact with cellular receptors. J Immunol 134:500-505.
12. Kantor FS, Cole RM (1959) A fibrinogen precipitating factor (FFP) of group A streptococci. Proc Soc Exp Biol Med 102:146-150.
13. Kantor FS (1965) Fibrinogen precipitation by streptococcal M protein. I. Identity of the reactants, and stoichiometry of the reaction. J Exp Med 121:849-859.
14. Whitnack E, Dale JB, Beachey EH (1984) Common protective antigens of group A streptococcal M proteins masked by fibrinogen. J Exp Med 159:1201-1212.
15. Whitnack E, Beachey EH (1985) Biochemical and biological properties of the binding of fibrinogen to M protein in group A streptococci. J Bacteriol 164:350-358.
16. Ludwicka A, Jeljaszewicz J (1978) Paracoagulation of fibrinogen in vitro and in vivo by protein T of Streptococcus pyogenes. Zentralbl Bakteriol Mikrobiol Hyg Ser A 241:301-307.
17. Schmidt KH, Köhler W (1981) Interaction of streptococcal cell wall components with fibrinogen. I. Adsorption of fibrinogen by immobilized T-proteins of Streptococcus pyogenes. Immunobiol 158:330-337.
18. Whitnack E, Dale JB, Beachey EH (1983) Streptococcal defenses against host immune attack: the M protein-fibrinogen interaction. Trans Assoc Am Physicians 96:197-202.
19. Poirier TP, Kehoe ME, Whitnack E, Beachey EH (1987) Surface expression of type 5 M protein of Streptococcus pyogenes in Streptococcus sanguis. In Ferretti JJ, Curtiss III R (eds) (1987) Streptococcal Genetics. American Society for Microbiology, Washington, D.C.
20. Poirier TP, Kehoe MA, Whitnack E, Dockter ME, Beachey EH (1987) Fibrinogen binding and resistance to phagocytosis of Streptococcus sanguis expressing cloned M protein of Streptococcus pyogenes. Submitted.
21. Kehoe MA, Poirier TP, Beachey EH, Timmis KN (1985) Cloning and genetic analysis of serotype 5 M protein determinant of group A streptococci: evidence for multiple copies of the M5 determinant in the Streptococcus pyogenes genome. Infect Immun 48:190-197.
22. Cunningham MW, Beachey EH (1974) Peptic digestion of streptococcal M protein. I. Effect of digestion at suboptimal pH upon the biological and immunochemical properties of purified M protein extracts. Infect Immun 9:244-248.
23. Whitnack E, Beachey EH (1985) Inhibition of complement-mediated opsonization and phagocytosis of Streptococcus pyogenes by D fragments of fibrinogen and fibrin bound to cell surface M protein. J Exp Med 162:1983-1997.
24. Runehagen A, Schönbeck C, Hedner V, Hessel B, Kronvall G (1981) Binding of fibrinogen degradation products to S. aureus and to beta-hemolytic streptococci group A,C and G. Acta Pathol Microbiol Scand [B] 89:49-55.

25. Vàradi A, Scheraga HA (1986) Localization of segments essential for polymerization and for calcium binding in the gamma-chain of human fibrinogen. Biochemistry 25:519-528.

26. Hui KY, Haber E, Matsueda GR (1983) Monoclonal antibodies to a synthetic fibrin-like peptide bind to human fibrin but not fibrinogen. Science 222:1129-1132.

27. Colvin RB, Dvorak HF (1975) Fibrinogen/fibrin on the surface of macrophages: detection, distribution binding requirements, and possible role in macrophage adherence phenomena. J Exp Med 142:1377-1390.

28. Sherman LH, Lee J (1977) Specific binding of soluble fibrin to macrophages. J Exp Med 145:76-85.

29. Gonda SR, Shainoff JR (1982) Adsorptive endocytosis of fibrin monomer by macrophages: evidence of a receptor for the amino terminus of the fibrin alpha chain. Proc Natl Acad Sci USA 79:4565-4569.

30. Phillips GN, Flicker PF, Cohen C, Manjula BN, Fischetti VA (1981) Streptococcal M protein: alpha-helical coiled-coil structure and arrangement on the cell surface. Proc Natl Acad Sci USA 78:4689-4693.

31. Hollingshead SK, Fischetti VA, Scott JR (1986) Complete nucleotide sequence of type 6 M protein of the group A Streptococcus [sic]. J Biol Chem 261:1677-1686.

32. Jones KF, Manjula BN, Johnston KH, Hollingshead SK, Scott JR, Fischetti VA (1985) Location of variable and conserved epitopes among the multiple serotypes of streptococcal M protein. J Exp Med 161:623-628.

33. Scott JR, Hollingshead SK, Fischetti VA (1986) Homologous regions within M protein genes in group A streptococci of different serotypes. Infect Immun 52:609-612.

A SURFACE PROTEIN OF <u>KLEBSIELLA</u> <u>PNEUMONIAE</u> AND <u>ESCHERICHIA</u> <u>COLI</u> THAT BINDS PHAGOCYTES AND INHIBITS PHAGOCYTOSIS

C. Pruzzo[1], S. Valisena[2], L. Baldi[1] and G. Satta[3]

[1]Institute of Microbiology
University of Genova
V. le Benedetto XV 10
16132 Genova Italy

[2]Institute of Microbiology
University of Padova
Padova Italy

[3]Institute of Microbiology
University of Siena
Siena Italy

Introduction

We have previously described a <u>Klebsiella</u> <u>pneumoniae</u> non-fimbrial adhesin (MIAT: mannose inhibitable adhesin/T7 receptor) which works as a receptor for phage T7 and mediates unencapsulated strain adherence to epithelial cells of the oral cavity, urinary tract and intestine (5-7).

This adhesin, which is made up of an outer membrane proteic fraction, also mediates attachment to inert human polymorphonuclear leucocytes (PMNs), but confers resistance to phagocytosis (6). In fact, MIAT-positive bacteria efficiently associate to PMNs either incubated at 4°C or fixed with glutaraldehyde at 37°C but are not efficiently phagocytized and killed by these cells incubated at 37°C. MIAT-positive bacteria pretreatment with compounds that specifically mask the MIAT (D-mannose, UV-inactivated T7 phages) makes these cells unable to associate to inert PMNs but enables phagocytes to ingest and kill such bacteria.

We now further analyze the mechanism by which the MIAT adhesin inhibits phagocytosis and we show that the MIAT-PMN binding is required to prevent phagocytosis of either MIAT-positive strains or MIAT-negative bacteria which are present in the same phagocytosis mixture.

NATO ASI Series, Vol. H24
Bacteria, Complement and the Phagocytic Cell
Edited by F.C. Cabello und C. Pruzzo
© Springer-Verlag Berlin Heidelberg 1988

In this paper, we also present a preliminary study of the presence and role of the MIAT adhesin in the closely related Escherichia coli. The results reported here show that in this bacterial species MIAT also mediates attachment to epithelial cells and protects bacteria from phagocytosis.

Materials and Methods

Bacterial strains

The MIAT-positive K. pneumoniae strain K59 and its MIAT-negative derivatives KRTT1 and KRTT2 were previously described (5). E. coli strains were isolated from midstream urine of patients with lower urinary tract infections. Phage T7 resistant mutants were selected as previously described (5). T7 phage was purchased from the American Type Culture Collection.

Media

Brain Heart Infusion broth and agar (Difco Laboratories) were used for culturing bacterial strains. Phosphate buffered saline (PBS, pH 7.2-7.4) and Hanks Balanced Salt Solution were used in adherence and phagocytosis experiments.

Adherence experiments

Adherence experiments to oral and urinary tract epithelial cells (EC) were performed as described (1,4).

Hemagglutination (HA) test

Agglutination of human and guinea pig red blood cells were evaluated as described (3). The effect of D-mannose on HA was evaluated by adding sugar to the red blood cell suspension at a final concentration of 20 mg/ml.

Phagocytosis experiments

Polymorphonuclear leucocytes (PMNs) were prepared as described by Boyum (2). Association to PMN monolayer at 4°C and 37°C was evaluated on coverslips as described by Mangan and Snyder (3), determining the percentage of PMNs with two or mo1:100.

Killing assay

Bacterial killing was assayed as described (3) using a PMN:bacteria ratio of 1:100. In mixed phagocytosis where the effect of PMN pretreatment with MIAT-positive bacteria (PMN:bacteria ratio 1:5) on killing of MIAT-negative cells was studied, bacteria having different antibiotic resistance markers were used and the rate of killing was evaluated using the duplicate pour-plate method on antibiotic containing agar.

Chemiluminescence (CL) assay

Luminol-induced CL responses were evaluated as previously described (8) using a PMN:bacteria ratio of 1:100.

Outer membrane proteins (OMPs) and lipopolysaccharide (LPS) purification

OMPs and LPS were isolated as described (7).

Phagocytosis experiments in the presence of different substrates

UV-inactivated T7 phages were prepared and added to the phagocytosis mixture as described (5,6). D-mannose, succinyl ConA, OMPs and LPS were added to the phagocytosis mixtures at final concentrations of 2mg/ml, 10mg/ml, 100μg/ml and 50μg/ml, respectively.

Results

The role of MIAT in resistance to phagocytosis by human PMNs of MIAT-positive strains was analyzed in several steps.

Binding of MIAT-positive strains to succinyl ConA-treated PMNs

To evaluate the possibility that resistance to phagocytosis shown by the MIAT-positive strain was due to the MIAT specific interaction with its receptor on the phagocyte membrane, we first masked the PMN mannose-containing receptors with the lectin succinyl ConA. As shown in Table 1, at 4°C PMN treatment with succinyl ConA reduced by 4-fold K59 adherence to phagocytes. At 37°C it was found that K59 was associated and killed (Table 1) by lectin-treated PMNs approximately 5-fold more efficiently than by untreated phagocytes. On the contrary, succinyl ConA treatment did not affect phagocytosis and killing of the MIAT-negative mutants, thus showing that succinyl ConA effects did not result from nonspecific stimulation of phagocytosis.

Table 1. Effect of PMN treatment with succinyl ConA (sConA) on phagocytosis and killing of MIAT-positive strain K59

Experim. conditions	% Association to PMNs (60 min)		% Killing by PMNs (60 min)
	4°C	37°C	
Control	81	22	18
+ sConA	24	86	91

Results represent an average of 3 experiments.

Effect of temperature shift on MIAT-positive strain association to phagocytes

As mentioned above, the MIAT-positive strain K59 does not associate to PMNs at 37°C and is resistant to phagocytosis despite its ability to adhere to phagocytes at 4°C. To confirm this phenomenon, we evaluated the effect of shifting the incubation temperature from 4°C to 37°C and vice versa on K59 binding to PMNs. As shown in Table 2, when the phagocytosis mixture was brought from 37°C to 4°C, bacteria which at 37°C did not efficiently associate with PMNs avidly bound to phagocytes. On the contrary, when the phagocytosis reaction was shifted from 4°C to 37°C, bacteria which at 4°C efficiently adhered to PMNs slowly detached from phagocytes. This further indicates that MIAT-positive strain inhibition of neutrophil phagocytic activity requires a metabolically active cell.

Table 2. Effect of temperature shift on association of MIAT-positive K59 bacteria with PMNs

Starting temperature	30'	Association with PMNs 60'		90'	120'
37°C	18	19 → no shift		21	20
		↘ shift to 4°C		76	90
4°C	74	84 → no shift		90	89
		↘ shift to 37°C		50	30

Results represent an average of 3 experiments.

CL response to MIAT-positive and MIAT-negative bacteria

The role of MIAT in MIAT-positive strain resistance to phagocytosis was further confirmed evaluating PMN luminol-enhanced CL, which is used as a quantitative measure of the interaction between PMNs and bacteria. For each preparation of PMNs, the degree of bacterial binding to phagocytes incubated at 4°C was also evaluated. As shown in Fig. 1, in the presence of K59 bacteria, which show a high level of binding to inert PMNs, a very low CL response was generated. On the contrary, a significant response was observed when PMNs were incubated with either the MIAT-negative bacteria KRTT1 and KRTT2 (Fig. 1) or the MIAT-positive K59 strain previously adsorbed with UV-inactivated T7 phages (not shown).

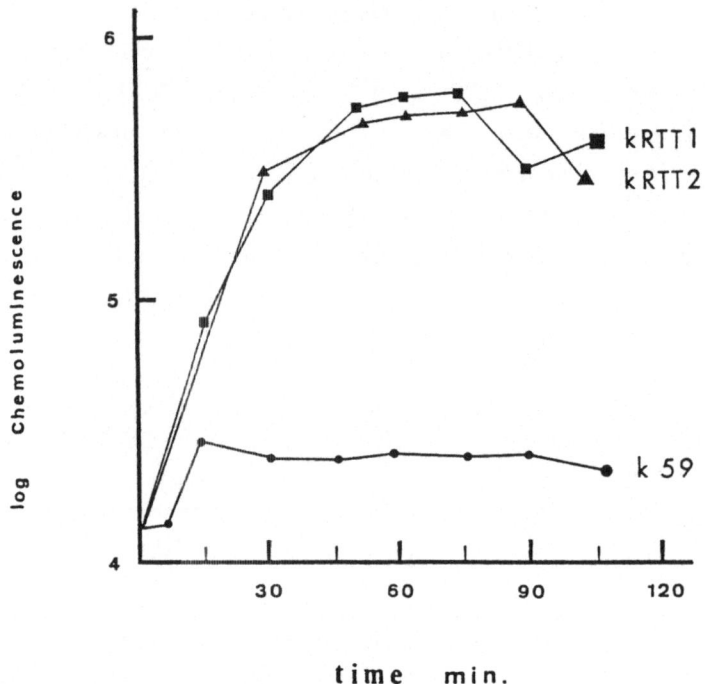

Fig. 1. PMN luminol-induced CL response to the MIAT-positive strain K59 and its MIAT-negative derivatives KRTT1 and KRTT2.

Effect of MIAT-positive bacteria on phagocytosis of MIAT-negative strains

If the suggestion that the interaction of MIAT with PMNs prevents them from binding and killing foreign particles is correct, PMNs preincubated with MIAT should also become unable to phagocytize MIAT-negative bacteria. To verify this hypothesis, we evaluated the effect of PMN pretreatment with the MIAT-positive Klebsiella K59 on the MIAT-negative KRTT1 strain sensitivity to phagocyte bactericidal activity. It was found that PMNs pretreated with K59, but not with KRTT1, showed a drastic reduction in their ability to kill KRTT1 bacteria (Table 3). When K59 bacteria were pretreated with either UV-inactivated T7 phages, or D-mannose, or anti-MIAT antibodies they lost their ability to inhibit phagocytosis. Since MIAT is made up of an outer membrane protein fraction, we tested the effect of K59 OMPs on the killing of KRTT1. It was found that in the presence of OMPs isolated from strain K59, but not from strain KRTT1, KRTT1 killing was reduced by 5-fold. LPS isolated from both K59 and KRTT1 had no effect on phagocytosis.

Table 3.   Effect of MIAT on KRTT1 sensitivity to PMN killing

Experimental conditions	% killing of KRTT1 nalR 60 min
PMNs + K59 nalS* + KRTT1 nalR*	15
PMNs + KRTT1 nalS* + KRTT1 nalR*	65

Killing was evaluated on Brain Heart Infusion agar containing nalidixic acid (final concentration 100μg/ml). Results represent an average of 3 experiments.
* nalS, nalR mean nalidixic acid sensitive and resistant bacteria, respectively.

Role of MIAT in E. coli adherence and resistance to phagocytosis

To analyze the role of MIAT in E. coli, we have first screened several clinical isolates for the presence of this adhesin. It was found that 70 of 250 strains isolated from human urines were MIAT-positive. On the basis of their HA pattern, they were subdivided into 3 groups. Group A: MS agglutination of guinea pig red blood cells and MR agglutination of human erythrocytes; group B: MS agglutination of guinea pig red blood cells only; group C: no agglutination. MIAT-negative spontaneous mutants were then selected from 1 strain of each HA group. They maintained the same HA pattern as the parent, but showed a 30% reduction in adherence efficiency to urinary tract and oral EC (Table 4).

When interactions with inert human PMNs were studied, it was found that the presence of MIAT did not affect this association (not shown). Despite this, all MIAT-positive strains were 2-4 fold more resistant to bactericidal activity by PMNs than their MIAT-negative mutants (Table 5). Pretreatment with UV-inactivated T7 phages made MIAT-positive strains as sensitive as their MIAT-negative mutants.

Table 4. Role of MIAT in E. coli adherence to epithelial cells

Strain	Agglut. of RBC		Mean No. of bacteria/EC from	
	guinea pig	human	Urinary tract	Oral cavity
12 MIAT+	-	-	74+/-10	80+/-9
12 MIAT-	-	-	27+/-9	26+/-3
56 MIAT+	+ Mann. sen	-	84+/-10	75+/-9
56 MIAT-	+ Mann. sen	-	53+/-5	44+/-5
77 MIAT+	+ Mann. sen	+ Mann. res	87+/-9	192+/-15
77 MIAT-	+ Mann. sen	+ Mann. res	51+/-13	82+/-11

Table 5. Role of MIAT in E. coli sensitivity to killing by PMNs

Strain	% killing by PMNs at 60 min
12 MIAT+	22
12 MIAT-	80
56 MIAT+	34
56 MIAT-	65
77 MIAT+	37
77 MIAT-	73

Results represent an average of 2 experiments.

Discussion

The results presented in this paper support our previous suggestion that the MIAT adhesin binding to a specific receptor in the PMN membrane, triggers changes in the cell surface receptors that inhibit further MIAT-mediated binding and phagocytosis. This hypothesis is supported by several data.

First, the fact that PMN treatment with succinyl ConA reduces K59 bacteria adherence to phagocytes at 4°C but enhances their sensitivity to killing, shows that when the receptors for the MIAT adhesin are masked, the interactions of MIAT-positive strains with PMNs are the same as with MIAT-negative bacteria.

It is not the presence of MIAT in the bacterial cell that <u>per se</u> makes bacteria unable to bind phagocytes and resistant to phagocytosis, but it is the interaction with a mannose containing receptor in the PMN membrane.

In phagocytosis mixtures incubated at 4°C and then shifted to 37°C the degree of association of MIAT-positive bacteria with PMNs rapidly decreased with time after shifting, while in mixtures incubated at 37°C and then shifted to 4°C the degree of K59 association was very low before shifting and then increased with time. These results suggest that binding of the MIAT adhesin to its receptor in the phagocyte triggers a change in the PMN surface which results in modification or removal of the MIAT receptors.

This is also supported by the fact that MIAT-positive bacteria, despite their high level of binding to PMNs at 4°C, do not induce a significant level of CL and PMNs pretreated with K59 <u>Klebsiella</u> become unable to engulf and kill the strains that do not carry the MIAT ligand.

The data presented in this paper also show that MIAT is present in a significant percentage of <u>E</u>. <u>coli</u> strains isolated form urinary tract infections. In these strains, as in <u>Klebsiella</u>, it confers bacteria with a higher adherence efficiency and makes them resistant to phagocytosis.

These results indicate that MIAT, which is an important virulence factor of <u>K</u>. <u>pneumoniae</u>, may also be important in the virulence of <u>E</u>. <u>coli</u> strains.

References

1. Beachey E.H., Bacterial adherence: adhesin receptor interactions mediating the attachment of bacteria to mucosal surfaces. J. Infect. Dis. 143:325-345 (1981).
2. Boyum A. Isolation of lymphocytes, granulocytes and macrophages. Scand. J. Immunol. 5:9-15 (1976).
3. Mangan D.F. and J. S. Synder, Mannose-sensitive interaction of Escherichia coli with human peripheral leukocytes in vitro. Infect. Immun. 26:520-527 (1979).
4. Ofek I., D. Mirelman, and N. Sharon, Adherence of E. coli to human mucosal cells mediated by mannose receptors. Nature (London) 265:623-625 (1977).
5. Pruzzo C., E. Debbia, and G. Satta, Identification of the major adherence ligand of Klebsiella pneumoniae in the receptor for coliphage T7 and alteration of Klebsiella adherence properties by lysogenic conversion. Infect. Immun. 30:562-571 (1980).
6. Pruzzo C., E. Debbia, and G. Satta, Mannose-inhibitable adhesins and T3-T7 receptors of Klebsiella pneumoniae inhibit phagocytosis and intracellular killing by human polymorphonuclear leukocytes. Infect. Immun. 36:949-957 (1982).
7. Pruzzo C., S. Valisena, and G. Satta, Laboratory and wild-type Klebsiella pneumoniae strains carrying mannose-inhibitable adhesins and receptors for coliphages T3 and T7 are more pathogenic for mice than are strains without such receptors. Infect. Immun. 39:520-527 (1983).
8. Svanborg-Eden C., L. M. Bjursten, R. Hull, K. E. Magnusson, Z. Moldavano, and H. Leppler, Influence of adhesins on the interaction of Escherichia coli with human phagocytes. Infect. Immun. 44:672-680 (1984).

ADHESINS, SERUM RESISTANCE AND CYTOLYSINS OF E. COLI- GENETIC STRUCTURE AND ROLE IN PATHOGENICITY

J. Hacker, W. Goebel, H. Hof, W. Konig[1], B. Konig[1], J. Scheffer[1], C. Hughes[2] and R. Marre[3]

Institut fur Genetik und Mikrobiologie, Institut fur Hygiene und Mikrobiologie
Univ. Wurzburg
Rontgenring 11
D-8700 Wurzburg
Fed. Rep. of Germany

[1]Arbeitsgruppe f. Infektabwehrmechanismen, Ruhr Univ. Bochum
D-4630 Bochum, Fed. Rep. of Germany
[2]Dept. of Pathology, Univ. of Cambridge, Tennis Court Road, Cambridge, U.K.
[3]Institut fur Med. Mikrobiologie, Univ. Lubeck, D2400 Lubeck, Fed. Rep. of Germany

1. Virulence factors of extraintestinal E. coli

Escherichia coli strains cause infections of the gut (intestinal infections), the urinary tract (urinary tract infections, UTI), the blood (sepsis) and are also the causative agents of new born meningitis (NBM). The infections which appear outside of the intestine have been termed as extraintestinal infections (Orskov and Orskov, 1985). It has been clear for several years that certain types of the O antigen (O1, O6, O18, O75, O83) and capsule (K1, K5, K12, K15) are strongly associated with such infections, as are special fimbrial adhesins, hemolysins (Hly) and iron binding chelators, such as aerobactin, a substance which takes up iron from the surrounding medium (Table 1).

2. General structure of hemolysin and adhesin genetic determinants

About 50% of uropathogenic E. coli are hemolytic and it has been shown that the E. coli hemolysin represents a protein which may act as cytolysin to normal tissue cells and to cells of the immune system (see below). The majority of Hly+ strains carry their hemolysin (hly) determinants on the chromosomes and only a few strains show transferable hemolysin plasmids (Hacker and Hughes, 1985). The hly gene clusters from different strains are highly homologous to each other. The hemolysin determinants consists of four genes, designated as hlyC,A,B,D.

NATO ASI Series, Vol. H24
Bacteria, Complement and the Phagocytic Cell
Edited by F.C. Cabello und C. Pruzzo
© Springer-Verlag Berlin Heidelberg 1988

Table 1: Virulence factors of extraintestinal E. coli isolates

Pathogenic E. coli	Adhesins	O-antigen	Toxins	Other virulence factors
Uropathogenic (UPEC)	P-fimbriae (S-fimbriae)	04,06,018 075	hemolysin	capsules serum resistance
Sepsis/ Meningitis	S-fimbriae	01,018,083		K1-antigen Aerobactin

While the HlyC and the HlyA gene products of 18 kilodalton (kd) and 110 kd in size form the intact hemolysin molecule, HlyB (82kd) and HlyD (54kd) are necessary for its transport across the outer membrane. A comparison of the entire nucleotide sequences of one plasmid-encoded hly determinant also discloses alterations in the composition of the hly coding region (Felmlee et al., 1985; Hess et al., 1986). The genes of the hemolysin determinants are transcribed from at least two promoter regions and are associated with flanking repeat sequences.

Fimbrial adhesins enable extraintestinal E. coli strains to attach to eukaryotic cells including phagocytes. They are distinguished by their receptor specificity. P and S fimbriae, predominantly found among extraintestinal isolates recognize the a-D-Gal-(1-4)-β-D-Gal-part of globosides and sialic acid containing receptors, respectively (Orskov and Orskov, 1983; Korhonen et al., 1984). Both belong to the mannose resistant hemagglutinating adhesins (Mrh). In contrast, type I fimbriae are associated with mannose sensitive hemagglutinating (Msh) adhesins and F1C fimbriae represent protein appendices without a detectable receptor specificity.

The genetic determinants of all types of fimbriae have been cloned (Uhlin et al., 1985). In Fig. 2 the general composition of the genetic determinant coding for the S fimbrial adhesin (sfa) is given. One has to distinguish between the gene sfaA which codes for the 16 kd fimbrillin subunit and a gene (sfaS), located near the 3' end of the determinant which codes for a 12kd protein identical to the S adhesin (Schmoll et al.,

1987; Moch et al., 1987). In addition, a region involved in the biogenesis of the fimbriae and sequences necessary for the control of transcription have been identified together with the presence of three promoter regions (p_A, p_S, p_C, see Fig. 1). The other fimbrial determinants mentioned above (type I, P, F1C) which all represent chromosomally encoded gene clusters show a rather similar composition to the sfa determinant. With the exception of S and FaC determinants, however, which show a high degree of homology to each other, the different fimbrial gene clusters mainly consist of non-homologous DNA sequences (Ott et al., 1987).

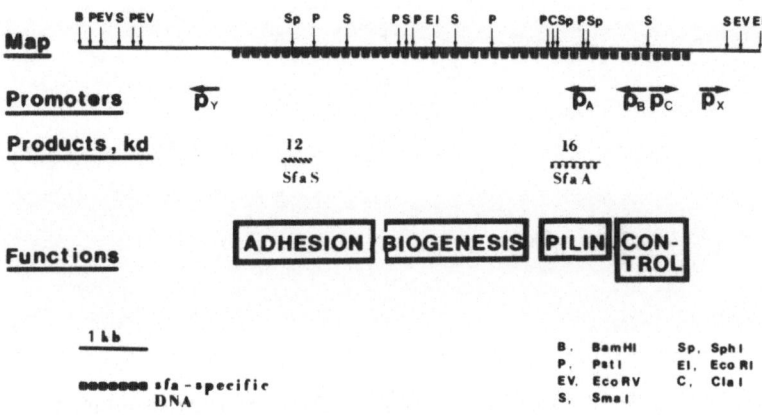

Fig. 1: Genetic map of the determinant coding for the S fimbrial adhesin

3.   Association of genetic virulence determinants

Virulent E. coli strains from extraintestinal sources can generate non-virulent mutants. For the uropathogenic strain 536 (06:K15) this phenomenon was analysed in detail (Knapp et al., 1986). In non-virulent isolates the hemolytic phenotype was abolished together with the expression of fimbrial adhesins. It has been shown that the two hemolysin determinants of that strain (hlyI, hlyII) which are localized on large chromosomal regions with 70 to 100 kilobases (kb) in size are deleted in non-virulent mutant strains like 536-21 (Table 2). In addition, mutants were generated which suffered deletion from one hly determinant only. As indicated in Table 2, mutants with deletions for both hly determinants or for hly determinant II had also lost the ability to produce fimbrial adhesins (Sfa, type I fimbriae). Using fimbria specific probes it was demonstrated that the corresponding determinants were still present on the chromosomes of the mutant strains. It seems, therefore, that a yet unknown transacting factor which may be encoded by the hly region II influences the expression of the adhesin determinants.

Table 2. Virulence properties of E. coli strain 536 and its mutants

Strain[1]	hlyI	hlyII	Sfa	Msh	22kd fim	Sre
536[2]	+	+	+	+	+	+
536-14	-	+	+	+	+	+
536-225	+	-	-	-	-	-
536-21	-	-	-	-	-	-
536-111	-	-	-	-	-	-
536-112	-	-	-	-	-	-

[1] hly means hemolysin determinant, Sfa means S fimbrial adhesin, Msh means mannose sensitive hemagglutination adhesin, 22kd fim means fimbriae with 22kd subunits, Sre means serum resistance
[2] all 536 strains exhibit the O6:K15 serotype

4.   Determination of serum resistance

A prerequisite for the capacity of virulent E. coli strains to cause extraintestinal infections is their ability to be resistant to the action of the complement system. As most of the extraintestinal pathogens strain 536 is serum resistant (Sre[+]). It has been shown that the mutant strains which had lost hemolysin production and adhesin formation are drastically reduced in resistance to normal human serum. Survival after 3 hours in undiluted serum was only 1-2% compared to over 400% survival in the case of the wild-type 536 (see Table 3).

Table 3. Serum resistance of E. coli strain 536 and its derivatives

Strain	Survival of strains (%) after growing in 95% human serum[1])		
	1h	2h	3h
536	200	650	430
536-14	185	600	420
536-225	160	75	2
536-21	140	60	2
536-21 (pANN5211)[2]	160	130	15
536-21 (pANN801-13)[3]	120	70	3
HB101 (K-12)	2	-	-

[1]nearyly 2x10^5 bacteria were set as 100%
[2]pANN5211 codes for hemolysin synthesis
[3]pANN801-13 codes for S fimbrial adhesin

To test whether hemolysin or adhesin determinants could influence the serum resistance cloned hly determinant (located on plasmid pANN5211) and gene clusters coding for S fimbrial adhesins (located on plasmid pANN801-13) were transformed into the mutant strain 536-21. Serum bactericidal assays with these transformants showed that hemolysin did not restore full serum resistance (Hughes et al., 1987). Nevertheless, the presence of multicopy plasmids carrying the fully functional hly determinants did invoke reproducible lower levels increases in 3 hour survival (15% instead of 1-2% of 536-21). The introduction of adhesin determinants had no effect on the serum resistance of the mutant strain. These data suggest a common expression of hemolysin, adhesin formation and serum resistance in strain 536. These different virulence factors seem to be organized in a functional "virulence block" (see also Table 2).

5.   Release of mediators of inflammation
It is clear from clinical and epidemiological studies that infections caused by extraintestinal E. coli very often show symptoms of inflammation

and fever. We have analysed the influence of different pathogenicity factors to these processes (Konig et al., 1986). For that reason, isogenic wild-type clones were constructed which differed in the expression of one single pathogenicity factor (hemolysin or adhesins). The release of leukotriene B_4, C_4 and D_4 and of histamine from human polymorphonuclear neutrophils (PMNs) and mast cells, respectively, was measured and used as an indication for inflammatory processes. As demonstrated in Figure 2 all hemolytic clones stimulated the release of leukotriene C_4 while non-hemolytic did not. Similar results were found for the release of the other leukotrienes and for histamine (Konig et al., 1986). Thus, it can be concluded that hemolysis is one factor involved in the generation of inflammatory processes. The degree of stimulation of leukocytes, however, depends on the presence of particular adhesin factors. It was shown that S fimbriae contribute to such processes to a high degree compared with P fimbriae. It can be summarized, therefore, that fimbriae in particular S fimbrial adhesins may act as co-factors in stimulation of release of mediators of infection.

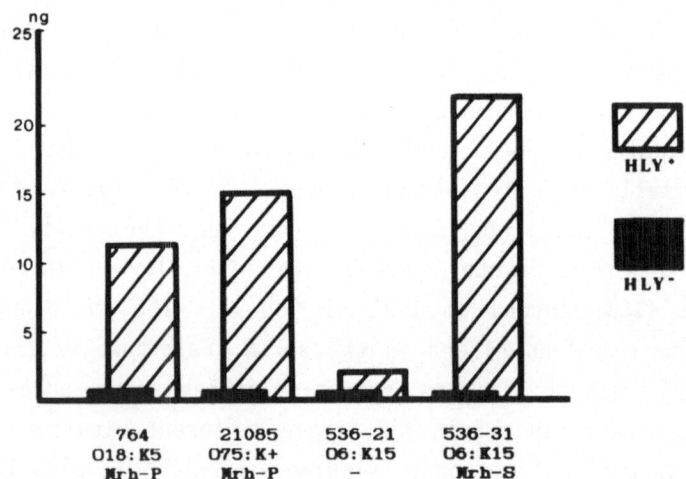

Fig. 2. Release of leukotriene C_4 from human PMNs measured after incubation of cells with isogenic E. coli strains which differ in the presence or absence of the hemolysin determinant

Table 4. Virulence properties of wildtype strain 536 and mutant strains which carry recombinant DNAs coding for different virulence factors

Strains[1]	cloned genes[2]	adhesion	Chicken embryo test[3]	Mouse LD_{50}[4]	Rat lethality[3]	Rat renal counts
536 (06,Sfa,Msh,Hly,Sre)	-	6.8	100	1×10^8	31	9.1×10^4
536-21 (06)	-	2	20	1×10^9	0	4.0×10^1
536-21 pANN801-4	sfa	10.1	20	1×10^9	0	1.0×10^3
536-21 pANN921	P-F8	13.5	25	8×10^9	0	5.0×10^1
536-21 pANN202-312	hly-p	2	30	8×10^9	0	1.2×10^2
536-21 pANN202-812	hly-p	2	100	1.2×10^8	42	nt
536-21	hly-c	2	90	1.3×10^8	0	2.0×10^2

[1] In the adhesion test bacteria per urinary epithelial cell are given, in the chicken embryo test and in the rat lethality test rate of killed animals are given
[2] hly-p means hemolysin determinat cloned from the plasmid pHly152, hly-c means hemolysin determinant cloned from the chromosome of an E. coli 018:K5 strain
[3] percent of killed animals
[4] number of bacteria

6. In vivo consequences

In order to find out whether the different cloned pathogenicity determinants have an influence on the in vivo virulence of strains they were assayed in different in vivo systems. As shown in Table 4 both types of adhesins under question, P and S, contribute to adherence of strains to uroepithelial cells. Hemolysin but not adhesins had a strong influence of in vivo toxicity, indicated by the use of a chicken embryo test and a mouse peritonitis model (Hacker et al., 1986). In a rat pyelonephritis model which more closely reflects the situation of a UTI in men (Marre et al., 1986) hemolysin and S fimbrial adhesins together with serum resistance contribute to nephrovirulence. These results together with the data of stimulation of mediators support the view that virulence of extraintestinal E. coli is a multifactorial phenomenon depending on different pathogenicity factors.

7.   References

Felmlee T, Pellett S, Welch RA (1985) Nucleotide sequence of an Escherichia coli chromosomal hemolysin. J Bacteriol 163:94-105.

Hacker J, Hughes C (1985) Genetics of Escherichia coli hemolysin. Curr Top Microbiol Immunol 118:139-162.

Hacker J, Hof H, Emody L, Goebel W (1986) Influence of cloned Escherichia coli hemolysin genes, S fimbriae and serum resistance on pathogenicity in different animal models. Microb Pathogenesis 1:533-557.

Hess J, Wels W, Vogel M, Goebel W (1986) Nucleotide sequence of a plasmid-encoded hemolysin determinant and its comparison with a corresponding chromosomal hemolysin sequence. FEMS Microbiol Lett 34:1-11.

Hughes C, Hacker J, Duvel H, Goebel W (1987) Chromosomal deletions and rearrangements cause coordinate loss of hemolysis, fimbriation and serum resistance in a uropathogenic strain of Escherichia coli. Microbiol Pathogenesis 2:227-230.

Knapp S, Hacker J, Jarchau T, Goebel W (1986) Large instable regions in the chromosome affect virulence properties of an uropathogenic Escherichia coli 06 strain. J Bacteriol 168:22-30.

Konig W, Konig B, Scheffer J, Hacker J, Goebel W (1986) On the role of Escherichia coli a-hemolysin and bacterial adherence in bacterial infection-requirements for the release of inflammatory mediators from granulocytes and mast cells. Infect Immun 54:886-892.

Korhonen TK, Vaisanen-Rhen M, Pere A, Parkkinen J, Finne J (1984) Escherichia coli fimbriae recognizing sialyl galactosides. J Bacteriol 159:762-766.

Marre R, Hacker J, Henkel W, Goebel W (1986) Contribution of cloned virulence factors from uropathogenic E. coli strains to nephro-pathogenicity in an experimental rat pyelonephritis model. Infect Immun 54:761-767.

Moch T, Hoschutzky H, Hacker J, Kronke KD, Jann K (1987) Isolation and characterization of the a-Sialyl-β-2.3-Galactosyl (S)-specific adhesin from fimbriated Escherichia coli. Proc Natl Acad Sci USA 84:3462-3466.

Orskov I, Orskov F (1983) Serology of Escherichia coli fimbriae. Prog. Allergy 33:80-105.

Orskov I, Orskov F (1985) Escherichia coli in extraintestinal infections. J Hyg Camb 95:551-575.

Ott M, Schmoll T, Goebel W, VanDie I, Hacker J (1987) Comparison of the genetic determinant coding for S fimbrial adhesin (sfa) of E. coli to other chromosomally encoded fimbrial determinants Infect Immun 55:1940-1943.

Schmoll T, Hacker J, Goebel W (1987) Nucleotide sequence of the sfaA gene coding for the S fimbrial protein subunit of Escherichia coli. FEMS Microbiol Lett 41:229-235.

Uhlin BE, Baga M, Goransson M, Lindberg FP, Lund B, Norgren M, Normark S (1985) Genes determining adhesin formation in uropathogenic <u>Escherichia coli</u>. Curr Top Microbiol Immunol 118:163-178.

PHAGOCYTOSIS AND INTRACELLULAR BIOLOGY OF LEGIONELLA PNEUMOPHILA

Marcus A. Horwitz

Division of Infectious Diseases
Department of Medicine
School of Medicine
University of California, Los Angeles
Center for the Health Sciences
Los Angeles, California, U.S.A.

Introduction

Legionella pneumophila is a gram-negative bacterium and the causative agent of two clinically distinct syndromes -- Legionnaires' disease, a severe and often fatal form of pneumonia, and Pontiac Fever, a mild non-fatal febrile illness. L. pneumophila is a facultative intracellular parasite that multiplies in human monocytes and alveolar macrophages (1,2).

Intracellular pathogens follow various phagocytic and intracellular pathways into and through the host cell. In this paper, I shall describe the phagocytic and intracellular pathways followed by L. pneumophila in human mononuclear phagocytes. I shall then note the similarities and differences between the pathways followed by L. pneumophila and other human intracellular pathogens.

Phagocytosis

L. pneumophila is phagocytized by an unusual mechanism termed coiling phagocytosis in which long monocyte pseudopods coil around the organism as it is internalized (3). Human monocytes, alveolar macrophages, and polymorphonuclear leukocytes all phagocytize L. pneumophila by coiling phagocytosis, and both live and dead L. pneumophila are phagocytized by this mechanism. Anti-L. pneumophila antibody neutralizes coiling phagocytosis; L. pneumophila coated with specific antibody are phagocytized by conventional phagocytosis, a process in which phagocyte

NATO ASI Series, Vol. H24
Bacteria, Complement and the Phagocytic Cell
Edited by F.C. Cabello und C. Pruzzo
© Springer-Verlag Berlin Heidelberg 1988

pseudopods move circumferentially and more or less symmetrically about the organism until their tips meet at its distal side and fuse. The end result of both coiling and conventional phagocytosis is the same -- the organism is enclosed in a membrane-bound vacuole or phagosome.

Receptors and Ligands Mediating Phagocytosis

Phagocytosis of L. pneumophila by human monocytes is mediated by monocyte membrane receptors for fragments of the third component of complement (4). This conclusion is supported by several lines of evidence. First, monoclonal antibodies against the type 3 complement receptor (CR3), which recognizes C3bi, and the type 1 complement receptor (CR1), which recognizes C3b, strongly inhibit adherence and consequently phagocytosis of L. pneumophila; monoclonal antibodies against other surface antigens (e.g. HLA-DR, Transferrin Receptor, CR2) have little or no effect on adherence. Second, as a consequence of their capacity to inhibit phagocytosis, monoclonal antibodies against CR1 or CR3 receptors inhibit L. pneumophila intracellular multiplication and protect monocyte monolayers from destruction. Third, monocytes plated on substrates of L. pneumophila membranes modulate their CR1 and CR3 receptors but not their Fc receptors.

Parallelling the above, phagocytosis of L. pneumophila is also mediated by fragments of complement component C3 fixed to the bacterial surface (4,5). This conclusion is supported by two major lines of evidence. First, L. pneumophila fixes C3 to its surface as demonstrated by an enzyme-linked immunosorbent assay (ELISA) and Western blot analysis. Fixation takes place by the alternative pathway of complement activation. Second, adherence of L. pneumophila takes place under conditions in which C3 is fixed to the bacterium, e.g. when the bacteria are incubated in fresh serum. Adherence is markedly reduced under conditions in which C3 is not fixed to the bacterium, e.g. when bacteria are incubated in heat-inactivated serum or in the absence of serum. In the absence of fresh serum, preopsonized L. pneumophila adhere strongly to monocytes but unopsonized L. pneumophila do not. Heat-treatment of preopsonized bacteria, which should not affect bacteria-C3 complexes already formed, has no effect on adherence.

Formation of the L. pneumophila Phagosome

After phagocytosis, the membrane-bound phagosome in which L. pneumophila resides undergoes a remarkable evolution involving the sequential interaction between the phagosomal membrane and three monocyte organelles (6). First, beginning a few minutes after ingestion, smooth vesicles cluster about the phagosome; these vesicles appear to fuse with and/or bud off from the phagosomal membrane. Second, beginning about 15 minutes after ingestion, mitochondria surround the phagosome, closely apposed to the phagosomal membrane; a majority of phagosomes are surrounded by mitochondria at 1 hour after ingestion. Third, several hours after ingestion, ribosomes and ribosome-lined vesicles line up around the phagosome at a distance of about 100 Angstroms from the phagosomal membrane; few smooth vesicles or mitochondria remain by this time. By 4-8 hours after ingestion, ribosome-lined phagosome has formed. At this point, L. pneumophila begins multiplying with a mid-log phase doubling time of about 2 hours. As the organism multiplies, the ribosome-lined phagosome enlarges to accommodate the increased numbers of bacteria. The bacteria multiply until the monocyte is chocked full of bacteria and ruptures.

Inhibition of Phagosome-Lysosome Fusion

L. pneumophila phagosomes do not fuse with monocyte primary or secondary lysosomes (7). This was determined by two methods. In the first method, monocytes were prelabelled with thorium dioxide, an electron-opaque marker that is concentrated in monocyte lysosomes. The monocytes were then infected with L. pneumophila and examined for the presence of the lysosomal marker in the phagosome, which would indicate that fusion had occurred. In the second method, monocytes were infected with L. pneumophila and then stained for acid phosphatase, a lysosomal enzyme. The first method measures fusion of the phagosome with only secondary lysosomes, i.e. lysosomes that have concentrated thorium dioxide as a consequence of fusion with pinocytic vesicles carrying the marker into the cell. The second method measures fusion of the phagosome with both primary and secondary lysosomes, both of which contain acid phosphatase. By both methods, phagosomes containing live L. pneumophila do not fuse with lysosomes. In contrast, phagosomes containing formalin-killed L. pneumophila do fuse with lysosomes; these dead bacteria are rapidly degraded within the phagolysosome.

Inhibition of Phagosome Acidification

Phagosomes containing L. pneumophila do not become acidified to the low levels characteristic of phagosomes containing intracellular pathogens which, in contrast to L. pneumophila, are rapidly killed by monocytes and degraded within phagolysosomes (8). This was determined by quantitative fluorescence microscopy (8). By this method, phagosomes containing live L. pneumophila have a mean pH of approximately 6.1. In contrast, phagosomes containing formalin-killed L. pneumophila have a mean pH of 5.4 and phagosomes containing either live or formalin-killed E. coli have a mean pH of 5.6; these two latter pH levels are typical of phagolysosomes.

The inhibition of acidification induced by L. pneumophila in monocytes is localized to the L. pneumophila phagosome. The presence of one phagosome containing L. pneumophila does not influence the pH of another phagosome containing an erythrocyte in the same monocyte, and vice versa. In such monocytes, phagosomes containing an erythrocyte are acidified to a level of approximately 5.0. Thus, phagosomes of different pH exist within the same cell.

Comparative Biology of Human Intracellular Pathogens

Intracellular pathogens enter mononuclear phagocytes by one of two general mechanisms -- coiling or conventional phagocytosis. In addition to L. pneumophila, Leishmania donovani appears to enter by coiling phagocytosis (9), and preliminary observations indicate that other intracellular pathogens enter by coiling phagocytosis as well. In contrast, Mycobacterium tuberculosis (10), Toxoplasma gondii (11), and Trypanosoma cruzi (12,13) enter mononuclear phagocytes by conventional phagocytosis.

Complement receptors appear to play a general role in mediating entry of intracellular pathogens into mononuclear phagocytes. In addition to L. pneumophila, complement receptors mediate ingestion of M. tuberculosis (10), L. donovani (14), and Histoplasma capsulatum (15). Since T. cruzi (16) and T. gondii (17) both fix C3 to their surface, it seems likely that complement receptors mediate uptake of these organisms as well.

The complement receptor pathway may be a protected route of entry for intracellular pathogens. Ligation of complement receptors for C3b and C3bi under conditions in which these receptors mediate phagocytosis does not result in the release of the oxidative metabolites hydrogen peroxide and superoxide or mediators of inflammation such as arachidonic acid (18,19,20).

Human intracellular pathogens follow at least three different pathways through the mononuclear phagocyte. L. pneumophila is prototypic of organisms that follow one such pathway. L. pneumophila forms a morphologically distinctive phagosome surrounded at specific times by smooth vesicles, mitochondria, and ribosomes, and this phagosome neither fuses with lysosomes nor becomes acidified to low levels (6-8). L. pneumophila shares this pathway with T. gondii and Chlamydia psittaci. All three organisms form a mitochondria-lined phagosome (6,11,22), all three inhibit phagosome-lysosome fusion (7,21,22), and at least two of them, L. pneumophila (8) and T. gondii (23) inhibit phagosome acidification.

L. donovani follows a second pathway through the mononuclear phagocyte. This organism does not form a phagosome with any of the morphologic features of the L. pneumophila phagosome, does not inhibit phagosome-lysosome fusion, and indirect evidence indicates that it does not inhibit phagosome acidification (24,25,26). Thus, L. donovani survives and multiplies within a phagolysosome.

T. cruzi follows yet a third pathway. This organism exits the phagosome shortly after ingestion and survives and multiplies within the cytoplasm of the mononuclear phagocyte (12,13).

Acknowledgements

Dr. Horwitz is a Gordon MacDonald scholar at UCLA and recipient of an American Cancer Society Faculty Research Award. This work is supported by grant DCB 8501458 from the National Science Foundation and grant AI-22421 from the National Institutes of Health.

References

1. M. A. Horwitz, S. C. Silverstein, The Legionnaires' disease bacterium (Legionella pneumophila) multiplies intracellularly in human monocytes. J. Clin. Invest. 66:441-450 (1980).
2. T. W. Nash, D. M. Libby, M. A. Horwitz, Interaction between the Legionnaires' disease bacterium (Legionella pneumophila) and human alveolar macrophages. Influence of antibody, lymphokines, and hydrocortisone. J. Clin. Invest. 74:771-782 (1984).
3. M. A. Horwitz, Phagocytosis of the Legionnaires' disease bacterium (Legionella pneumophila) occurs by a novel mechanism: Engulfment within a pseudopod coil. Cell 36:27-33 (1984).
4. N. R. Payne and M. A. Horwitz, Phagocytosis of Legionella pneumophila by human monocytes is mediated by membrane receptors for the third component of complement and monoclonal antibodies against these receptors inhibit intracellular multiplication. In: Program of the 1987 Annual Meeting of the American Society of Microbiology, Atlanta, GA, March 1-6, p. 86 (1987).
5. C. G. Bellinger-Kawahara, and M. A. Horwitz, Legionella pneumophila fixes complement component C3 to its surface - demonstration by ELISA. In: Program of the 1987 Annual Meeting of the American Society of Microbiology, Atlanta, GA, March 1-6, p. 86 (1987).
6. M. A. Horwitz, Formation of a novel phagosome by the Legionnaires' disease bacterium (Legionella pneumophila) in human monocytes. J. Exp. Med. 158:1319-1331 (1983).
7. M. A. Horwitz, The Legionnaires' disease bacterium (Legionella pneumophila) inhibits phagosome-lysosome fusion in human monocytes. J. Exp. Med. 158:2108-2126 (1983).
8. M. A. Horwitz, F. R. Maxfield, Legionella pneumophila inhibits acidification of its phagosome in human monocytes. J. Cell. Biol. 99:1936-1943 (1984).
9. K. P. Chang, Leishmania donovani promastigote-macrophage surface antigens in vitro. Exp. Parasitol. 48:175-189 (1979).
10. N. R. Payne, C. G. Bellinger-Kawahara, and M. A. Horwitz, Phagocytosis of Mycobacterium tuberculosis by human monocytes is mediated by receptors for the third component of complement. Clinical Research 35:617A (1987).
11. T. C. Jones, S. Yeh, J. G. Hirsch, The interaction between Toxoplasma gondii and mammalian cells. I. Mechanisms of entry and intracellular fate of the parasite. J. Exp. Med. 136:1157-1172 (1972).

12. N. Nogueira, Z. A. Cohn, Trypanosoma cruzi: Mechanism of entry and intracellular fate in mammalian cells. J. Exp. Med. 143:1402-1420 (1976).

13. H. Tanowitz, M. Wittner, Y. Kress, and B. Bloom, Studies of in vitro infection by Trypanosoma cruzi. I. Ultrastructural studies on the invasion of macrophages and L-cells. Am. J. Trop. Med. Hyg. 25:25-33 (1975).

14. J. M. Blackwell, R. A. B. Esekowitz, M. B. Roberts, J. Y. Channon, R. B. Sim, and S. Gordon, Macrophage complement and lectin-like receptors bind Leishmania in the absence of serum. J. Exp. Med. 162:324-331 (1985).

15. W. E. Bullock and S. D. Wright, Role of the adherence-promoting receptors CR3, LFA-1, and p150,95, in binding of Histoplasma capsulatum by human macrophages. J. Exp. Med. 165:195-210 (1987).

16. K. A. Joiner, S. Hieny, L. V. Kirchhoff, and A. Sher, gp72, the 72 kilodalton glycoprotein, is the membrane acceptor site for C3 on Trypanosoma cruzi epimastigotes. J. Exp. Med. 161:1196-1212 (1985).

17. S. A. Fuhrman and K. A. Joiner, Binding of the third component of complement (C3) to Toxoplasma gondii. Clinical Research 35:475A (1987).

18. K. R. Yamamoto and B. Johnston, Jr., Dissociation of phagocytosis from stimulation of the oxidative metabolic burst in macrophages. J. Exp. Med. 159:405 (1984).

19. S. D. Wright and S. C. Silverstein, Receptors for C3b and C3bi promote phagocytosis but not the release of toxic oxygen from human phagocytes. J. Exp. Med. 158:2016-2023 (1983).

20. A. A. Aderem, S. D. Wright, S. C. Silverstein, and Z. A. Cohn, Ligated complement receptors do not activate the arachidonic acid cascade in resident peritoneal macrophages. J. Exp. Med. 161:617-622 (1985).

21. T. C. Jones and J. G. Hirsch, The interaction between Toxoplasma gondii and mammalian cells. II.The absence of lysosomal fusion with phagocytic vacuoles containing living parasites. J. Exp. Med. 136:1173-1194 (1972).

22. R. R. Friss, Interaction of L-cells and Chlamydia psittaci -- entry of the parasite and host response to its development. J. Bacteriol. 110:706-721 (1972).

23. L. D. Sibley, E. Weidner, and J. L. Krahenbuhl, Phagosome acidification blocked by intracellular Toxoplasma gondii. Nature 315:416-419 (1985).

24. J. Alexander and K. Vickerman, Fusion of host cell secondary lysosomes with the parasitophorous vacuoles of Leishmania mexicana-infected macrophages. J. Protozool. 22:502-508 (1975).

25. K. P. Chang and D. M. Dwyer, Multiplication of a human parasite (Leishmania donovani) in phagolysosomes of hamster macrophages in vitro. Science (Washington, D.C.) 193:678-690 (1976).

26. A. J. Mukkada, J. C. Meade, T. A. Glaser and P. F. Bonventre, Enhanced metabolism of Leishmania donovani amastigotes at acid pH: an adaptation for intracellular growth. Science 229:1099-1101 (1985).

INDUCTION OF INFLAMMATION BY ESCHERICHIA COLI AT A MUSCOSAL SITE: REQUIREMENT FOR ADHERENCE AND ENDOTOXIN

Catharina Svanborg Edén, Inga Engberg, Henrik Linder

Department of Clinical Immunology, University of Göteborg
Guldhedsgatan 10, S-413 46 Göteborg, Sweden

Introduction

Most infectious agents enter the body via the mucosal surfaces. The host defense mechanisms at these sites combine to dislodge pathogens from the surface. Microorganisms resist elimination by attaching to components of the mucosal lining (1). The attachment can result from specific interactions of bacterial surface lectins with receptors consisting of oligosaccharide sequences in epithelial glycoconjugates (2).

Uropathogenic Escherichia coli attach to uroepithelial cells via the globoseries of glycolipid receptors with the common disaccharide Galα1→4 Galβ. The bacteria can recognize other receptor specificities e.g. mannosides (3). The Galα1→4Gal receptor specificity promotes the localization of E. coli to the kidney (4).

Once bacteria reach the tissues, inflammation and immunity are activated. The link between surface attachment and the tissue reactions has not been defined. This report demonstrates a mucosal inflammatory response within 24 h after local challenge with E. coli, and suggests that two components are required:

1. Ability of the infected host to react to the lipid A moiety of endotoxin.

2. Bacterial attachment to the globoseries of glycolipid receptors.

Endotoxin reactivity as a prerequisite for the acute local inflammatory response

The congenic mouse strains C3H/HeN (Lps^n/Lps^n) and C3H/HeJ (Lps^d/Lps^d)

NATO ASI Series, Vol. H24
Bacteria, Complement and the Phagocytic Cell
Edited by F. C. Cabello und C. Pruzzo
© Springer-Verlag Berlin Heidelberg 1988

differ at the _Lps_ locus on chromosome four, which determines the reactivity with the lipid A portion of lipopolysaccharide, LPS, from gram-negative bacteria (5). C3H/HeJ mice are refractory to the many actions of lipid A, including the toxic, mitogenic and immunomodulatory effects (6). The role of LPS responsiveness for the mucosal inflammatory response was analyzed here using a model for ascending urinary tract infection, UTI (4). LPS responder and non-responder mice were challenged intravesically with _E. coli_ Hu734 (075:K5:H-) (7), and the acute inflammatory response was monitored as the influx of leucocytes into the urine (8). Within 24 h of infection as increase in urinary leucocytes occurred in C3H/HeN mice; >90% were identified as polymorphonuclear leucocytes by Giemsa-stained cytocentrifuge preparations. In contrast, the LPS non-responder mice did not show a significant increase in urinary leucocyte counts (Table 1). These results suggested that the lipid A moiety of Lps played a crucial role among the many inflammatogenic bacterial surface structures, in the triggering of the acute inflammatory response in this model. This was in contrast to previous _in vivo_ studies

Table 1. Inverse relationship between bacterial persistence and leucocyte excretion.

Mouse strain	LPS genotype	Urinary leucocytes Mean x 10^{-4}	Bacterial recovery Kidneys
C3H/HeN	(Lps^n, Lps^n)	232	132
C3H/HeJ	(Lps^d, Lps^d)	8	18197

Female C3H/HeJ mice (Jackson Laboratories, Bar Harbor, ME) and C3H/HeN mice (Charles River, UK) were used at 6-10 weeks of age. Bacteria were cultured in nutrient broth to end logarithmic phase. Mice were challenged intravesically with 0.1 ml of a suspension containing about 10^9 _E. coli_ Hu734. Urine was cultured prior to challenge to avoid contamination. Leucocyte excretion was quantitated microscopically in urine obtained 24 h after infection. The bacterial concentration in the inoculum and in the tissues after sacrifice, was quantitated by viable counts of homogenized tissues (8). The mean represents two experiments, 10 mice per experiment and mouse strain.

of the inflammatory response to systemically administered LPS, and in vitro studies showing normal chemotaxis of PMNs from C3H/HeJ mice towards LPS (9,10). The chemotactic signal normally induce by LPS at the mucosal site appeared to be deficient in the non-responder mouse.

Inverse relationship between bacterial persistence and the inflammatory response

The susceptibility to local gram-negative infection of C3H/HeJ and C3H/HeN mice was determined as the persistence of E. coli Hu734 in kidneys and bladders 24 h after infection and later (4,7). The LPS responder mice cleared the infection within a few days (Table 2). In contrast, the non-responder mice had about 1000-fold higher E. coli kidney counts than CeH/HeN mice 24 h after infection, and remained infected for at least a month. The resistance of the mice to gram-negative infection thus paralelled the LPS responder phenotype, and the ability to mount an inflammatory response. The dramatic difference in bacterial counts within 24 h of infection suggested that bacterial clearance from mucosal sites involves host resistance factors which are activated much faster than it would take to induce specific mucosal immune responses, which previously have been emphasized as the first line of defense.

The inflammatory signal can be triggered by dead bacteria

The inflammatory signal might have been delivered subsequent to bacterial invasion through interaction of LPS with, e.g. tissue macrophases, or through a direct effect at the mucosal surface. Live and formalin-killed bacteria were compared for their ability to elicit the inflammatory response (Table 2). In LPS responder mice a significant increase in urinary leucocytes occured within 24 h of exposure to both live and formalin-killed E. coli Hu734 and Hu824, but not Hu742. This demonstrated that the inflammatory signal can be delivered without the aid of metabolic activity in the bacteria.

Table 2. Recruitment of leucocytes by live and killed whole bacteria

Inoculum	Receptor specificity of adhesins	Urinary leucocytes Mean (range) x 10⁻⁴	
		C3H/HeN	C3H/HeJ
Hu734	Galα1→4Gal		
Live	"Mannosides"	98 (20- 400)	6 (0- 30)
Killed		165 (12- 600)	8 (0- 32)
Hu824	Galα1→4Gal		
Live		205 (28- 600)	3 (0- 7)
Killed		127 (12- 500)	8 (0- 38)
Hu742	"Mannosides"		
Live		58 (8 - 175)	8 (4- 12)
Killed		8 (0 - 19)	0 (0- 0)

Mice were challenged with live or formalin-fixed whole bacteria and the influx of urinary leucocytes quantitated after 24 h (see footnote Table 1). The expression of adhesins on the bacteria was controlled prior to challenge by the ability to agglutinate human and guinea pig erythrocytes, as well as latex beads covalently coupled with Galα1→4Gal.

Bacteria adherence required for adequate delivery of the LPS signal

The difference in ability to trigger an inflammatory response between E. coli Hu824 and Hu742, was attributed to the difference in adhesive properties. Hu824 and Hu742 were mutants of the E. coli strain Hy734 (11). They retained the parent strain LPS and capsular polysaccharide, were non-motile and hemolysin negative, but differed in the receptor specificity of their adhesins. The parent strain expressed adhesins binding the globoseries of glycolipid receptors Galα1→4Gal and mannose. The Hu824 mutant retained the specificity for Galα1→4Gal, and attached well to mouse and human uroepithelial cells. The Hu742 mutant retained the mannose-sensitive adhesins, and attached poorly to mouse and not to human uroepithelial cells. The ability of the two mutants to deliver the LPS signal was analysed using formalin-killed bacteria (Table 2). The Hu824 mutant induced a significant influx of leucocytes, Hu742 did not. These results suggested that bacterial attachment was required for the LPS signal to be delivered to the mucosal surface.

Inhibition of inflammation by receptor analogues

The previous results demonstrated that the LPS-induced inflammation could be aborted if the host was unresponsive to lipid A. Provided that specific adherence was required for optimal delivery of the LPS signal, blocking of adherence would similarly abort the inflammatory response. This hypothesis was tested by pretreatment of $\underline{E.}$ $\underline{coli}$ Hu824 with globotetraosylceramide, globotetraose or $Gal\alpha1{\rightarrow}4Gal$ prior to injection into C3H/HeN mice. The urinary leucocyte response was compared to the saline-treated control (Table 3). A significant inhibition of the inflammatory response occurred in mice receiving bacteria pretreated with receptor analogue.

Table 3. Abrogation of the inflammatory response by competitive inhibition of adherence with receptor analogues

Pretreatment	Attachment bacteria/cell	Urinary leucocytes Mean (range) x 10^{-4}	
PBS	42	121	(12-500)
Globoside	0	22	(0-103)

C3H/HeN mice (Lps^n, Lps^n) were challenged with formalin-killed $\underline{E.}$ $\underline{coli}$ Hu824, expressing adhesins specificed for $Gal\alpha1{\rightarrow}4Gal$-containing receptors. Bacteria were incubated at 37^0C for 30 min in PBS, 10 mg/ml globoside ($GalNac\beta1{\rightarrow}3Gal\alpha1{\rightarrow}4Gal\beta1{\rightarrow}4Glc$-ceramide) sonicated in PBS[11]. An aliquot was tested for adherence to mouse uroepithelial cells[4]; attachment is expressed as the mean no. of bacteria attached to 40 epithelial cells; two experiments per inhibitor. Urinary leucocytes were quantitated in urine obtained 24 h after infection.

We propose the following interpretation of the results: E. coli expressing adhesins specific for Galα1→4Gal containing receptors attached to the mucosal lining of the kidneys, and delivered the endotoxin to the tissues. The attachment probably served as a means of accumulating sufficient concentrations of toxin at the surface. The poor inflammatory response to the mutant Hu742, with the same endotoxin but adhesins specific for mannose, probably reflected its lower attachment. Alternatively, the crosslinking of Galα1→4Gal-containing receptors induced the epithelial cells to take up the bacteria, thus permitting the toxin to be transported to the tissues. This remains to be investigated.

LPS can trigger chemotaxis by interaction with tissue macrophages, complement components, etc. (10). The demonstration here that inflammation was triggered by killed bacteria delivered at the mucosal surface raised the possibility that the epithelial cells themselves were active in this process. This is contrary to the generally accepted hypothesis that attachment is followed by active bacterial invasion followed by activation of host responses.

Results similar to those reported here were recently obtained from clinical studies (12). In infants with urinary tract infections, there was a relationship between the intensity of the inflammatory response and the attachment of the infecting E. coli strain. Infants infected with bacteria which expressed adhesins specific for Galα1→4Gal-containing receptors, had higher urinary leucocyte counts than infants infected with other bacteria. Blocking of adherence, thus, has the potential of both reducing bacterial colonization and of aborting the inflammatory response at mucosal surfaces.

Acknowledgements

This study was supported by grants from the Swedish Medical Research Council (Grant no. 215 and no. 7934), the BACH project of SSA-KABI, the Swedish Board for Technical Development and the Lundberg Foundation. Globotetraose was kindly provided by B. Nilsson and collaborators, the Swedish Sugar Company.

References

1. Beachey, E.H. J. Infect. Dis. 143, 325-345 (1981)
2. Leffler, H.., Svanborg-Edén, C. in Microbial Lectins and Agglutinins: Properties and Biological Activity (ed. Mirelman, D.) 83-111 (John Wiley and Sons, 1986)
3. Ofek, I., Mirelman, D., and Sharon, N. Nature 265, 623-625 (1977)
4. Hagberg, L., Hull, R., Hull S., Falkow, S., Freter, R., Svanborg-Edén, C. Infect. Immun. 40, 265-272 (1983)
5. Watson, J. and Riblet, R. J. Exp. Med. 140, 1147-1161 (1974)
6. Vogel, S.N., Weinblatt, A.C. and Rosenstreich, D.L. Inherent macrophage defects in mice. in Immunological Defects in Laboratory Animals. 327-357 (Plenum Press, New York 1981).
7. Hagberg, L., Briles, D.E., Edén, C.S. J. Immunol. 134, 4118-4122 (1985)
8. Svanborg-Edén, C. et al. in Genetic Control of Host Resistance to Infection and Malignancy (ed. Shamene) 385-391 (Alan Liss Inc. 1985)
9. Sultzer, B.M. Nature 219, 1253 (1968)
10. Verghese, M.W. and Snyderman, R. J. Immunol. 127, 288-293 (1981)
11. Marild, S. et al. IVth Int. Symp. on Pyelonephritis. (Chicago University Press) (in press)
12. Svanborg-Edén, C., Freter, R., Hagberg, L., Hull, R., Hull, S., Leffler, H., Schoolnik, G. Nature 298, 560-562 (1982)

IgA- ARMED T LYMPHOCYTES AS ANTI- S. TYPHI EFFECTORS IN HUMANS

L. Nencioni, L. Villa, M.T. De Magistris, M. Romano, D. Boraschi and
A. Tagliabue

Sclavo Research Center
Via Fiorentina
53100 Siena
Italy

Typhoid fever still remains one of the major causes of death in several developing countries. Furthermore, cases imported from these endemic regions continue to cause problems also in industrialized areas of the world where typhoid fever is no longer endemic (Edelman & Levine, 1986).

Most data support the concept that the critical defence elicited by the host to typhoid fever is cell-mediated. However, until recently knowledge about the cells involved and the antigens responsible for the specificity of the cellular immune response against Salmonella typhi (S. typhi) was rather scanty. In particular, it was always thought that the cells mainly involved in the antibacterial activity belonged to the monocyte-macrophage series. Despite this common belief, the evidence that lymphocytes also play an important role in protecting the host against bacterial microorganisms came from studies in the mouse system. In fact, by using a previously described short term in vitro assay (Nencioni et al., 1983), murine lymphoid cells from different anatomical sites, including the gut associated lymphoid tissues (GALT), were demonstrated to express natural antibacterial (NA) activity against enteropathogenic bacteria (Nencioni et al., 1983; 1985). As in the mouse system, human peripheral blood mononuclear cells (PBMC) from normal donors were also shown to reduce the viability of S. typhi when used as targets in the in vitro assay, whereas cell populations highly enriched in monocytes do not have this capacity (Fig. 1). The phenotypic characterization of the effector cell of the NA activity revealed that this cell is a CD2, CD3, CD4, CD28 positive, and CD8, CD11b, CD16, CD21 negative T lymphocyte as resulted from the treatment of nylon wool nonadherent peripheral blood lymphocytes (PBL) with a panel of cytotoxic monoclonal antibodies plus complement (Table 1).

NATO ASI Series, Vol. H24
Bacteria, Complement and the Phagocytic Cell
Edited by F.C. Cabello und C. Pruzzo
© Springer-Verlag Berlin Heidelberg 1988

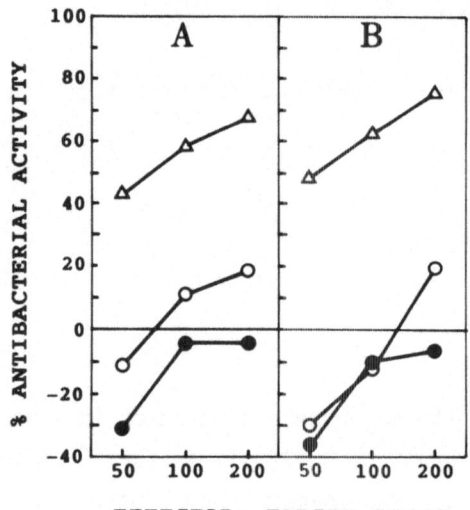

Fig. 1. NA activity against S. typhi of PBMC from normal donors before (O——O) and after enrichment in monocytes (●——●) or lymphocytes (▲——▲). Enriched cell populations were obtained by adherence on plastic (panel A) or by one-step discontinuous gradient of Percoll as previously described (Tagliabue et al., 1985).

Moreover, as shown in Figure 2, an enriched population of CD4 positive cells obtained from PBL after FACS purification expresses a strong NA activity, even at low effector to target ratios, whereas no activity was observed with a CD4 negative enriched population. Finally, but no less important, it was found that cells pretreated with fragments of anti-human IgA but not anti-IgG antibodies were no longer able to exert NA activity (Fig. 3).

Thus, a preliminary conclusion we can draw from these results is that, as found in the mouse system (Tagliabue et al., 1983; 1984), the NA activity observed in vitro might be due to an antibody-dependent cellular cytotoxicity (ADCC) mechanism. This antibacterial ADCC against S. typhi and other enteropathogenic bacteria, is expressed by CD4 positive T lymphocytes in the presence of naturally preexisting IgA antibodies bound to the cell surface by means of Fcα receptors. The origin of naturally occurring IgA antibodies still remains, however, to be clarified. Several hypotheses can be advanced to provide an explanation on this issue. The most likely is that these antibacterial IgA are the result of subclinical infections. Alternatively, according to studies in neonates (Mellander et al., 1986), natural antibodies might be due to an anti-idiotypic network induced by maternal antibodies.

Table 1. Characterization of the phenotype of cells exerting NA
activity against S. typhi

Treatment	Cluster of differentiation	% antibacterial activity		
		50[a]	100	200
None		10	20	26
C		9	18	27
OKT3+C	CD3	-6	-2	1[b]
OKT4+C	CD4	-4	-12	-8[b]
OKT8+C	CD8	13	20	29
OKT11+C	CD2	-5	-3	0[b]
OKM1+C	CD11	27	37	43[b]
OKB7+C	CD21	20	23	27
AB8-28+C	CD16	21	26	29
Leu8+C	CD28	-2	-3	-1[b]

[a]Effector:target ratio

[b]$P \leq 0.5$ versus corresponding C group.

Monomeric, polymeric and secretory IgA antibodies were previously shown
to drive antibacterial ADCC with murine lymphoid cells from GALT
(Tagliabue et al., 1983, 1984). Thus, this antibacterial mechanism might
be considered the major expression of immunosurveillance against
infections at the mucosal level.

It was, therefore, expected that an oral stimulation of the mucosal
immune system could potentiate cell-mediated immunity against enteric
pathogens. Indeed, mice orally immunized with S. typhimurium became
resistant to oral reinfection with the same bacteria and cells from
spleens and Peyer's patches of surviving mice expressed a higher in vitro
NA activity than normal mice (Tagliabue et al., 1985).

We, thus, followed the same approach, employing in humans the widely
used live oral GalE mutant S. typhi vaccine, strain Ty 21a (Germanier &
Furer, 1975), which was shown to be safe and effective in field trials
carried out in endemic countries such as Egypt and Chile (Wahdam et al.,
1982; Levine et al., 1987).

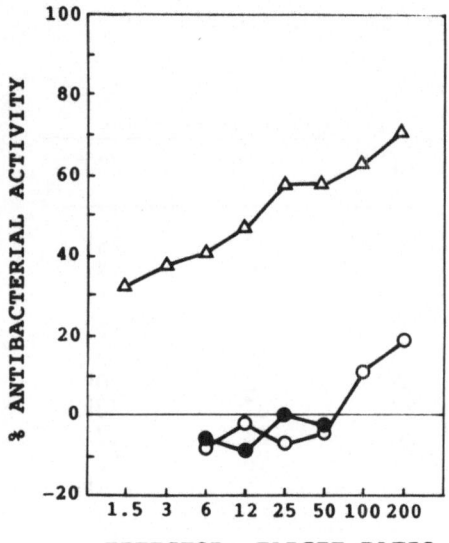

Fig. 2. NA activity against $\underline{S.}$ $\underline{typhi}$ of PBL (O——O) from normal donors and of CD4+ (●——●) or CD4- (▲——▲) enriched populations separated by FACS according to these surface markers.

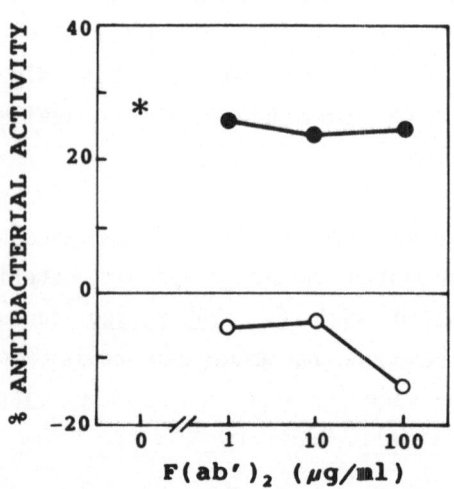

Fig. 3. Effect of pretreatment with different concentrations of $F(ab')_2$ fragments against human IgA (O——O) and IgG (●——●) of PBL exerting NA activity (✳) against $\underline{S. \ typhi}$.

Indeed, PBMC of healthy adult volunteers from areas not endemic for typhoid who had never had the disease showed a significant increase of the in vitro antibacterial activity against S. typhi, S. paratyphi A and B at different times after vaccination with Ty21a (Fig. 4). The enhancement of the activity observed in our in vitro system perfectly correlates with the long lasting protection in vivo, not only against S. typhi but also against S. paratyphi B as determined during the Chilean field trial (Levine et al., 1987), whereas no data are available so far in vivo against S. paratyphi A. Furthermore, the increased cell-mediated immune response was strain-specific because, as shown in Fig. 4, there was no corresponding increase in reactivity of PBMC to S. paratyphi C or S. tel-aviv, which do not share common epitopes with S. typhi within surface antigens.

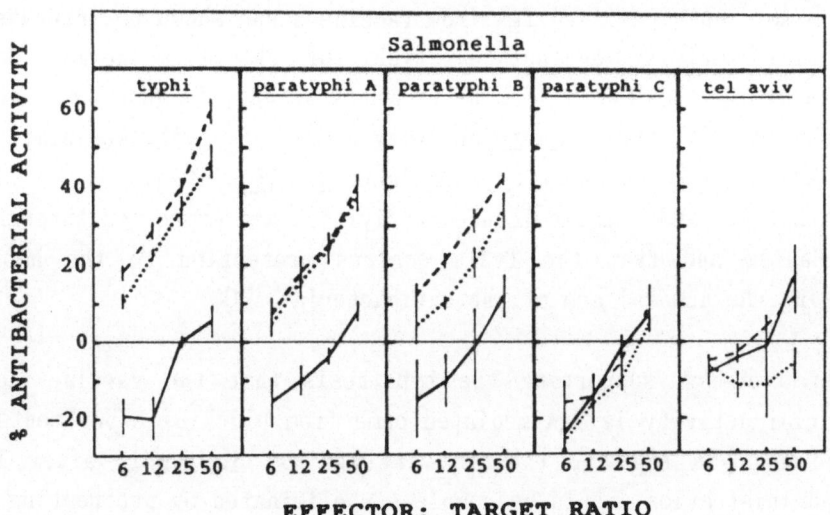

Fig. 4. Cell-mediated in vitro antibacterial activity of PBMC from volunteers before (——), 15 days (– – –) and 120 days (·····) after the last administration of live oral typhoid vaccine Ty21a.

Indeed, both somatic (0) and flagellar (H) but not Vi antigens are the targets of the cellular immune response of vaccines, as demonstrated by inhibition studies on the antibacterial activity of postvaccination PBMC by killed bacteria selected for 0, H and Vi antigens (Tagliabue et al., 1986). In contrast, only 0 polysaccharide exhibited an inhibitory effect on the natural immunity against S. typhi before vaccination. The fact that PBMC from our healthy volunteers never expressed cell-mediated responses to the Vi polysaccharide capsular antigen, which is recognized as a virulence factor of S. typhi (Robbins & Robbins, 1984) is in agreement with humoral studies by other investigators (Tacket et al., 1986). In fact, these authors found that only one-third of patients with typhoid fever have high levels of serum antibodies, whereas this situation is present in most chronic S. typhi carriers. On the other hand, Ty21a vaccine lacks Vi (Germanier & Furer, 1975), yet it has been shown to induce protection in the host against wild-type strains of S. typhi.

As far as humoral responses are concerned, the Ty21a vaccine was capable of inducing an enhancement of IgA and IgG but not IgM serum antibodies specific for both 0 and H antigens. However, when added in the in vitro assay, only purified IgA from vaccinees was shown to increase the antibacterial activity against S. typhi of PBL from normal donors (Tagliabue et al., 1986). Since, after vaccination, the phenotype of the cell responsible for the in vitro activity against S. typhi was similar to the effector of the NA activity (Tagliabue et al., 1986), it is likely that both natural and vaccine-induced activities are expressed through the same mechanism and that the Ty21a confers protection to the host by potentiating the humoral arm of the antibacterial ADCC.

Further evidence supporting the hypothesis that the vaccine-induced antibacterial activity is IgA-mediated came from blocking experiments. In fact, as shown in Fig. 5, the activity against S. typhi, after Ty21a vaccine administration, could be completely eliminated by pretreating PBMC from vaccines with anti-human IgA antibodies before use in the in vitro test.

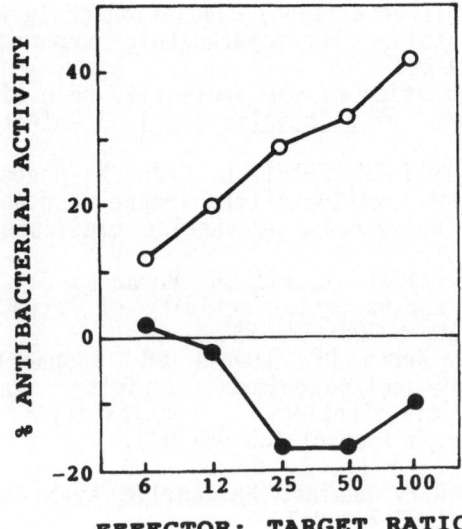

Fig. 5. Antibacterial activity of PBMC from a volunteer vaccinated with the live oral Ty21a strain before (O—O) and after (●—●) pretreatment with anti-human IgA antibodies.

In conclusion, the results reported herein show the presence in normal volunteers of an antibody dependent cell-mediated immunity against S. typhi which can be potentiated by the administration of an attenuated oral typhoid vaccine. This in vitro antibacterial activity involving IgA-armed CD4 positive T lymphocytes and manifested at the peripheral level could reflect a situation occurring also at the GALT level where enteric pathogens first come into contact with the host.

References

Edelman R, Levine MM (1986) Summary of an international workshop on typhoid fever. Rev Infect Dis 8:329-349.

Germanier R, Furer E (1975) Isolation and characterization of GalE mutant Ty21a of Salmonella typhi: a candidate for a live oral typhoid vaccine. J Infect Dis 131:553-558.

Levine MM, Ferreccio C, Black RE, Germanier R, Chilean Typhoid Committee (1987) Large-scale field trial of Ty21a live oral typhoid vaccine in enteric-coated capsule formulation. Lancet i:1049-1052.

Mellander L, Carlsson B, Hanson LA (1986) Secretory IgA and IgM antibodies to E. coli O and poliovirus type I antigens occur in amniotic fluid, meconium and saliva from newborns. A neonatal immune response without antigen exposure: a result of anti-idiotypic induction? Clin Exp Immunol 63:555-561.

Nencioni L, Villa L, Boraschi D, Berti B, Tagliabue A (1983) Natural and antibody-dependent cell-mediated activity against Salmonella typhimurium by peripheral and intestinal lymphoid cells in mice. J Immunol 130:903-907.

Nencioni L, Villa L, Boraschi D, Tagliabue A (1985) Modulation of in vitro natural cell-mediated activity against enteropathogenic bacteria by simple sugars. Infect Immun 47:534-539.

Robbins JD, Robbins JB (1984) Reexamination of the protective role of the capsular polysaccharide (Vi antigen) of Salmonella typhi. J Infect Dis 150:436-449.

Tacket CO, Ferreccio C, Robbins JB, Tsai CM, Schulz D, Cadoz M, Godeau A, Levine MM (1986) Safety and characterization of the immune response to two Salmonella typhi Vi capsular polysaccharide vaccine candidates. J Infect Dis 154:342-345.

Tagliabue A, Nencioni L, Villa L, Keren DF, Lowell GH, Boraschi D (1983) Antibody dependent cell mediated antibacterial activity of intestinal lymphocytes with secretory IgA. Nature 306:184-185.

Tagliabue A, Boraschi D, Villa L, Keren DF, Lowell GH, Rappuoli R, Nencioni L (1984) IgA-dependent cell mediated activity against enteropathogenic bacteria: distribution, specificity and characterization of effector cells. J Immunol 133:988-992.

Tagliabue A, Nencioni L, Caffarena A, Villa L, Boraschi D, Cassola G, Cavalieri S (1985) Cellular immunity against Salmonella typhi after live oral vaccine. Clin Exp Immunol 62:242-247.

Tagliabue A, Villa L, De Magitris MT, Romano M, Silvestri S, Boraschi D, Nencioni L (1986) IgA-driven T cell mediated antibacterial immunity in man after live oral Ty21a vaccine. J Immunol 137:1504-1510.

Wahdam MH, Serie C, Cerisier Y, Sallam S, Germanier R (1982) A controlled field trial of live Salmonella typhi strain Ty21a oral vaccine agianst typhoid: three year results. J Infect Dis 145:292-296.

INTERFERENCE OF A STAPHYLOCOCCUS AUREUS BACTERIOLYTIC ENZYME WITH POLYMORPHONUCLEAR LEUCOCYTE FUNCTIONS

S. Valisena[1], C. Pruzzo[2], P. E. Varaldo[3], and G. Satta[4]

[1]Istituto di Microbiologia
dell'Universita
Via A. Gabelli 63
35121 Padova, Italy

[2]Institute of Microbiology
University of Genova
Italy

[3]Institute of Microbiology
University of Ancona
Italy

[4]Institute of Microbiology
University of Siena
Italy

Introduction

Staphylococcus aureus is one of the most important human pathogens responsible for serious infections (1). Infections caused by this microorganism are often difficult to treat because of the high frequency with which S. aureus strains are resistant to the most common antibiotics. A way to overcome this problem could be the development of methods for enhancing host defenses against S. aureus. To this purpose, the main virulence factors of this microorganism have been investigated and almost all of the biologically active compounds produced by this microorganism have been analyzed (2). Despite this, an enzyme, the endo-β-N acetylglucosaminidase (SaG) which hydrolyzes peptidoglycan and is produced by all S. aureus strains (3), was never considered as a possible virulence factor.

Previous studies on the properties of the bacteriolytic enzymes produced by the different pathogenic and non pathogenic staphylococcal species (4,5), suggested to us the possibility that the S. aureus glucosaminidase could be involved in virulence. More recently, we have

NATO ASI Series, Vol. H24
Bacteria, Complement and the Phagocytic Cell
Edited by F. C. Cabello und C. Pruzzo
© Springer-Verlag Berlin Heidelberg 1988

purified the bacteriolytic enzymes produced by three different staphylococcal species (6) and have shown that the S. aureus purified SaG interferes with several physiological properties of human fibroblasts (7,8). On the other hand, experiments performed by others with two enzymes (hen egg white and human lysozymes) sharing with SaG the mechanism of action and the property of being cationic proteins, have shown that the former can remove reducing groups from vertebrate cells (9) and the latter represses the activated state of human polymorphonuclear cells (10). These data have encouraged us to evaluate the effects of S. aureus purified SaG on human polymorphonuclear leucocytes (PMNs) function. We found that this enzyme inhibits S. aureus and Klebsiella pneumoniae phagocytosis and intracellular killing by human PMNs, and highly increases staphylococcal pathogenicity for mice.

Materials and Methods

Strains

We used the strains: S. aureus A8 (11), Staphylococcus epidermidis AH-15 (12), Staphylococcus simulans 0-12 (6), Staphylococcus saprophyticus BO-3 (6) and K. pneumoniae RR4 (13). S. aureus A8 mutants defective in SaG production were isolated by methylmethanesulfonate treatment (14) and replica plating on Brain Heart Infusion Agar (DIFCO Laboratories) containing Micrococcus luteus. The mutant MA8a produced a very reduced amount of enzyme, while the mutant MA8b did not produce any detectable amount of it.

Media and buffers

Brain Heart Infusion Broth (BHIB) and Brain Heart Infusion Agar (BHIA) (DIFCO Laboratories) were used for culturing all bacterial strains. Phosphate buffered saline (0.1 M Na_2KHPO_4, 0.15 M NaCl, 0.1 M KH_2PO_4, pH 7.2-7.4) (PBS) and Hank's Balanced Salt Solution were utilized.

Animals

Male Swiss albino mice weighing about 20g and 3-Kg white New Zealand rabbits were employed.

SaG purification and assay of lytic activity

The endo-β-N-acetylglucosaminidase excreted by S. aureus (SaG) has been purified and characterized as previously described (6). Criteria for lytic enzyme purity, definition and determination of the arbitrary lytic units have also been previously described (6,15).

Preparation and source of the other tested substances and chemicals

Tri-N-acetylglucosamine (chitotriose) was obtained by partial acid hydrolysis of chitin (INC Pharmaceutical, Plainview, N.Y.) followed by charcoal column fractionation as described by Rupley (16). Lysostaphin and poly-L-lysine were obtained from SIGMA. Heat inactivation of SaG was as previously described (17).

Preparation of immune sera

Specific anti-SaG antisera were obtained in rabbits. The initial dose of antigen was given intramuscularly as a water-in-oil emulsion with complete Freund's adjuvant (DIFCO Laboratories). Two other doses were given subcutaneously as emulsions with incomplete Freund's adjuvant (DIFCO Laboratories) 10 and 30 days later. Blood for serum was collected 5 to 7 days after the second booster.

Polymorphonuclear leucocyte (PMN) preparation

PMNs were obtained from healthy adult donors as described by Boyum (18). PMN counts were then performed using standard methods and the final leucocyte pellet was resuspended at a concentration of about 5×10^6 PMNs/ml.

Bacterial-leucocyte association on coverslips and bacterial intracellular survival

Bacterial association to the PMN monolayer was examined as described by Mangan and Snyder (19). Each PMN was scored as positive if 2 or more bacteria could be seen attached to or internalized into the cells. Intracellular survival was determined as described by Verhoef et al. (20). In all experiments bacteria/PMNs ration was 100/1.

Chemoluminescence (CL) response

Luminol-induced CL responses were evaluated as previously described (21) at a bacteria/PMNs ratio of 100/1.

LD_{50} determination

The 50% lethal dose (LD_{50}) values were calculated according to Reed and Muench (22). Four different doses of overnight broth-cultures were injected intraperitoneally to a group of five mice for each dose. The calculations were based on the numbers of survivals on day 15.

Passive protection studies in mice

Mice were intraperitoneally inoculated with 0.2ml of anti-SaG antiserum. The first administration included anti-SaG antiserum and bacteria at two different sites.

Results

Effect of SaG on phagocytosis and intracellular killing by human PMNs

The effect of SaG on human PMN functions was studied evaluating capability of these cells to phagocytize and kill S. aureus A8 and K. pneumoniae RR4 strains. It is evident from Tables 1 and 2 that SaG inhibits both microorganisms' association to PMNs and resistance to intracellular killing. The inhibitory effect is proportional to the SaG concentration; in the presence of 100 μg/ml of enzyme the number of Klebsiella and S. aureus bacteria associated with PMNs decreased by 67 and 50%, respectively. In the same experimental conditions, the number of survivors among ingested microorganisms increased by 4 and 1.5 fold, respectively for Klebsiella and S. aureus. The described inhibition of PMN functions did not depend either on bacterial surface alterations or on a SaG aspecific interference with surface interactions between phagocytes and bacteria, but depended on a SaG direct effect on PMNs. In fact, very similar results were obtained when phagocytes were pretreated with the enzyme for 30 minutes before the addition of bacteria and then washed. Under these conditions, the inhibitory action was reversible. It was fully expressed for 60 min and then the PMN functions gradually resumed and appeared normal after 90 minutes (not shown).

Table 1. Effect of SaG on <u>K</u>. <u>pneumoniae</u> RR4 and <u>S</u>. <u>aureus</u> A8
association to human PMNs

Experimental conditions	Challenged strain	Association to PMNs (%)[a] 30 min	60 min
Untreated PMNs	RR4	41	72
	A8	28	50
PMNs + SaG (10 μg/ml)	RR4	21	37
	A8	16	29
PMNs + SaG (100 μg/ml)	RR4	12	24
	A8	10	25
PMNs + heat inactivated SaG[b] (100 μg/ml)	RR4	39	79
	A8	24	49
PMNs + SaG (100 μg/ml) + CT[c] (100 μg/ml)	RR4	43	82
	A8	29	54
PMNs + SaG (100 μg/ml) + anti-SaG ab	RR4	44	68
	A8	19	47
PMNs pretreated with SaG[d] (100 μg/ml)	RR4	17	29
	A8	16	17

[a]Results are expressed as percentage of 200 PMNs examined having
two or more bacteria associated. Values represent an average of
two experiments.
[b]The enzyme was heat inactivated by exposure to 120°C for 10 min.
[c]CT (chetotriose) alone had no effect on phagocytosis.
[d]PMNs were pretreated with SaG for 30 min at 37°C then washed
and added to bacteria.

Several control experiments were then performed both to exclude the
possibility of a SaG a specific toxic effect on granulocytes and to
evaluate if the observed interferences with PMN functions were due to
either the enzyme activity or the basic nature of the SaG protein. It was
observed that in presence of SaG (100, 200, 300 μg/ml, final
concentration) PMN viability was always greater than 90% as determined by
trypan blue exclusion. In addition, it was found that all observed effects
were caused by the protein component of the purified SaG preparation and
required an intact enzyme activity. In fact, the antiphagocytic effects
were completely abolished by 10 min. autoclaving, by the specific
glucosaminidase inhibitor chetotriose, and by anti-SaG rabbit antibodies

(Tables 1 and 2). In contrast to this, three other cationic polypeptides (RNAse, lysostaphin and polylysine) did not interfere with PMN capability to ingest and kill either S. aureus or K. pneumoniae cells.

Table 2.   Effect of SaG on K. pneumoniae RR4 and S. aureus A8
survival to bactericidal activity by human PMNs

Experimental conditions	Challenged strain	Intracellular survival (%)[a]	
		30 min	60 min
Untreated PMNs	RR4	32	12
	A8	56	55
PMNs + SaG (10 μg/ml)	RR4	49	45
	A8	74	71
PMNs + SaG (100 μg/ml)	RR4	62	53
	A8	82	79
PMNs + heat inactivated SaG[b] (100 μg/ml)	RR4	34	14
	A8	58	47
PMNs + SaG (100 μg/ml) + CT[c] (100 μg/ml)	RR4	34	11
	A8	54	49
PMNs + SaG (100 μg/ml) + anti-SaG ab	RR4	39	17
	A8	49	47
PMNs pretreated with SaG[d] (100 μg/ml)	RR4	59	56
	A8	78	72

[a]Values represent an average of two experiments.
[b]The enzyme was heat inactivated by exposure to 120°C for 10 min.
[c]CT (chetotriose) alone had no effect on phagocytosis.
[d]PMNs were pretreated with SaG for 30 min at 37°C then washed
 and added to bacteria.

Sensitivity to phagocytosis of mutants with altered production of SaG

   To further confirm the role of SaG in the interactions between PMNs and S. aureus cells we isolated two mutants from S. aureus A8: one produced a very reduced amount of SaG (MA8a); the other did not show any detectable SaG activity (MA8b). Except for SaG production, these strains demonstrated properties identical to those of the parent. As A8 they were unencapsulated, carried protein A, grew with the same generation time and

Table 3. Sentitivity of S. aureus A8 and its mutants MA8a and
MA8b to phagocytosis by human PMNs

Strains and experimental conditions	Association to PMNs (%)[a]		Intracellular Survival (%)[b]	
	30 min	60 min	30 min	60 min
A8	30	58	49	47
MA8a[c]	44	69	36	27
MA8a + SaG (100 μg/ml)	27	49	51	48
MA8b[c]	80	87	32	15
MA8b + SaG (100 μg/ml)	18	32	54	50

[a]Results are expressed as percentage of 200 PMNs examined having
two or more bacteria associated. Values represent an average of
two experiments.
[b]Values represent an average of two experiments.
[c]MA8a produces a 3-fold reduced amount of SaG; MA8b does not
produce any detectable SaG activity.

produced the same amount of all analyzed extracellular enzymes. Moreover,
antisera prepared against the parental strain after absorption with either
mutant, completely lost capability to agglutinate the bacteria against
which they were prepared. Table 3 shows that both mutants were more
sensitive than the parent to phagocytosis and intracellular killing by
PMNs.

In particular, the mutant not producing the enzyme was much more
sensitive than the one producing a reduced amount of it. If SaG was added
to the phagocytosis mixtures, both mutants acquired the same phagocytosis
resistance level as the parent.

Effect of SaG on PMN CL response

The role of SaG in enhancing resistance to phagocytosis was further
confirmed evaluating PMN luminol-enhanced CL, which is used as a
quantitative measure of the interactions between PMNs and bacteria. As
shown in Fig. 1, A8 bacteria induced a CL response lower than the MA8b
mutant, not producing SaG. When SaG was added to the phagocytosis

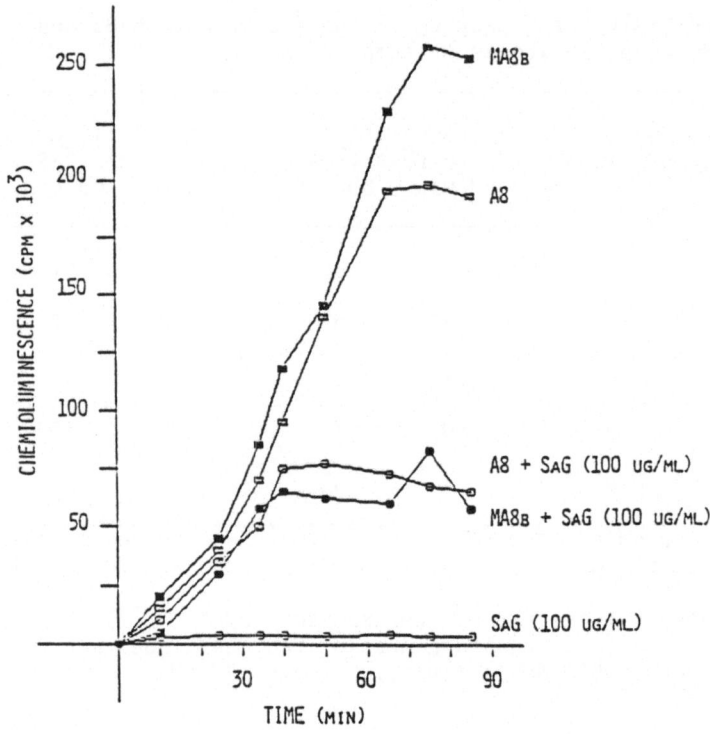

Fig. 1. Effect of SaG on PMN Luminol-induced CL response to S. aureus A8 and its mutant (MA8b) which does not produce the bacteriolytic enzyme.

Table 4. LD$_{50}$ of S. aureus strain A8 and its mutants MA8a and MA8b for intraperiotoneally injected mice

Strain	LD$_{50}$ (no. of cells)	Fold difference[a]
A8	3.2×10^5	-
MA8a[b]	2.3×10^7	(+) 71.8
MA8b[b]	4.2×10^7	(+)131.2

[a]Fold differences were calculated by comparing the LD$_{50}$ values of two S. aureus A8 mutants with that of the parental strain. (+) positive difference.
[b]MA8a produces a 3-fold reduced amount of SaG; MA8b does not produce any detectable SaG activity.

Table 5. Effect of SaG and specific anti-SaG antiserum on LD_{50} of
Staphylococci from different species for intraperitoneally-
injected mice

Strain and experimental condition[a]	LD_{50} (no. of cells)	Fold difference[b]
S. aureus A8	8.3×10^6	-
" A8 + SaG	2.7×10^5	(-) 30.7
" A8 + anti-SaG ab	5.9×10^8	(+) 71
" MA8a[c]	1.3×10^8	-
" MA8a + SaG	4.5×10^6	(-) 28.1
" MA8a + anti-SaG ab	3.9×10^8	(+) 3.1
" MA8b[c]	6.2×10^8	-
" MA8b + SaG	2.4×10^7	(-) 25.8
" MA8b + anti-SaG ab	1.9×10^8	(-) 3.2
S. epidermidis AH-15	1.1×10^6	-
" + SaG	6.1×10^4	(-) 18
" + anti-SaG ab	9.4×10^5	(-) 1.1
S. simulans 0-12	3.1×10^7	-
" + SaG	1.2×10^6	(-) 25.8
" + anti-SaG ab	4.2×10^7	(+) 1.3
S. saprophyticus BO-3	1.8×10^6	-
" + SaG	8.5×10^4	(-) 21.1
" + anti-SaG ab	1.2×10^6	(-) 1.5

[a]SaG was intraperitoneally given at day 1,2 and 3 at dose of 100 µg/
mouse, the first administration included SaG and bacteria at two
different sites. Anti-SaG antisera were given as described in Materials and Methods section.
[b]Fold differences were calculated by comparing LD_{50} of each strain in
SaG or serum-treated mice with that of the same strain in saline-
treated mice. (+) positive difference; (-) negative difference.
[c]MA8a produces a 3-fold reduced amount of SaG; MA8b does not produce
any detectable SaG activity.

mixtures, CL responses to both strains were drastically reduced being CL
values 2-4 fold lower in comparison to the untreated samples. No CL
response was observed in the presence of SaG alone.

Effect of SaG on S. aureus pathogenicity for mice

The possible role of SaG in S. aureus pathogenicity was then evaluated
by analyzing its effect on the LD_{50} for mice. In a first group of
experiments, the virulence of strain A8 was compared with that of the

mutants with SaG altered production. Table 4 shows that both inability to produce SaG and a reduced production of it are associated with reduced virulence. In fact, MA8b and MA8a were respectively 130- and 70-fold less pathogenic than the parent. In other experiments we have evaluated the effect of SaG on the LD_{50} for mice utilizing different staphylococcal species. Table 5 shows that enzyme administration caused a 30-fold reduction in S. aureus A8 LD_{50} and reduced by 18-25-fold the LD_{50} of three other staphylococcal strains belonging to different species (S. epidermidis, S. simulans and S. saprophyticus). In contrast, anti-SaG specific antisera caused a 70-fold increase in S. aureus A8 LD_{50}, but did not influence the LD_{50} of either the S. aureus mutant not producing SaG or strains of the other species (Table 5).

Discussion

The data presented in this paper clearly show that the main bacteriolytic enzyme excreted by all S. aureus strains and by no strain of other staphylococcal species (3) inhibits phagocytic activity of human PMNs and acts as a major virulence determinant in S. aureus pathogenecity for the mouse. This is a completely new finding since none of the bacterial enzymes that hydrolyze peptidoglycan and cause bacteriolysis were previously proposed to play a role in pathogenicity of any bacterial species.

It is likely that such a possibility was previously excluded on the theoretical ground that enzymes destroying a polymer essential for bacterial survival seemed more likely to be potentially involved in protection against infections rather than in virulence. However, particularly in the case of exosaminidase, a theoretical ground exists for suggesting a possible role in pathogenicity which has been overlooked so far. Some exosaminidases have been shown to be capable of partially hydrolyzing chitin (16); this indicates that such a group of enzymes may also split bonds between other exosamines aside from covalent bonds between muramic acid and N-acetylglucosamine. Exosamines are frequently present in the polysaccharidic portion of glycoproteins which, in turn, constitute major surface receptors of vertebrate cells (23-25). It appears not unlikely that bacteriolytic glucosaminidases can also interact with surface glycoproteins of vertebrate cells, thus interfering in their physiology. This suggestion is supported by the fact that hen egg white lysozyme has been shown to remove reducing groups from chick fibroblast (9) and that human lysozyme represses the activated state of human monocytes (10). Moreover, several effects of various exosaminidases on vertebrate cells have been demonstrated (4,5,7-9,17,26-28). In addition, we have recently found that SaG releases reducing groups from human PMNs. This makes likely that the modification of some surface receptors by this glucosaminidase triggers inhibition of both phagocytosis and oxygen consumption.

These observations appear important for several reasons. They have led to the identification of a new pathogenicity determinant of one of the most important human pathogens and have raised the possibility that bacteriolytic enzymes of other microbial species may also be major pathogenicity determinants. In this respect, it is interesting to mention that we have recently shown that both in the Enterococcus and the Pseudomonas genera a clear correlation exists between pathogenicity of the various species and production of bacteriolytic enzymes (29). Exosaminidases of various types (muramidases and glucosaminidases) are produced in vertebrates by cells involved in host defenses such as macrophages, monocytes and PMNs. The observation that an exogenous glucosaminidase heavily interferes with PMN functions indicates that these enzymes are probably involved in regulation of host defense activity. As mentioned before, a similar possibility was previously proposed by others for lysozyme. Our findings extend this possibility to N-acetyl-glucosaminidase which is also produced by monocytes and macrophages and raises the possibility that all enzymes of this group may share a similar role.

In previous papers we described an original system for the separation of staphylococci into species based on the identification of the specific bacteriolytic enzymes produced (3). The findings described in this paper confirm that this method of species identification is more valid for clinical purposes than other methods, since it is based on the detection of properties of a bacterial product that has an important function in the pathogenicity of staphylococci.

Acknowledgement

This work was supported by grant no. 8601631.52 from the Consiglio Nazionale delle ricerche.

References

1. J. O. Cohen (ed.), The staphylococci. Wiley-Interscience, New York (1972).
2. J. Jeljaszewicz (ed.), Staphylococci and staphylococcal diseases. Gustav Fischer Verlag, Stuttgart (1976).
3. G. Satta, P. E. Varaldo, G. Grazi, and R. Fontana, Bacteriolytic activity in staphylococci. Infect. Immun. 16:37-42 (1977).
4. P. E. Varaldo, F. Biavasco, C. Pruzzo, and G. Satta, Interference of Staphylococcus aureus glucosaminidase with mitogen responsiveness by human lymphocytes. In Staphylococci and staphylococcal infections (J. Jeljaszewicz, ed.) Gustav Fischer Verlag Publisher, Stuttgart, p. 895-899 (1981).
5. B. Azzarone, P. E. Varaldo, S. Valisena, R. Fontana, and G. Satta, Effect of Staphylococcus aureus (biogroup I) glucosaminidase on vertebrate cells. In Staphylococi and staphylococcal infections (J. Jeljaszewicz, ed.) Gustav Fischer Verlag Publisher, Stuttgart, p. 403-406 (1981).
6. S. Valisena, P. E. Varaldo, and G. Satta, Purification and characterization of three separate bacteriolytic enzymes excreted by Staphylococcus aureus, Staphylococus simulans, and Staphylococcus saprophyticus. J. Bacteriol. 151:636-647 (1982).
7. G. Satta, P. E. Varaldo, B. Azzarone, and C. A. Romanzi, Effects of Staphylococcus aureus lysozyme on human fibroblasts. Cell. Biol. Int. Rep. 3:525-533 (1979).
8. G. Satta, B. Azzarone, P. E. Varaldo, R. Fontana, and S. Valisena, Stimulation of spreading of trypsinized human fibroblasts by lysozymes from Staphylococcus aureus, hen egg white and human urine. In Vitro 16:738-750 (1980).
9. H. Ausdourian, L. Chu, K. Lan, and H. Amos, Lysozyme: evidence for effects on chick fibroblasts, HeLa cells, and their products. Biochim. Biophys. Res. Commun. 64:1144-1151 (1975).
10. L. I. Gordon, S. D. Douglas, N. E. Kay, O. Yamada, E. F. Osserman, and H. S. Jacob, Modulation of neutrophil function by lysozyme. Potential negative feedback of inflammation. J. Clin. Invest. 64:226-232 (1979).
11. C. Pruzzo, S. Valisena, P. E. Varaldo, and G. Satta, Interference of Staphylococcus aureus bacteriolytic enzyme with phagocytosis by human polymorphonuclear leukocytes. Abst. 13th Int. Congr. Chemother. 90:82-85, Vienna, (1983).
12. S. Valisena, L. Radin, P. E. Varaldo, R. Fontana, and G. Satta, Purification and properties of the lytic enzyme excreted by staphylococci of different biogroup (or species). In Staphylococci and staphylococcal infections (J. Jeljaszewicz, ed.) Gustav Fischer Verlag Publisher, Stuttgart, p. 398-401 (1981).

13. C. Pruzzo, E. Debbia, and G. Satta, Mannose-inhibitable adhesin and T3-T7 receptors of Klebsiella pneumoniae inhibit phagocytosis and intracellular killing by human polymorphonuclear leukocytes. Infect. Immun. 36:949-957 (1982).

14. J. H. Miller, Experiments in molecular genetics. Cold Spring Harbor Laboratory (1977).

15. S. Valisena, P. E. Varaldo, and G. Satta, Biochemical and physical properties of the endo-β-N-acetylglucosaminidase from Staphylococcus aureus, Staphylococcus simulans and Staphylococcus saprophyticus. Microbiologica 6:277-291 (1983).

16. J. A. Rupley, The hydrolysis of chitin by concentrated hydrochloric acid and the preparation of low-moleuclar-weight substrates for lysozyme. Biochim. Biophys. Acta 83:245-255 (1964).

17. G. Satta, P. E. Varaldo, and B. Azzarone, Stimulation of DNA synthesis and cell proliferation of transformed human fibroblasts by pure Staphylococcus aureus lysozyme. Microbios. Lett. 6:55-60 (1978).

18. A. Boyum, Isolation of lymphocytes, granulocytes and macrophages. Scand. J. Immunol. 5:9-15 (1976).

19. D. F. Mangan, I. S. Snayder, Mannose-sensitive interaction of Escherichia coli with human peripheral leukocytes in vitro. Infect. Immun. 26:520-527 (1979).

20. J. Verhoef, P. K. Petterson, and P. G. Quie, Human polymorphonuclear leukocytes receptors for staphylococcal opsonins. Immunology 33:231-239 (1977).

21. C. Svanborg-Eden, L. M. Bjurstern, R. Hull, K. E. Magnusson, Z. Moldovano, and H. Leppler, Influence of adhesin on the interaction of Escherichia coli with human phagocytes. Infect. Immun. 44:672-680 (1984).

22. L. J. Reed and H. Muench, A simple method for estimating fifty percent endpoint. Am. J. Hyg. 27:493-499 (1938).

23. A. G. Athely, B. J. Barnhart, and P. M. Kramer, Growth and biochemical characteristic of a detachment variant of CHO cells. J. Cell. Physiol. 90:375-386 (1977).

24. S. Roseman, Sugar of the cell membrane. Hosp. Pract. 10:61-71 (1975).

25. K. M. Yamada and K. Olden, Fibronectins-adhesive glycoproteins of cell surface and blood. Nature 275:179-184 (1978).

26. E. F. Osserman, M. Klockars, J. Halper, and R. E. Fischel, Effects of lysozyme on normal and transformed mammalian cells. Nature 243:331-335 (1973).

27. J. J. Rinehart, J. G. Cerilli, H. S. Jacob, and E. F. Osserman, Lysozyme stimulates lymphocyte proliferation in monocyte-depleted mixed lymphocyte cultures. J. Lab. Clin. Med. 99:370-381 (1982).

28. T. Tokaoka, H. Kotsuta, and T. Ishiki, Effects of lysozyme on the proliferation of fibroblasts in tissue culture. Japan. J. Exp. Med. 42:221-232 (1972).

29. G. Satta, F. Palmas, P. E. Varaldo, G. Grazi, O. Soro, and R. Pompei, Analysis of the bacteriolytic pattern as a new tool for species separation and identification in Micrococcaceae, Streptococci and Pseudomonadaceae. Abst. 13th Int. Congr. Chemother., Vienna (1983).

THE OPSONIC CAPACITY OF MONOCLONAL ANTIBODIES AGAINST ESCHERICHIA COLI 0111 AND ITS ROUGH MUTANT E. COLI J5

(Complement activating and opsonic capacity of MAb)

J. Verhoef, R. W. Vreede*, and A. S. Bouter

Laboratory of Microbiology
State University of Utrecht
Catharijnesingel 59
3511 GG Utrecht
The Netherlands

Abstract

Six monoclonal antibodies raised against Escherichia coli 0111 and against its rough mutant (chemotype Rc) were studied. One IgG2A against E. coli J5, one IgM anti-J5 and one IgG2a anti-0111 monoclonal antibody did not bind to lipopolysaccharides of the homologous strain, but cross-reacted with heterologous gram-negative rods in an enzyme-linked immunosorbent assay. These three monoclonal antibodies activated complement when incubated with homologous or heterologous strains, but were opsonic neither in the presence nor in the absence of complement. The other three monoclonal antibodies were directed against lipopolysaccharide of the homologous strain, but showed no cross-reactivity. However, these antibodies cross-reacted when bacteria were exposed to low concentrations of antibiotics (monobactams). The IgG3 and one IgM anti-J5 monoclonal antibodies activated complement and were opsonic only in the presence of complement. The IgM anti-0111 monoclonal antibody activated complement and was opsonic both in the presence and absence of complement.

Introduction

Studies in animals and humans have shown that antisera prepared against rough (R) mutants of Escherichia coli or Salmonella species (1) are cross-protective against a variety of gram-negative microorganisms (2-8). Also in humans antiserum against an R mutant of Escherichia coli 0111:B4, also known as E. coli J5, appears to prevent the development of endotoxic shock in patients with gram-negative bacteremia (9). Although it seems that protection is afforded by immunoglobulins directed against the core region of the lipopolysaccharides (LPS) (10), the exact nature of the protective factor is still not understood. Antibodies in such sera might neutralize the endotoxic effects of LPS (4) or enhance opsonization and subsequently phagocytosis of bacteria leading to enhanced clearance from the blood-stream (11).

Recently, monoclonal antibodies raised against LPS derived from the J5 mutant of the smooth (S) strain Escherichia coli 0111 have been described (12). These antibodies react with purified LPS from mutants of Escherichia coli and Salmonella typhimurium and also with LPS of other both smooth- and rough-phenotype, gram-negative bacilli, and appear to bind determinants present in the LPS core region (12).

We were interested in the immunodominant group being responsible for the induction of protective antibodies. Therefore, monoclonal antibodies (MAbs) were raised, and selected not only on the basis of their reactions with bacteria but also based on their functional capacities (13). The results presented in this paper deal with the complement activating and opsonic capacities of MAbs. Six samples of ascites fluid containing different MAbs raised against Escherichia coli 0111 and its J5 mutant (chemotype Rc) are described. It was found that cross-reactivity of anti-R mutant MAbs was enhanced when bacteria were first treated with monobactam antibiotics.

Materials and Methods

Antibiotics. The Monobactam antibiotics Aztreonam (Squibb, Princeton NJ, USA) and Carumonam (Roche, Basle, Switzerland) were used. MIC's were determined by standard methods.

Bacteria. The smooth (S) strain Escherichia coli 0111 and its rough J5 mutant, chemotype Rc, and Salmonella typhimurium M206 (0 type 1.4.12) and its rough mutants Ra (SF 1592), Rc (SF 1195), and Re (SF 1398) were used. Escherichia coli J5 was a gift from Dr. M. P. Glauser (Centre Hospitalier

Universitaire Vaudois, Lausanne, Switzerland). This strain lacks both the enzyme uridine-diphosphate-galactose 4-epimerase and the ability to incorporate exogenous galactose into its LPS (9). Escherichia coli 0111 and the Salmonella typhimurium strains were kindly provided by Dr. P. A. M. Guinée (National Institute of Public Health and Environmental Hygiene (RIVM), Bilthoven, The Netherlands). The latter strains, whose LPS structure was outlined previously (14), were originally selected in the Max Planck Institut fur Immunobiologie, Freiburg im Breisgau, FRG, and were provided by the Institut Pasteur, Paris, France. After conservation in 15% glycerol solution, these strains were kept at -70°C until use. In addition strains isolated from blood samples of patients with bacteremia hospitalized at the University Hospital of Utrecht, The Netherlands, and Klebsiella K21 (NTCC 9141) were used. These strains were kept in lyophilized state until use.

Monoclonal Antibodies. Bacteria were grown overnight at 37°C on Iso-sensitest-agar (Oxoid Ltd., UK), washed once, adjusted to a concentration of 2.5 x 10^8 CFU/ml phosphate buffered saline (PBS), pH 7.4, by a spectrophotometric method, and boiled for 1 h. Spleen cells from adult BALB/c mice immunized with heat-killed Escherichia coli 0111 or Escherichia coli J5 were fused with the mouse myeloma cell line Sp2/0-Ag14 (15) according to techniques described previously (16,17). Hybridomas were screened for antibody activity by an enzyme-linked immunosorbent assay (ELISA) on undiluted culture supernatants. All hybrids selected were subcloned twice by limiting dilution. Ascites fluids were obtained from BALB/c mice, which had been treated with pristane (Janssen Chimica, Belgium) and subsequently injected with 5 x 10^5 cells of selected clones.

Lipopolysaccharides. LPS of smooth strains were isolated according to the phenol/water method described by Westphal et al. (18), those of rough strains according to the phenol-chlorophorm-petroleum ether method of Galanos and co-workers (19).

Enzyme-Linked Immunosorbent Assay. The assay was performed as described previously (13). Briefly, heat-killed bacteria or LPS were coated on wells of a 96-well microplate. After washing, the wells were subsequently incubated with ascites fluids, horseradish peroxidase-conjugated goat-anti-mouse immunoglobulins and substrate solution.

The enzyme reaction was stopped and absorbance measured at 450 nm. Optical density (OD) below 0.200 was considered negative.

Sera. Sera from venous blood obtained from ten healthy, adult donors were pooled and stored in 1.0 ml aliquots at -70°C. This normal human serum (NHS) served as a standard source of complement. NHS was thawed prior to use.

Haemolytic Complement Assay. Veronal buffer containing calcium and magnesium (VBS^{++}) was used as diluent. Heat-inactivated bacteria were adjusted to a concentration equivalent to 5×10^8 CFU/ml VBS^{++}. Bacterial suspension (160 μl), ascites fluid (40 μl of a 1:20 dilution in VBS^{++}), and VHS (600 μl of a 5% dilution in VBS^{++} were mixed. After incubation for different periods of time at 37°C, the remaining complement activity in the supernatant of the incubation mixture was measured with optimally sensitized sheep erythrocytes in a microtitre assay as described by Klerx et al. (21). Complement consumption was expressed as percentage of the total amount of complement added.

Opsonization Procedure. Ascites fluids and NHS were diluted in Hank's Balanced Salt Solution containing 0.1% gelatin (Gibco Bio-cult Ltd., UK) (gHBSS). In polypropylene vials (Beckman Instruments, Ireland), 0.1 ml of a prewarmed (37°) bacterial suspension (2.5×10^8 CFU/ml gHBSS) was added to 0.4 ml of diluted ascites fluid, fresh serum, or both. The suspensions were incubated at 37°C in a shaking waterbath (150 rpm), and opsonization was stopped by adding 2.0 ml ice-cold PBS. After centrifugation for 15 min at 1600 g at 4°C, opsonzied bacteria were resuspended in 0.5 ml gHBSS and kept at 0°C until use. Care was taken that low concentrations of serum were used. No lysis of bacteria in the serum was observed.

Leucocytes. Polymorphonuclear granulocytes (PMN) were isolated according to a modification of a method of Boyum (21) as described previously (22). In brief, venous blood samples from healthy adult donors were heparinized and allowed to settle by gravity in 6% dextran (mol. wt. 70,000; Pharmacia Fine Chemicals, Sweden) in normal saline. Leucocyte-rich plasma was withdrawn and centrifuged for 10 min at 160 g. The pellet was resuspended in gHBSS, carefully layered onto Ficoll-Paque (Pharmacia Fine

Chemicals) and centrifuged for 35 min at 160 g. Residual erythrocytes in the pellet were lysed with ice-cold 0.87% NH_4Cl (w/v) in sterile water. After centrifugation for 10 min at 160 g, PMN were washed twice and adjusted to a concentration of 5 x 10^6/ml gHBSS.

Phagocytosis Assay. Phagocytosis was measured according to a method described previously (23,24). After being cultured in 5 ml Mueller-Hinton broth (Difco Laboratories, USA) containing 0.02 mCi ^3H-thymidine (specific activity 5.0 Ci/mmol; Radiochemical Centre, UK), bacteria were washed three times and adjusted to a concentration of 2.5 x 10^8 CFU/ml gHBSS. Opsonization was performed as described above.

Equal volumes of opsonized bacterial and PMN suspensions were added to each of four polypropylene vials (Beckman), and the vials were shaken (150 rpm) in a waterbath at 37°C. After 2, 6 and 12 min, phagocytosis was stopped by adding ice-cold PBS to each of three vials. Leucocyte-associated bacteria were then washed free of non-leucocyte-associated bacteria by three cycles of differential centrifugation (160 g at 4°C). Pellets were resuspended in 2.5 ml scintillation liquid (Aqua Luma Plus, Lumac/3M, The Netherlands). Leucocyte-associated radioactivity was measured in a liquid scintillation counter (Philips, The Netherlands) and expressed as percentage of the total radioactivity added, as determined in the pellet of the fourth vial resuspended in scintillation liquid (Lumac/3M) after centrifugation for 15 min at 1600 g. In an electron microscopic study from our laboratory it has been shown that the leucocyte-associated radioactivity can be attributed primarily to intracellular bacteria (25).

Results

Reactivity of Monoclonal Antibody in Immunoassays. All samples of ascites fluid reacted with the homologous, heat-killed strain in ELISA (Table 1). Three samples of ascites (4-9A1, 4-7B5, and 3-12A1) contained IgG3 anti-J5, IgM anti-J5, IgM anti-J5, and IgM anti-0111 MAb, respectively, which were directed against LPS of the homologous strains (Table 1).

Table 1.

Reciprocal titres of monoclonal antibodies raised against Escherichia coli J5 and Escherichia coli 0111 incubated with heat-killed bacteria or LPS (ELISA)

Monoclonal antibodies [a]		Strains		LPS of	
		J5	0111	J5	0111
Anti-J5					
IgG2A	(2-4B3)	⩾ 1,024,000		-	
IgG3	(4-9A1)	⩾ 1,024,000		16,000	
IgM	(2-5B2)	16,000		-	
IgM	(4-7B5)	64,000		32,000	
Anti-0111					
IgG2A	(5-5A1)		512,000		-
IgM	(3-12A1)		64,000		128,000

[a] Numbers in parentheses indicate clone numbers

The two anti-J5 LPS MAbs (4-9A1 and 4-7B5) reacted with heat-killed bacteria of <u>Escherichia coli</u> J5 and <u>Salmonella typhimurium</u> Rc in ELISA (Table 2). Both samples of ascites fluid also reacted with LPS of the latter strain, showing titres of 16,000 and 64,000, respectively. The IgM anti-0111 LPS MAb (3-12A1) reacted only with the homologous strain (Table 2). The three other samples of ascites fluid (2-4B3, 2-5B2, and 5-5A1) contained IgG2A anti-J5, IgM anti-J5, and IgG2A anti-0111 MAbs, respectively. These MAbs did not bind to LPS of the homologous strains (Table 1), but reacted with heat-killed bacteria of all mutant and parent strains tested in ELISA (Table 2).

Table 2.

Reactivity of anti-J5 and anti-0111 monoclonal antibodies to heat-killed bacteria of two smooth and four rough strains as measured by enzyme-linked immunosorbent assay. Ascitic fluid was tested in 1:1000 dilutions.

Monoclonal antibodies [a]		Strains					
		J5	0111	Ra	Rc	Re	M206
Anti-J5							
IgG2A	(2-4B3)	+	+	+	+	+	+
IgG3	(4-9A1)	++	-	-	+	-	-
IgM	(2-5B2)	++	++	+++	+++	+++	+++
IgM	(4-7B5)	+++	-	-	+++	-	-
Anti-0111							
IgG2A	(5-5A1)	+	+	+	+	+	++
IgM	(3-12A1)	-	+++	-	-	-	-

[a] Numbers in parentheses indicate clone numbers.
-     = optical density (OD) below 0.200
+   = OD between 0.200 and 0.500
++   = OD between 0.500 and 1.000
+++   = OD between 1.000 and 1.500

Because the anti-LPS MAbs only reacted with the homologous strains or homologous LPS, we decided to treat bacteria first with subinhibitory concentrations of carumonam or aztreonam. Bacteria were grown overnight in concentrations of 1/4 MIC of the drugs carumonam or aztreonam. MAbs 4-9A1 and 4-7B5 cross-reacted with E. coli 0111 and other gram-negative strains that were pretreated with one of the monobactams. Thus, culture of gram-negative bacilli in broth containing 1/4 MIC of monobactams changed the cell surface in a way that they reacted with MAb against E. coli J5 LPS (26).

Complement Consumption by Monoclonal Antibody. Each sample of ascites fluid containing MAb was tested for complement activating capacity in a haemolytic complement (CH50) assay. Heat-killed bacteria were incubated in 1% dilutions of ascites fluid and 5% dilutions of NHS, as complement source, and then centrifuged, and the haemolytic activity of the supernatants was measured. Results were compared to those of the sample of control ascites fluid produced by the cell line Sp2/0-Ag14.

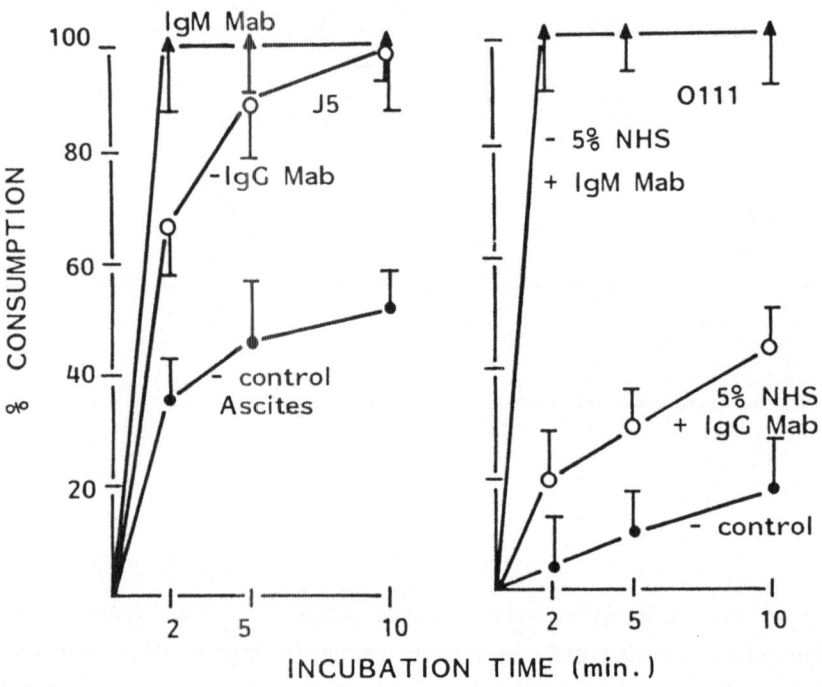

Complement consumption by anti -LPS and non LPS Mabs

Fig. 1. Kinetics of complement consumption expressed as percentage of the total amount of complement added. Heat-killed bacteria of <u>Escherichia coli</u> J5 (left) or <u>Escherichia coli</u> O111 (right) were incubated for various periods of time in a 5% dilution of normal human serum and 1% dilution of control ascites (●), 5% NHS and 1% dilution of ascites containing homologous IgG2A MAb (o), or 5% NHS and 1% dilution of ascites containing homologous IgM MAb (▲). Results are means of three independent observations ± standard deviation.

Figure 1 presents the kinetics of complement consumption by two anti-J5 (2-4B3 and 2-5B2) and two anti-0111 MAbs (5-5A1 and 3-12A1) after incubation with the homologous strains. Within 2 min, all complement (100%) was activated by both IgM MAbs, whereas the complement consumption by the two IgG MAbs occurred at a slower rate.

Complement consumption by each sample of ascites fluid alone - i.e. without bacteria - varied between 13% and 23%, whereas that of control ascites fluid amounted to 11.1% ± 8%.

With the results from ELISA as a guide, the MAbs were tested for their ability to activate complement after incubation with either the homologous bacteria only or heterologous bacteria, too. A sample of ascites fluid, normal human serum, and bacteria were incubated for 10 min. Values differing less than 20% of that of control ascites fluid were considered negative.

A sample of ascites fluid (4-9A1) containing IgG3 and another (4-7B5) containing IgM anti-J5 LPS MAb activated more complement than control ascites fluid when incubated with Escherichia coli J5 or Salmonella typhimurium Rc (Figure 2).

A sample of ascites fluid (3-12A1) containing IgM anti-0111 MAb directed against LPS consumed 100% of the available complement after incubation with Escherichia coli 0111, whereas control ascites fluid activated 15% only (results not shown).

The IgG2A (2-4B3) and IgM anti-J5 MAbs (2-5B2), directed against other cell wall components besides LPS, activated more complement than control ascites when incubated with either homologous or each of the heterologous strains (Fig. 3). The sample of ascites fluid (5-5A1) with IgG2A anti-0111 MAb activated more complement than control ascites fluid when incubated with all strains tested with the exception of Escherichia coli 01 (Figure 3).

Complement consumption by <u>E. coli</u> and <u>S. typhimurium</u>
Rc in the presence of anti-LPS Mabs

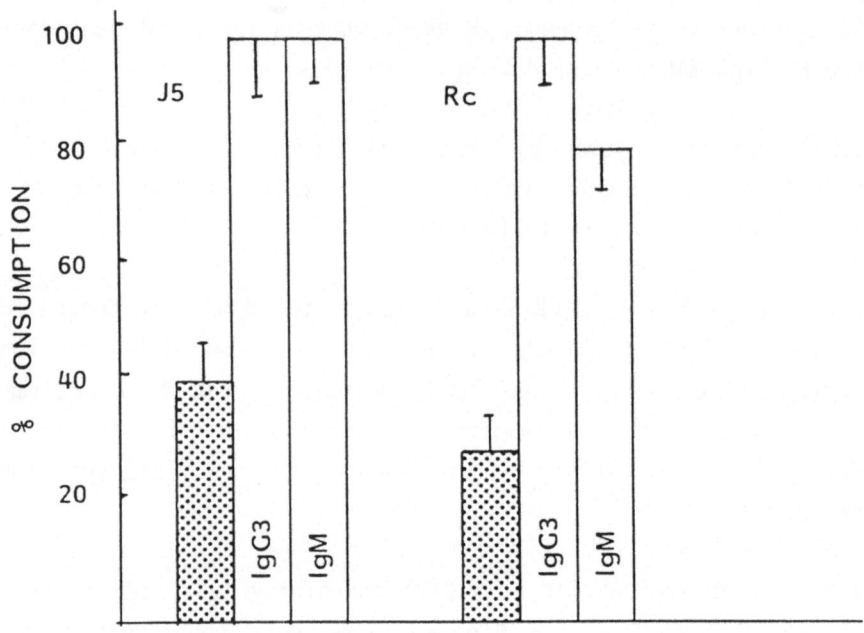

Fig. 2. Complement consumption (percentage) by IgG3 and IgM anti-J5 LPS
MAbs (4-9A1 and 4-7B5) after incubation of heat-killed bacteria
of <u>Escherichia coli</u> J5 or <u>Salmonella typhimurium</u> Rc in 1%
dilutions of the corresponding ascites and 5% NHS. Results were
compared to those of 1% dilutions of control ascites and 5% NHS
presented as hatched bars. Results are means of three independent
observations ± standard deviation.

Complement consumption by <u>E. coli</u> in the presence
of anti - non LPS Mabs

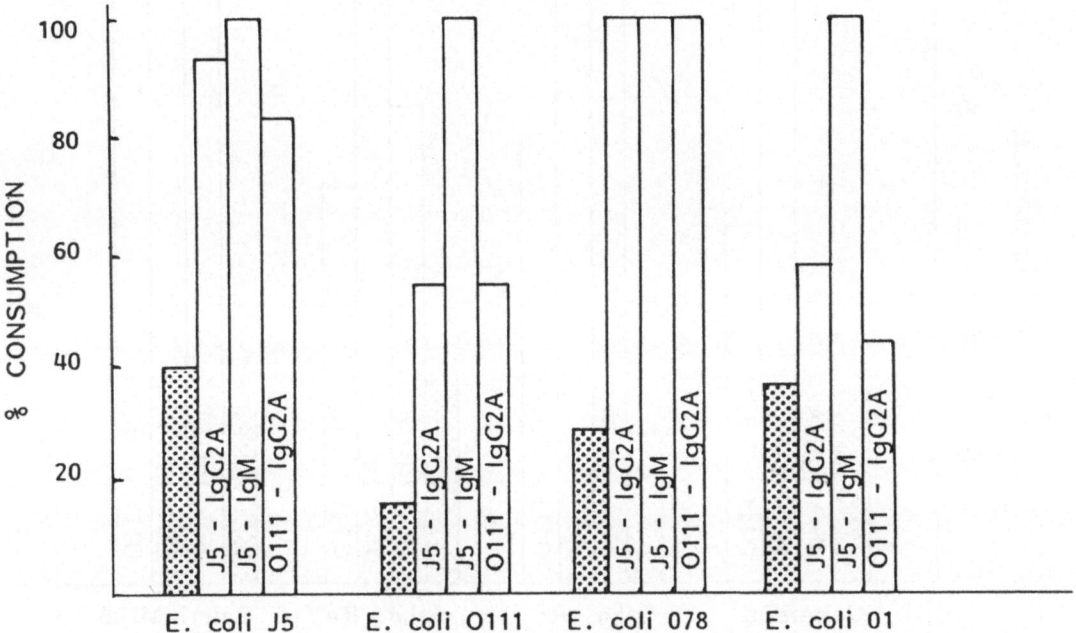

Fig. 3a

Complement consumption by bacteria in the presence
of anti - non LPS Mabs

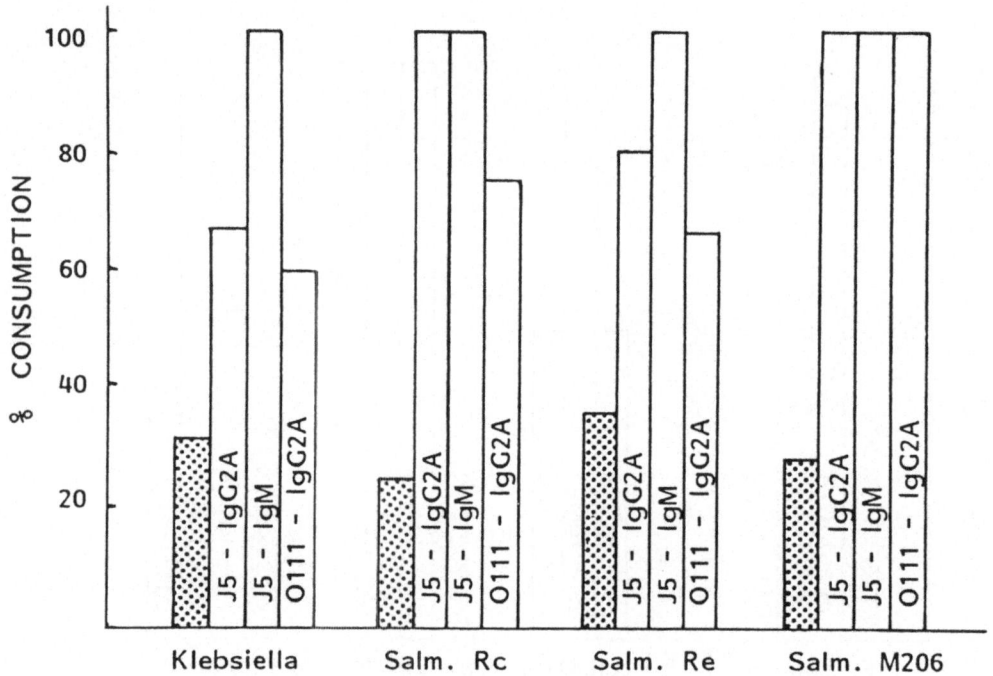

Fig. 3b

Fig. 3. Complement consumption (percentage) by anti-J5 or anti-O111 MAbs
not directed against LPS after incubation of heat-killed bacteria
of homologous and heterologous strains in 1% dilutions of ascites
and 5% NHS. Bars represent, successively, control ascites
(hatched) and the samples of ascites (2-4B3), (2-5B2), and
(5-5A1) containing, respectively, IgG2A anti-J5 MAb (J5-IgG2A),
IgM anti-J5 MAb (J5-IgM), and IgG2A anti-O111 MAb (O111-IgG2A).

The positive results of the latter three samples of ascites incubated with Salmonella typhimurium Ra are not shown, because the patterns were quite similar to those found when the two other mutants of Salmonella typhimurium were tested.

Opsonic Activity of Monoclonal Antibody. To assess the opsonic activity of MAbs, radio-actively- labeled bacteria were incubated in 1% dilutions of ascites fluid with or without appropriate dilutions of NHS as complement source, centrifuged, resuspended in buffer, and added to PMN.

Since the IgG2A MAb (2-4B3) and IgM MAb (2-5B2) produced against Escherichia coli J5 and the IgG2A MAb (5-5A1) against Escherichia coli O111 showed no opsonic activity for the homologous strain, either in the presence or absence of complement, these MAbs were not tested further with regard to opsonic activity for heterologous strains.

The sample of ascites fluid (4-9A1) containing IgG3 anti-J5 LPS MAb showed indirect opsonic activity for Escherichia coli J5 and Salmonella typhimurium Rc, i.e. in the presence of complement, as did the sample of ascites fluid (4-7B5) with the IgM anti-J5 LPS MAb (Figure 4). But, neither MAb directed against LPS had opsonic activity in the absence of complement. Dilutions of ascites fluid higher or lower than 1% had no influence on the percentage of bacterial phagocytosis (data not shown). 20%-30% uptake of bacteria incubated in the samples of the two latter ascites fluids was a non-specific effect, because pre-incubation of Rc mutants in only buffer also resulted in a similar percentage of phagocytosis by PMN (Figure 4).

Uptake of <u>E. coli</u> J5 and <u>S. typhimurium</u> Rc
opsonized with anti LPS Mab

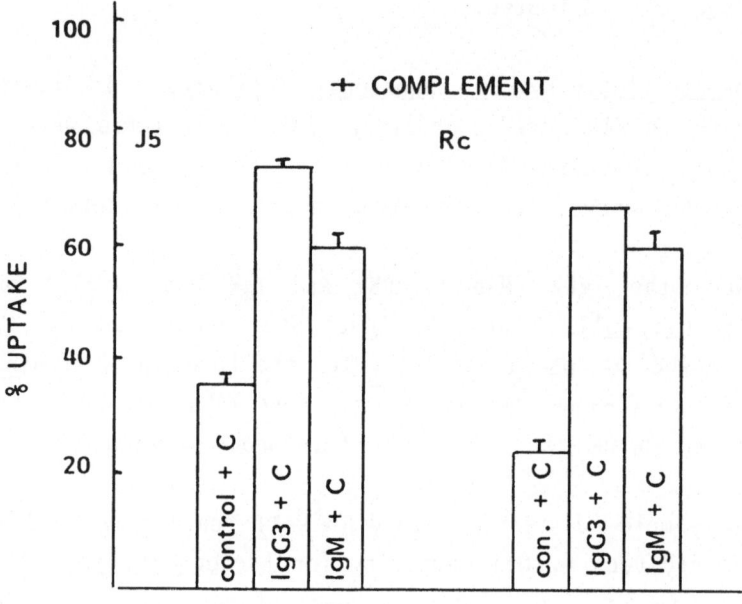

Fig. 4a

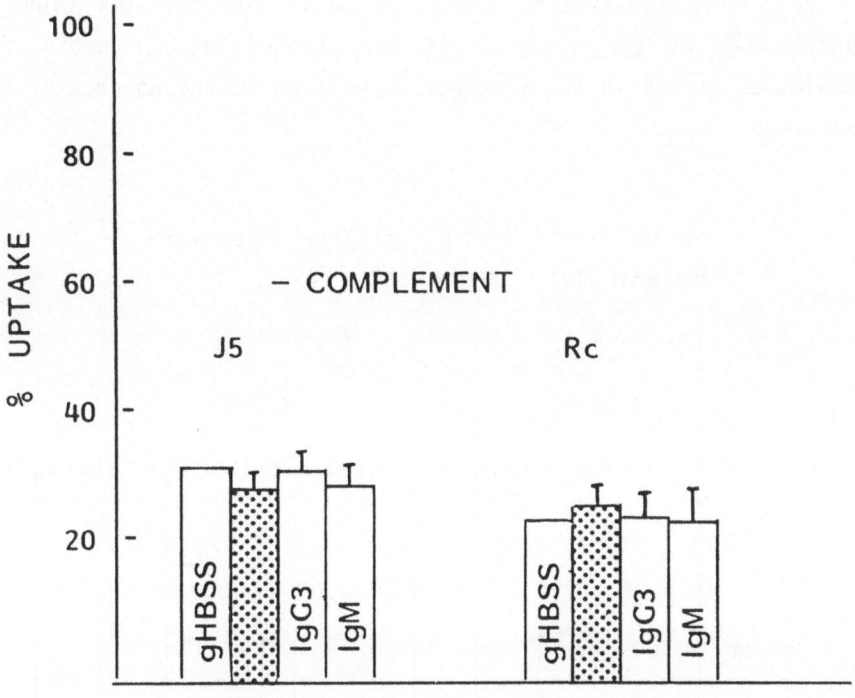

Fig. 4b

Fig. 4.   Fig. 4a: Phagocytosis (percentage) of <u>Escherichia</u> <u>coli</u> J5 or
<u>Salmonella</u> <u>typhimurium</u> Rc opsonized first in 1% dilutions of
ascites containing IgG3 anti-J5 LPS MAb (4-9A1) or IgM anti-J5
LPS MAb (4-7B5), and secondly in 1% NHS (+C), compared to that of
bacteria opsonized in equal dilutions of control ascites and NHS.
Fig. 4b: Phagocytosis (percentage) of <u>Escherichia</u> <u>coli</u> J5 or
<u>Salmonella</u> <u>typhimurium</u> Rc pre-incubated in 1% dilutions of
ascites containing IgG3 anti-J5 LPS MAb (4-9A1) or IgM anti-J5
LPS MAb (4-7B5) compared to that of bacteria opsonzied in 1%
dilutions of control ascites (hatched bars) or buffer (gHBSS).
Opsonized bacteria were incubated for 12 min with PMN. Results
are means of three independent observations ± standard deviation.

The sample of ascites fluid (3-12A1) from the IgM clone directed against LPS of <u>Escherichia</u> <u>coli</u> 0111 showed opsonic activity for the homologous strain in the presence as well as in the absence of complement (Figure 5).

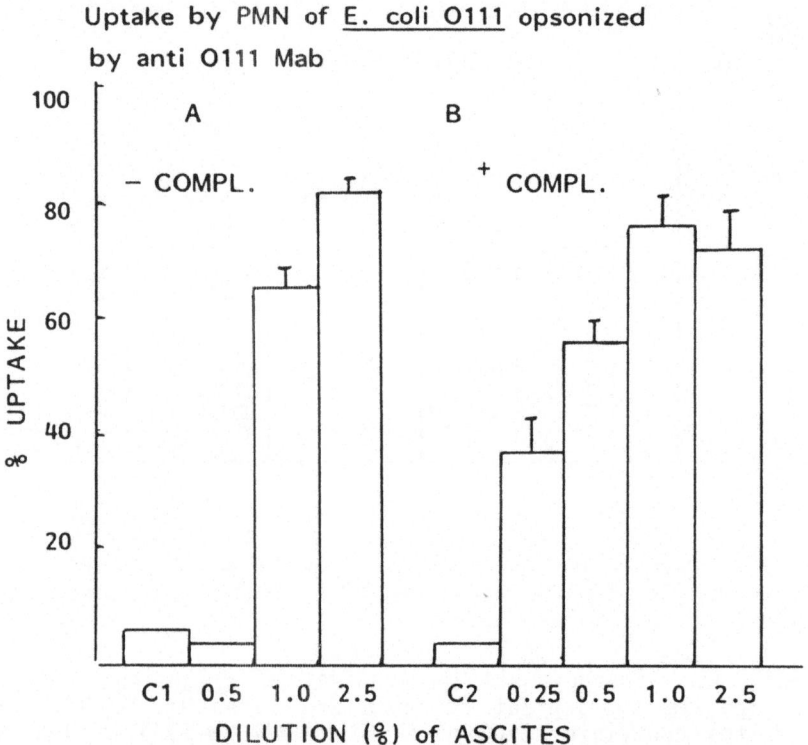

Fig. 5. Phagocytosis (percentage) of <u>Escherichia</u> <u>coli</u> 0111 opsonized in various dilutions of ascites (3-12A1) containing anti-0111 IgM MAb in the absence (A) or presence (B) of 2% NHS as complement source. Controls comprised of bacteria opsonized in control ascites (C1) or NHS plus control ascites (C2). Opsonized bacteria were incubated for 12 min with PMN. Results are means of three independent observations ± standard deviation.

Discussion

As part of our investigations of the immunodominant groups of the endotoxin molecule, which are able to induce cross-protective antibodies, and of the mechanism underlying this protection, we raised monoclonal antibodies in mice. Firstly, we decided to screen the MAbs for their complement activating, opsonic, and endotoxin neutralizing capacities. Data on the opsonic and complement activating capacities are presented in this paper. We realize that the study at hand has some inherent limitations because MAbs raised in mice, human complement, and human leucocytes were used. Despite this heterologous system some conclusions can be drawn.

First, in contrast to the anti-J5 MAbs described by Nelles and Niswander (12), our IgG2A and IgM anti-J5 MAbs, which cross-reacted with heterologous strains, were not directed against LPS of Escherichia coli J5. These two MAbs activated complement but were not opsonic, even when complement was present. Second, the other two anti-J5 MAbs, which reacted with the two Rc mutants used in this study, bound to LPS isolated from Escherichia coli J5. The latter MAbs, belonging to the IgG and IgM classes, respectively, activated complement after incubation with the homologous strain and the Rc mutant of Salmonella typhimurium. Both MAbs were also indirectly opsonic for these two Rc mutants, i.e. in the presence of complement.

Our finding that the IgG anti-J5 MAb had no direct opsonic activity is in agreement with a previous study in which we found that in the absence of complement IgG antibodies in polyclonal antisera raised against rough mutants of Escherichia coli and Salmonella spp. do not enhance phagocytosis of homologous bacteria (27).

Our three IgG MAbs were able to activate complement as measured in a CH50 microtitre assay. It is important to note that the choice of 10 min for incubation time in the complement consumption studies favours the classical over the alternative pathway. Couderc et al. also showed that an IgG MAb - directed against trinitrophenol - was able to activate complement via the classical pathway (28).

When the consumption of complement by the IgG2A anti-J5 and IgG2A anti-O111 MAbs incubated with the homologous strains is followed in time (Figure 1), it is clear that a higher level is reached in tests with the anti-J5 MAbs than with the anti-O111 MAbs. This might be due to the presence of more antibodies within the samples of ascites fluid containing anti-J5 MAbs than those containing anti-O111 MAbs, leading to a higher degree of binding of complement molecules. Another explanation might be that Escherichia coli J5 activates complement via the classical pathway of complement independently of antibody (29, 30), whereas Escherichia coli O111 hardly does (29). Here, it is assumed that not all LPS molecules have bound antibodies.

Although the IgG3 anti-J5 MAb showed indirect opsonic activity, i.e. in the presence of complement, the IgG2A anti-J5 and anti-O111 MAbs did not. It might be that the complement factors activated by these IgG2A MAbs play a role in serum bactericidal properties instead of opsonic activity. This is being further investigated.

When considered together our monoclonal antibodies produced against Escherichia coli J5 and Escherichia coli O111 reveal several patterns. IgG2A and IgM MAbs directed against cell wall components other than LPS showed cross-reactivity to heterologous strains in an ELISA using heat-killed bacteria. They were also able to activate complement when incubated with heat-killed bacteria of homologous and heterologous strains, but they had no opsonic activity for homologous and heterologous gram-negative rods. The IgG3 and IgM MAbs directed against LPS of the strains used for immunization showed no cross-reactivity, only activated complement when incubated with the homologous strain, and were opsonic for bacteria of the homologous strain in the presence of complement. In the assays mentioned no differences were found between the two Rc mutants as regards interactions with IgG3 and IgM anti-J5 LPS MAbs.

Because the anti-coreglycolipid MAbs 4-9A1 and 4-7B5 did not react with heterologous bacteria, no complement consumption or uptake by phagocytes could be detected when heterologous bacteria were incubated with complement or with complement and phagocytes respectively. When these bacteria were treated with 1/4 MIC of the monobactams, incubation with

MAbs and complement may enhance complement consumption. Moreover, incubation with complement and phagocytic cells could lead to phagocytosis of drug-treated bacteria exposed to these MAbs (to be published).

Finally, it is remarkable that one of the IgM MAb showed indirect as well as direct opsonic activity for the homologous strain. It is generally accepted that IgM antibodies are not capable of opsonic activity without complement. Further studies are needed on this IgM MAb, e.g. to clarify if the antibodies are comprised of true pentameres. The binding of MAbs, directed against cell wall components other than LPS, e.g. to outer membrane proteins, are also currently being studied.

Acknowledgements

Part of this work was supported by the "Preventiefonds", project number 28-964.

References

1. Lüderitz O, Staub AM, Westphal O (1966) Immunochemistry of O and R antigens of Salmonella and related Enterobacteriaceae. Bacteriological Reviews 30:192-255.
2. Braude AI, Douglas H (1972) Passive immunization against the local Schwartzman reaction. J Immunol 108:505-512.
3. Davis CE, Ziegler EJ, Arnold KF (1978) Neutralization of meningococcal endotoxin by antibody to core glycolipid. J Exp Med 147:1007-1017.
4. Johns M, Skehill A, McCabe WR (1983) Immunization with rough mutants of Salmonella minnesota. IV. Protection by antisera to O and rough antigens against endotoxin. J Infect Dis 147:57-67.
5. Marks MI, Ziegler EJ, Douglas H, Corbeil LB, Braude AI (1982) Induction of immunity against lethal Haemophilus influenzae type b infection by Escherichia coli core lipopolysaccharide. J Clin Invest 69:742-749.
6. McCabe WR (1972) Immunization with R mutants of Salmonella minnesota. I. Protection against challenge with heterologous gram-negative bacilli. J Immunol 108:601-610.
7. Young LS, Stevens P, Ingram J (1975) Functional role of antibody against "core" glycolipid of Enterobacteriaceae. J Clin Invest 56:850-861.
8. Ziegler EJ, Douglas H, Sherman JE, Davis CE, Braude AI (1973) Treatment of E. coli and Klebsiella bacteremia in agranulocytic animals with antiserum to a UDP-gal epimerase-deficient mutant. J Immunol 111:433-438.

9. Ziegler EJ, McCutchan JA, Fierer J, Glauser MP, Sadoff JC, Douglas BS, Braude AI (1982) Treatment of gram-negative bacteremia and shock with human antiserum to a mutant Escherichia coli. New Eng J Med 307:1225-1230.
10. Wolff SM (1982) The treatment of gram-negative bacteremia and shock. Editorial. New Eng J Med 307:1267-1268.
11. Kirkland TN, Ziegler EJ (1984) An immunoprotective monoclonal antibody to lipopolysaccharide. J Immunol 132:2590-2592.
12. Nelles MJ, Niswander CA (1984) Mouse monoclonal antibodies reactive with J5 lipopolysaccharide exhibit extensive serological cross-reactivity with a variety of gram-negative bacteria. Infect Immunity 46:677-681.
13. Vreede RW, Leuvenink J, Bouter AS, Brouwer EC, Marcelis JH, Verhoef J (1986) Complement activating and opsonic capacity of monoclonal antibodies raised against Escherichia coli 0111 and its rough mutant J5. Eur J Clin Microbiol 5:141-147.
14. Ivanoff B, André C, Fontanges R, Jourdan G (1982) Secondary immune response to oral and nasal rough mutant strains of Salmonella typhimurium. Annales d'Immunologie (L'Institut Pasteur) 133 D:61-70.
15. Shulman M (1978) A better cell line for making hybridomas secreting specific antibodies. Nature (London) 276:269-270.
16. Kohler G, Milstein C (1975) Continuous culture of fused cells secreting antibody of predefined specificity. Nature (London) 256:495-497.
17. Fazekas de St. Groth S, Scheidegger D (1980) Production of monoclonal antibodies: strategy and tactics. J Immunol Methods 35:1-21.
18. Westphal O, Jann K (1965) Bacterial lipopolysaccharides. Methods of Carbohydrate Research 5:83-91.
19. Galanos C, Luderitz O, Westphal O (1969) A new method for the extraction of R lipopolysaccharides. Eur J Biochem 9:245-249.
20. Bos ES, van der Doelen AA, van Rooy N, Schuurs AHWM (1981) 3,3', 5,5'-Tetramethylbenzidine as a ames test negative chromogen for horse-radish-peroxidase in enzyme-immunoassay. J Immunoassay 2:187-204.
21. Klerx JPAM, Beukelman CJ, van Dijk H, Willers JMN (1983) Microassay for colorimetric estimation of complement activity in guinea pig, human and mouse serum. J Immunol Methods 63:215-220.
22. Boyum A (1968) Isolation of mononuclear cells and granulocytes from human blood. Scandinavian J Clin Invest 97:S77-89.
23. Verbrugh HA, Peters R, Peterson PK, Verhoef J (1978) Phagocytosis and killing of staphylococci by human polymorphonuclear and mononuclear leukocytes. J Clin Path 31:539-545.
24. Verhoef J, Peterson PK, Quie PG (1977) Kinetics of staphylococcal opsonization, attachment, ingestion and killing by human polymorphonuclear leukocytes: a quantitative assay using ^{3}H-thymidine labeled bacteria. J Immunol Methods 14:303-311.
25. Rosenberg-Arska M, Salters MEC, van Strijp JAG, Geuze JJ, Verhoef J (1985) Electron microscopic study of phagocytosis of Escherichia coli by human polymorphonuclear leukocytes. Infect Immun 50:852-859.
26. Overbeek BP, Schellekens JFP, Lippe W, Dekker HAT, Verhoef J (1987) Carumonam enhances reactivity of E. coli with mono- and polyclonal antisera to rough mutant E. coli J5. J Clin Microbiol 25:1009-1013.

27. Vreede RW, Marcelis JH, Verhoef J (1986) Antibodies raised against rough mutants of E. coli and Salmonella strains are opsonic only in the presence of complement. Infect Immun 52:892-896.
28. Couderc J, Kazatchkine MD, Ventura M, Thien Duc H, Maillet F, Thobie N, Liacopoulos P (1985) Activation of the human classical complement pathway by a mouse monoclonal hybrid IgG1-2A monovalent anti-TNP antibody bound to TNP-conjugated cells. J Immunol 134:486-491.
29. Betz J, Isliker H (1981) Antibody-independent interactions between Escherichia coli J5 and human complement components. J Immunol 127:1748-1754.
30. Tenner AJ, Ziccardi RJ, Cooper NR (1984) Antibody-independent C1 activation by E. coli. J Immunol 133:886-891.

A MUTATION IN SALMONELLA TYPHIMURIUM THAT ENHANCES RESISTANCE TO OXYGEN-INDEPENDENT ANTIMICROBIAL NEUTROPHIL PROTEIN

John K. Spitznagel[1], William M. Shafer[1], Monica M. Farley[2]

[1]Department of Microbiology and Immunology
Emory University
Atlanta, Georgia   30322 USA

[2]Division of Infectious Diseases
Veterans Administration Medical Center
Atlanta, Georgia   30033 USA

Introduction

Our experience with the cationic antimicrobial proteins (CAPs) of human neutrophil leukocytes and the experiences of other investigators have led us to believe that resistance to them in Salmonella typhimurium is under the control of at least two distinct genes or clusters of genes. We think the matter of importance for, although these bacteria can survive intracellularly in neutrophils, the basis for their capacity to survive is poorly understood. Perhaps enhanced understanding of the in vitro resistance of Salmonella to the CAPs will shed light on Salmonella strategies for intracellular survival.

It is known that Salmonella at risk of being killed by neutrophils potentially have two defensive strategies. They can avoid being phagocytized or they can foil neutrophil microbicidal mechanisms after being phagocytized. To a substantial degree the Salmonella 0 antigen serves both purposes. It is antiphagocytic, but it also inhibits intracellular killing (Okamura and Spitznagel, 1982).

Table 1.   Intraphagosomal Killing Systems

      O_2 Independent

      O_2 Dependent

NATO ASI Series, Vol. H24
Bacteria, Complement and the Phagocytic Cell
Edited by F. C. Cabello und C. Pruzzo
© Springer-Verlag Berlin Heidelberg 1988

What is responsible for the complex process of intracellular killing (Table 1) (Spitznagel, 1984)? We now recognize two mechanisms, each involving several mediators. One of these, the oxygen-independent killing mechanism, functions whether oxygen is present in the neutrophil or not. The other mechanism, the oxygen-dependent mechanism, only functions in the presence of molecular oxygen and essential reducing enzymes. What are the mediators of oxygen-independent killing? What are the known genetic controls of microbial resistance to them, controls with which bacterial cells might resist killing after being phagocytized by neutrophils?

Table 2. Antimicrobial Mechanisms of PMN

Independent of O_2

Hydrogen ion shift
Organic acids
Lysozyme
Lactoferrin
Cationic proteins

Oxygen-independent antimicrobial action (Spitznagel, 1984) is mediated by several things (Table 2): low pH, low molecular weight fatty acids, and several distinctly different antimicrobial proteins associated with the granules of the neutrophil. Convincing evidence for the importance of the cationic proteins in this process is steadily growing (Hirsch, 1956; Hovde and Gray, 1986; Modzrakowski et al, 1979; Odeberg and Olsson, 1975; Shafer et al, 1984; Skarnes and Watson, 1956; Spitznagel and Chi, 1963; Weiss et al, 1978; Zeya and Spitznagel, 1968). Clearly, Salmonella that find themselves in the phagolysosomes of neutrophils would need to be able to deal with many of these mediators. Since our own work on antimicrobial resistance mechanisms has been greatly concerned with one such protein, which we call CAP57, this is what we shall discuss here.

Table 3. The 57kD cationic antimicrobial protein
CAP57 has a defined range of activity

Resistance or sensitivity of gram-negative bacteria	%[a] Survival at pH:	
	5.5	7.5

Sensitive

Acinetobacter lwoffii	34	80
Escherichia coli	4	60
Neisseria gonorrhoeae	38	53
Pseudomonas aeruginosa	36	100
Pseudomonas cepacia	39	72
Salmonella typhi	14	40
Salmonella typhimurium (LT-2)	42	93
Shigella sonnei	29	82

Resistant

Proteus mirabilis	100	95
Proteus vulgaris	100	99
Serratia marcescens	100	100
Bacillus subtilis	100	100
Staphylococcus aureus (Wood)	100	100
Staphylococcus epidermidis	100	92
Streptococcus pneumoniae	98	98
Streptococcus pyogenes	100	98

[a] % bacteria surviving after one hour of 5 μg CAP57/ml.

CAPS57 is bactericidal for a considerable spectrum of gram-negative bacteria (Table 3): Salmonella, Escherichia coli, Neisseria, Pseudomonas, etc. However, CAP57 has different minimal inhibitory concentrations for different species and for different strains of the same species. What are the determinants of resistance?

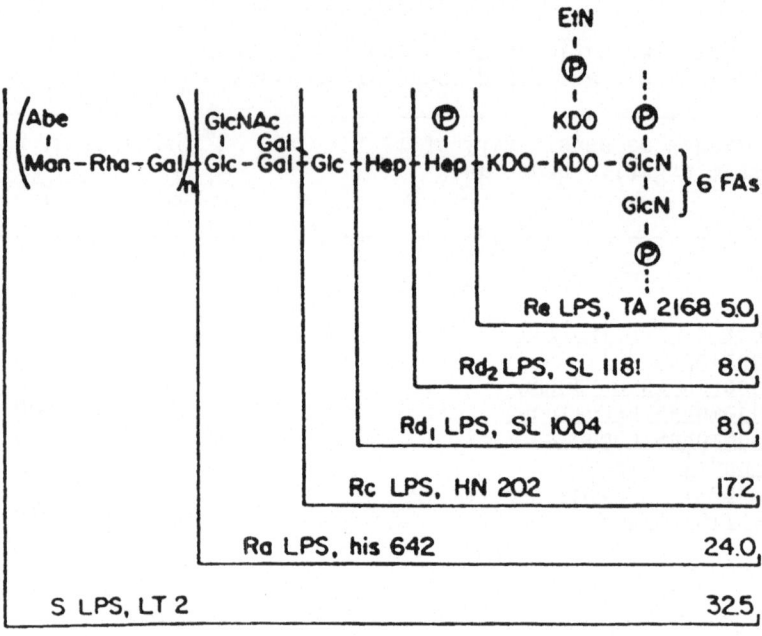

Fig. 1. <u>Relation of LPS chemotype to resistance to CAP</u>. LPS chemotypes (Ra, Rc, Rd_1, Rd_2, Re) of <u>Salmonella typhimurium</u> LT2 and its OM mutants (<u>his642</u>, HN202, SL1004, SL1181, TA2168) were used in these experiments. The numbers to the right of the figure indicate the concentration per milliliter of azurophil granule protein required to kill 50% of a standard inoculum of <u>Salmonella typhimurium</u> and several OM mutants in 60 min. Abe, Abequose; Man, mannose; Rha, rhamnose; Gal, galactose; GlcNAc, N-acetylglucosamine; Glc, glucose; Hep, heptose; EtN, ethanolamine; KDO, ketodeoxyoctonate; GlcN, glucosamine; P, phosphate.

It has long been known that smooth <u>Salmonella</u> and related species lose their resistance to CAPs as these bacteria mutate to roughness. In Figure 1 you can see the percent of bacteria killed by various concentrations of granule protein in relation to the chemotype of the <u>Salmonella</u> lipopolysaccharide (from Okamura and Spitznagel, 1982). As the bacteria lose first their O-side chains and then their core polysaccharides, they become less and less able to resist CAP. This phenotypic behavior defines certain gene clusters as influential in this form of antimicrobial resistance. They include the rfa and rfb gene clusters, which map at 79 and 42 minutes, respectively. Since they direct the synthesis of these outer membrane polysaccharide antigens, they control one form of resistance to CAPs. The mutant alleles are rough and have lower antimicrobial resistance. They would not be expected to enhance the survival of Salmonella in neutrophils. Reversion to smooth chemotypes tends to increase antimicrobial resistance in the rough mutants.

Are there any mutations that can increase the resistance of Salmonella to CAP57? Recently, Bill Shafer (1983) discovered that the pmrA mutation, described and mapped by Makela (1978) at 94 minutes, increases the resistance of the rough mutants to CAP57. These rough forms are usually less resistant to CAP57 than in the wild-type parents. With pmrA they gain substantial resistance to it. In other words, Shafer found the pmrA strains were cross resistant to CAP57 and polymyxin B (Shafer et al, 1984). Figure 2 shows the concentration dependency of CAP57 killing of SH9178, an Rb strain of Salmonella, and SH7426, a pmrA mutant of SH9178 hyperresistant to polymyxin B. The abscissa shows the concentrations of CAP57 used and the ordinate shows the percentage of bacteria killed. More CAP57 (three- to four-fold) was required to kill 50% of the SH7426 than was required to kill 50% of SH9178.

To better understand the effects of the pmrA mutation, we studied the binding of iodinated (but still biologically active) CAP57 to the pmrA mutant SH7426 and to the parent SH9178. Constant amounts of iodinated CAP57 were added in the presence of increasing excess of cold CAP57. The results (not shown) revealed that the pmrA mutant bound substantially less iodinated protein. Moreover, the binding with both organisms was saturable, suggesting that it was specific. Degree of binding correlated with killing capacity of CAP57 for the bacteria.

These results raised several questions. Could a mutation to CAP resistance appear spontaneously in Salmonella? What might be the frequency of the spontaneous mutation? Would the mutation be expressed in smooth organisms? Where would the mutation map? We thought that if such events could happen with fair frequency they might have some bearing on pathogenesis. At the very least, we might find out something more about how CAP kill <u>Salmonella</u>. Because polymyxin is readily available and cheap we decided to select for spontaneous mutations to resistance to polymyxin B. Then, we planned to examine such mutants for their resistance to CAPs. Mutations with cross resistance to CAPs were to be mapped and then the nature of the genes characterized.

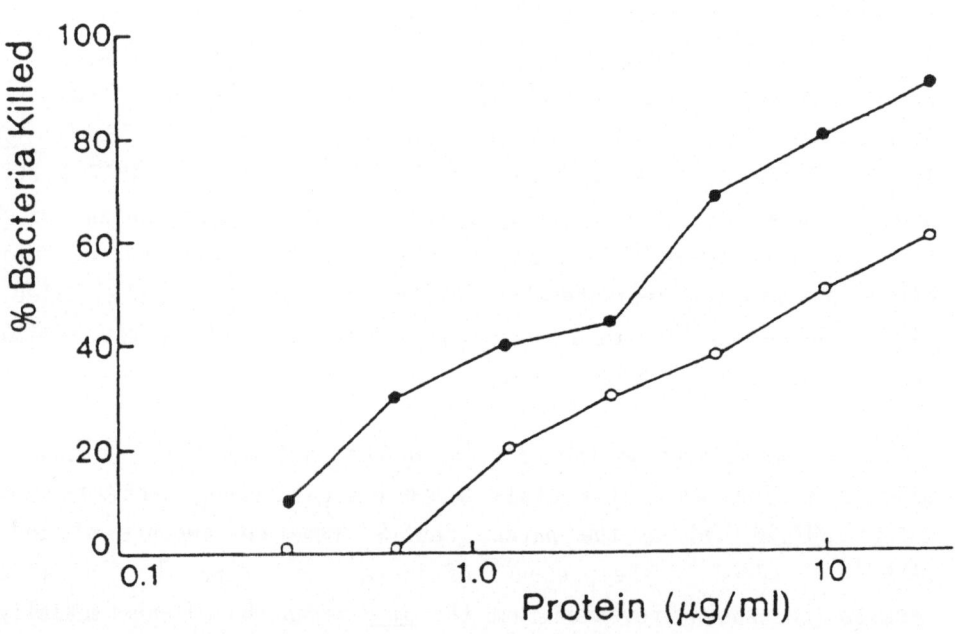

Fig. 2. <u>Salmonella typhimurium resistance to CAP57 is enhanced by the pmrA mutation</u>. o----o depicts concentration dependence for killing SH7426 <u>rfa</u>J <u>pmr</u>A163 (R471a); LD$_{50}$, ca. 10 μg/ml. ●----● depicts concentration dependence for killing SH9178 <u>rfa</u>J4041 (R417a); LD$_{50}$, ca. 3 μg/ml.

Table 4.   Frequency of Spontaneous Mutation to Pb[r]

Polymyxin b[a] $(\mu g/ml)$

Cells Plated	0	1	2	4	8	16
3×10^7	--	CG[b]	CG	1.1[c]	.28	.15
3×10^6	--	CG	29	1.5	.33	.15
3×10^5	--	CG	30	0	1.5	0
3×10^2	289	--	--	--	--	--

[a]μ/ml of LB agar
[b]Confluent growth
[c]Colonies X 10^{-6} cells

The spontaneous mutation rate to Pb[r] was 3×10^{-7}. An overnight culture of S. typhimurium LT-2 was plated on LB agar. The number of cells plated is shown in the first column. Resistance to polymyxin B was defined as the capacity to grow on ≥ 8 μg of polymyxin B per ml LB agar.

Table 5.   Properties of Spontaneous Mutants with Pb[r] Phenotype

Strain	Polymyxin B $(\mu g/ml)$	MIC Granule Protein $(\mu g/ml)$	P22 Sensitive	Remarks
JKS0001	50	95	Yes	Spontaneous mutant of LT-2
JKS0002	50	80	Yes	"
JKS0003	12.5	52	Yes	"
JKS0004	25.0	57	Yes	"
JKS0005	50.0	>100	Yes	"
JKS0007	50	40	Yes	"
LT-2 WT	12	25	Yes	---
SH9178	3	5	No	Rb chemotype LPS LT-2 mutant

Construction of Generalized Transducing Phage P22

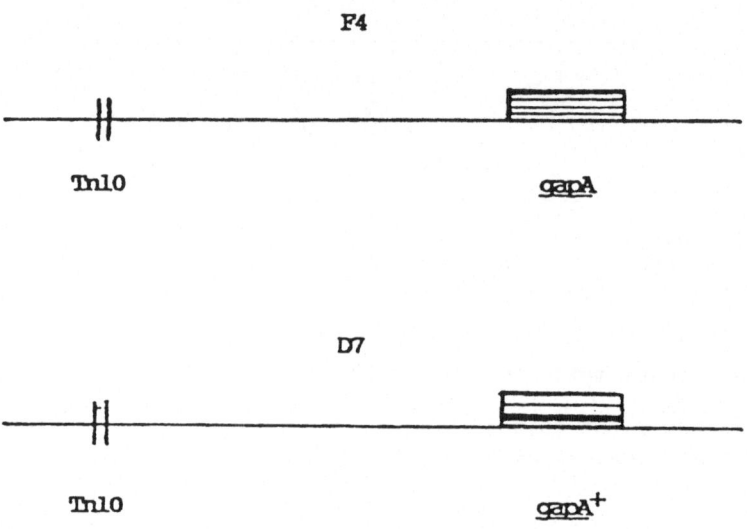

Fig. 3.   We have constructed two kinds of transducing phage in order to
characterize the gene(s) responsible for the CAP57^r phenotype.
First, we inserted the Tn10 transposon into the JKS5 strain with
the phage NK337 hisC527leu414supE and selected on agar plates
containing tetracycline. The high frequency transducing phage
P22(HT) int$^-$ was grown on this library and the transducing lysate
was   used   to   transduce   LT-2.   We   selected   for
tetracycline-resistant   transductants   and   screened   these   for
polymyxin B resistance. Several of the Tet^{r}Pb^r transductants were
used to prepare transducing lysates. We then examined these for
their capacity to cotransduce Tet^{r}Pb^r, gapA::Tn10, with high
frequency. F4 was a lysate that transduced Tet^{r}Pb^r and CAP57^r
with frequency of 48%. We also prepared a transducing lysate, D7,
that transduces Tet^{r}Pb^s into JKS5, a Pb^r strain with a frequency
of 90%. The gapA$^+$ allele was derived from a Tn10 insertion
library of LT-2. These strains were used to transduce Hfr strains
to Tet^{r}Pb^r or Tet^{r}Pb^s. Thus, we had the advantages of Tn10 marker
for use in mapping with the marker inserted near the gene of
interest.

We found that spontaneous mutants appeared in populations of LT-2 with a frequency of about 10^{-7} (see Table 4). We then proceeded to characterize several of these mutants (Table 5). As you can see, the MIC for polymyxin B increased two- to four-fold for several of the mutants, compared to the MIC for LT-2. It increased from four- to almost sixty-fold when compared with the Rb chemotype 9178. Comparable increases were found in the MICs for CAP57 (not shown). The sensitivity of all of the mutants to phage P22 showed that the mutants were still fully smooth. The electrophoretic behavior of their LPS was identical with the LPS of LT-2 (not shown). In addition, Monica Farley also compared saturable binding of CAP57 with the mutant JKS5 and the parent LT-2. The mutation to resistance resulted in a definite reduction in binding. The binding was saturable (data not shown). We found that overall the smooth <u>Salmonella</u> bound less protein than the rough ones. However, even in LT-2 the mutation to polymyxin B resistance and CAP57 resistance correlated with a further decrease in binding by LT-2.

We have begun experiments with mapping. For this purpose we have constructed two generalized transducing phages F4 and D7 with P22(HT)int⁻ (Fig. 3). These bacteriophage were prepared with standard techniques (Davis <u>et al</u>, 1980). F4 transduces a <u>Tn</u>10 transposon inserted near the mutation of interest, which we tentatively call <u>gapA</u> (granule antibacterial protein). It cotransduces LT-2 to Tet^{r}Pb^r at a rate of 48%. D7 has the <u>Tn</u>10 inserted near <u>gapA</u>⁺. It cotransduces JKS5 (Table 5) to Tet^{r}Pb^r with a frequency of 90%.

We have transduced several Hfrs (obtained from Dr. Kenneth Sanderson) and performed both interrupted and uninterrupted matings (Fig. 4). Interrupted and uninterrupted matings were done as described in Miller (1972). The figure shows the results of interrupted matings done with HfrKA and HfrK17. The recombinants were selected on tetracycline and counterselection was done with streptomycin. HfrKA inserts clockwise and HfrK15 inserts counterclockwise. As shown in Figure 4, the <u>Tn</u>10 entered the recipient at 98 minutes from HfrKA and at 3 minutes from HfrK17. We are making further efforts to refine this mapping so we can proceed to characterize the gene. The recombinants are Tet^{r}Pb^{r}Str^r.

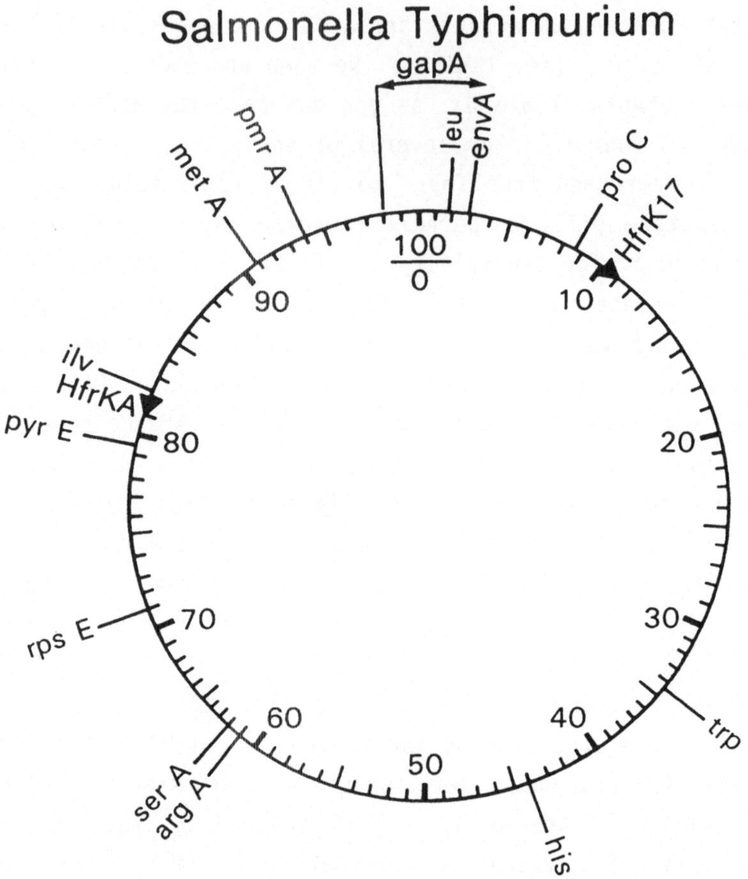

Salmonella Typhimurium

Fig. 4. The tentative location of gapA on the Salmonella chromosome lies within a region, represented here as an arc, extending clockwise from 98 min to 3 min. We use gapA to signify the locus for a mutant allele with phenotype Pb^r and CAP57^r because the locus maps away from the pmrA locus established by Makela (1978). Shown by arrowheads are the points of origin and direction of transfer of two Hrf strains transduced to Tet^{r}Pb^r with phage F4 and used for conjugal mapping. The markers for envA and pmrA are shown, as are the auxotrophic mutations of F$^-$ strain SA1145, which we used as recipient in these crosses. Recombinants were selected on LB agar with tetracycline and streptomycin. Uninterrupted matings (gradient of transfer) led to similar results.

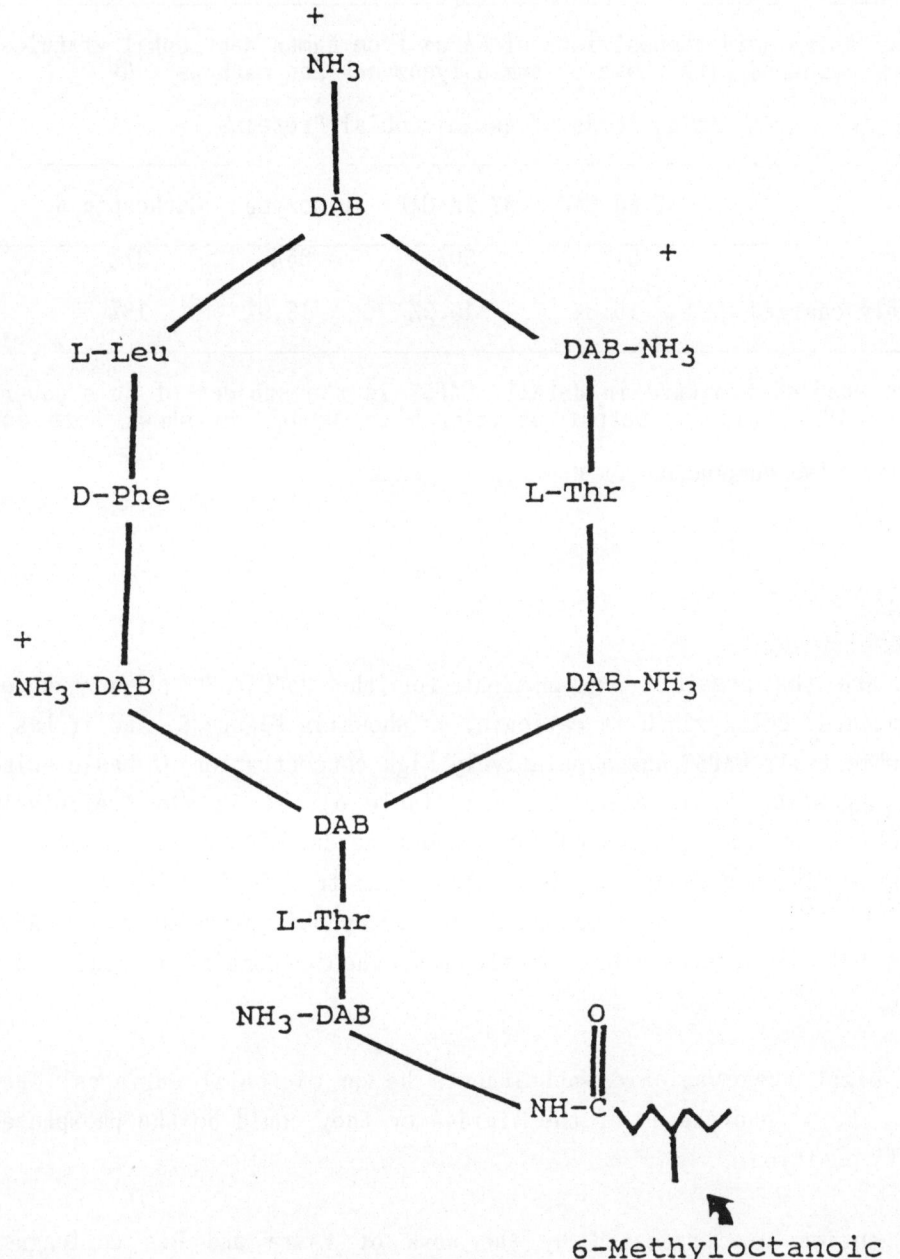

Fig. 5. <u>Structure of polymyxin B</u>. The molecule comprises a hydrophilic,
strongly cationic head group that is a cyclic peptide of
L-leucine (L-Leu), L-threonine (L-thr), and diaminobutyric acid
(DAB), and a hydrophobic tail (indicated by the wavy line) that
is 6-methyl octanoic acid. The other polymyxins have different
hydrophobic tail groups (not shown).

Table 6. Amino acid compositions of CAPs[a] from human neutrophil granules compared with those of human lysozyme[b] and cathepsin G[b].

Amino Acids of Antimicrobial Proteins

	57 Kd CAP	37 Kd CAP	Lysozyme	Cathespin G
Nonpolar	51%	50%	35%	37%
Positively charged	15.8%	15.2%	13.9%	18%

[a]We have studied two CAPs in detail. CAP57 is the subject of this paper. However, CAP37, also a potent antimicrobial agent, is shown here for comparison.
[b]These are also components of neutrophil granules.

Discussion

What are the possible explanations for the Pb[r]CAP57[r] phenotypes of these mutants? Polymyxin B is cationic, as shown in Figure 5, and it has a hydrophobic tail. CAP57 has a relatively high concentration of basic amino acids, suggesting it is also cationic (Table 6). It is also relatively hydrophobic due to its hydrophobic amino acids. It is reasonable to suppose the cationic groups of Pb or CAP57 bind to anionic substituents on the microbial outer surface, bringing the hydrophobic parts of Pb or CAP57 close enough to intercalate into the hydrophobic domains of the outer membrane.

What might these anionic substituents be on microbial surfaces? They could be the 4' phosphates of the lipid A or they could be the phosphates in the 1' position.

This notion was suggested by the work of Vaara and his colleagues (1981) who found that polymyxin B-sensitive strains of Salmonella typhimurium have less than 20% of their lipid A 4' phosphates blocked with 4-amino-arabinose. Thus, 80% or more of these phosphates bear two negative charges. Substitution with 4-amino-arabinose converts the $PO_4^=$ groups to Zwitterions, resulting in fewer negative charges. In contrast to the Pb-sensitive strains, the polymyxin B-resistant pmrA mutants have 100% of their lipid A 4' phosphates blocked. It is also important that the 1' phosphates of lipid A tend to show more extensive substitution with ethanolanine in the pmrA mutants. This would also tend to reduce the anionic charges due to these phosphates. Unfortunately, our gapA mutation maps at a distance from pmrA. Therefore, the biochemistry of our gapA mutants must be compared with that of their $gapA^+$ isogenic parents before we can do more than speculate on these interactions. We cannot assume that the situation will be the same in gapA mutants as it is in the pmrA mutants.

Conclusions

Smooth Salmonella do mutate to cross resistance to polymyxin B and CAP57. They do this with a frequency of 10^{-7}. The mutation in JKS5 is tentatively mapped between 98 and 3 minutes. Resistance is associated with reduced binding of the CAP57 protein to gapA strains and to pmrA strains when they are compared to their parent $gapA^+$ and $pmrA^+$ strains. The binding is saturable with all strains, smooth or rough, $gapA^+$ or $pmrA^+$.

Acknowledgments

This work was supported by NIH Grant #AI17662 to JKS and an NIH Training Grant # AI07265. We wish to thank Joy Royer for forbearance and good humor in the preparation of this manuscript.

References

Davis RW, Botstein D, Roth JR (1980) Advanced Bacterial Genetics. Cold Spring Harbor Laboratory, Cold Spring Harbor, New York

Ganz T, Selsted ME, Szklarek D, Harwig SS, Daher D, Bainton DF, Lehrer RI (1985) J Clin Invest 76:1427-1435

Hirsch JG (1956) Phagocytin: A bactericidal substance form polymorphonuclear leukocytes. J Exp Med 103:589-611

Hovde CJ, Gray HH (1986) Characterization of a protein from normal human polymorphonuclear leukocytes with bactericidal activity against Pseudomonas aeruginosa. Infect Immun 54:142-148

Makela PH, Sarvas M, Calcagno S, Lounatmaa K (1978) Isolation and genetic characterization of polymyxin-resistant mutants of Salmonella. FEMS Microbiol Letters 3:323-326

Makela PH, Stocker BAD (1984) Genetics of lipopolysaccharide, in Handbook of Endotoxin VI, Chemistry of Endotoxin. Rietschel ET (ed) Elsevier Amsterdam, pp 59-119

Miller JH (1972) Experiments in Molecular Genetics. Cold Spring Harbor Laboratory, Cold Spring Harbor, New York

Modzrakowski MC, Cooney MH, Martin LE, Spitznagel JK (1979) Bactericidal activity of fractionated granule contents from human polymorphonuclear leukocytes. Infect Immun 23:587-590

Odeberg H, Olsson I (1975) Antibacterial activity of cationic proteins from human granulocytes. J Clin Invest 56:1118-1124

Okamura N, Spitznagel JK (1982) Outer membrane mutants of Salmonella typhimurium LT2 have lipopolysaccharide-dependent resistance to the bactericidal activity of an aerobic human neutrophils. Infect Immun 36:1086-1095

Sanderson KE, Ross H, Ziegler L, Makela PH (1972) F+, Hfr, and F' strains of Salmonella typhimurium and Salmonella abony. Bact Rev 36:608-637

Shafer WM, Casey SJ, Spitznagel JK (1983) Lipid A and resistance of Salmonella typhimurium to antimicrobial granule protein of human neutrophil granulocytes. Infect Immun 43:834-838

Shafer WM, Martin LE, Spitznagel JK (1984) Cationic antimicrobial proteins isolated from human neutrophil granulocytes in the presence of diisopropyl fluorophosphate. Infect Immun 45:29-35

Skarnes RC, Watson DW (1956) Characterization of leukin: An antimicrobial factor from leukocytes active against gram-positive pathogens. J Exp Med 104:829-845

Spitznagel JK (1984) Nonoxidative antimicrobial reactions of leukocytes, in regulation leukocyte function. Snyderman R (ed) Plenum, New York and London, pp. 284-343

Spitznagel JK, Chi H-Y (1963) Cationic proteins and antibacterial properties of infected tissues and leukocytes. Am J Pathol 43:697-711

Vaara M, Vaara T, Jensen M, Helander I, Nurminen M, Rietschel ET, Makela PH (1981) Characterization of the lipopolysaccharide from the polymyxin-resistant pmrA mutants of Salmonella typhimurium. FEBS Letters 129:145-149

Weiss J, Elsbach P, Olsson I, Odeberg H (1978) Purification and characterization of a potent bactericidal and membrane active protein from the granules of human polymorphonuclear leukocytes. J Biol Chem 253:2664-2672

Zeya HI, Spitznagel JK (1963) Antibacterial and enzymatic basic proteins from leukocyte lysosomes: Separation and identification. Science 142:1085-1087

Zeya HI, Spitznagel JK (1968) Arginine-rich proteins of polymorphonuclear leukocyte lysosomes. Antimicrobial specificity and biochemical heterogeneity. J Exp Med 127:927-941 van der Meulen, J. H., Woldendorp, J. W. ... Ammonia oxidation and nitrification ... Soil Biol. Biochem. ...

PLASMID-MEDIATED COMPLEMENT AND PHAGOCYTOSIS RESISTANCE IN E. COLI

M.E. Fernández-Beros[1], C. González[1], V. Kissel[1], M.E. Agüero[1],
M. Binns[2], G. de la Fuente[3], E. Vivaldi[3] and F.C. Cabello[1]

[1] Department of Microbiology and Immunology
New York Medical College
Valhalla, NY 10595 USA

[2] Houghton Poultry Research Station
Houghton, Huntingdon, Cambs.
England

[3] Facultad de Ciencias Biológicas
Universidad de Concepciòn
Concepciòn, Chile

Introduction

The pioneering work of W. H. Smith and others such as the Orskovs, more than fifteen years ago, indicated the relevance of plasmids to the pathogenesis of bacterial diseases (1). They were able to show the involvement of plasmid genes in the ability of E.coli to adhere and produce toxins in the intestine, and the role of the ColV plasmid in the ability of bacteria to produce extraintestinal infection in animals, and probably in humans (2). Since those days many different plasmid encoded properties have been recognized as virulence factors in bacteria.

Plasmid coded properties are involved in all the steps necessary for successful infection of the host, from colonization of the host to the ability to survive in the intracellular milieu, underlining the evolutionary importance of these so called dispensable genetic structures. While the role of plasmid genes in the ability of bacteria to produce intestinal infection has been studied extensively, their role in the ability of bacteria to produce extraintestinal infections such as bacteremia and urinary tract infections has been less well characterized.

NATO ASI Series, Vol. H24
Bacteria, Complement and the Phagocytic Cell
Edited by F.C. Cabello und C. Pruzzo
© Springer-Verlag Berlin Heidelberg 1988

Table 1 contains a list of plasmid-coded properties that may be involved in the ability of Enterobacteriaceae to invade tissues and blood. The roles of many of these plasmid coded properties in virulence remains to be confirmed by more experimental and epidemiological work.

Table 1

Some Plasmid Mediated Invasion and Dissemination Traits in Bacteria

Extracellular survival

Complement resistance:	TraT protein (5)
	pomp Y. enterocholitica (6)
	60 kdal OMP S. typhimurium (7)
	11 kdal polypeptide of the
	cryptic plasmid S. typhimurium (8)
	lipopolysaccharide? (9)
	iss gene (10)
Phagocytosis resistance:	TraT protein (5)
	pomp Y. enterocholitica (6)
	lipopolysaccharide? (9)
	iss gene (this paper)
Iron uptake:	ColV plasmids (11)
	R plasmids (12)

Intracellular survival

Non-oxidative bactericidal activity resistance	TraT protein?
	lipopolysaccharide?
	GAPA and PMRA mutations
	(This book)
Oxidative bactericidal activity resistance	
Iron-deprivation resistance:	ColV?
	R plasmids?
Low Ca^{2+} response:	PCD1, Y. pestis (13)

Table 2

Influence of the K1 Capsule and the ColV Plasmid on Susceptibility
of E. coli to Phagocytosis and Killing by Serum and Virulence

Property[a]	Phagocytosis[b] %	Survival in serum[c] (%)	LD$_{50}$[d]	Bacteriemia from Kidney[e] (%)	Gut[f]	Mortality Kidney[e] (%)	Gut[f]
K1$^+$ ColV$^+$	26	170	10^5	62.6	25.0	33.3	86.0
K1$^+$ ColV$^-$	36	120	10^6	16.6	15.5	3.0	0
K1$^-$ ColV$^+$	90	30	<10^9	ND	ND	ND	ND
K1$^-$ ColV$^-$	90	5	<10^9	0	0	0	0

[a]The K1$^+$ phenotype and carriage of the ColV plasmid of E. coli FC001 (018 ab, ac:K1:H7) and its K1$^-$ and ColV$^-$ derivatives obtained as K1-specific phage resistant mutants and ColV$^-$ derivatives obtained by SDS curing.

[b]Percentage of mouse adult peritoneal macrophages having phagocytosed the indicated bacteria.

[c]Percentage of bacteria surviving incubation for 90 min in presence of 10% human serum. Values greater than 100% indicate that bacterial multiplication occurred during the incubation period.

[d]The 50% lethal dose obtained by intraperitoneal injection of Swiss-Webster adult mice with bacteria plus hog gastric mucin.

[e]Sprague-Dowley rats were inoculated intravesically with x10^6 bacteria. Cultures were taken by cardiac puncture. Animals were observed for 12 days.

[f]New-born Sprague-Dowley rats were fed 1x10^8 bacteria through an oral gastric tube. Blood cultures were taken by puncturing the tail vein at the 2nd, 5th and 7th day post inoculation. Animals were observed for 2 weeks.

Plasmid ColV and virulence

Plasmid ColV is one of the plasmids found more frequently among E. coli strains isolated from extraintestinal infection (3). We chose to study a

ColV plasmid harbored by an E. coli K1 strain isolated from a case of neonatal meningitis. This conjugative large plasmid of approximately 120 Kb, as it can be seen in Table 2, has the ability to increase the virulence of E. coli in at least three animal models (4).

The loss of the plasmid increases the LD50 of E. coli for mice after intraperitoneal injection, and reintroduction of the plasmid by conjugation restores the LD50 to a value quite similar to that of the original parental strain (Table 2). It can be also seen that E. coli harboring the plasmid produces, in the new born rat model of neonatal meningitis, significantly more bacteremia and mortality than the isogenic strain without the plasmid (25.4% vs 15.5% and 15.6% vs 0%). We believe that these effects may reflect both a better colonization of the intestine by the ColV+ strains and an increased ability to invade the blood stream. At similar levels of colonization the ColV+ strains still produce more bacteremia and mortality in the model (data not shown).

Table 2 also shows that in a model of urinary tract infection the E. coli strain harboring this plasmid generates more bacteremia and mortality at similar levels of kidney infection (62,6% vs 16.6% in bacteremia, 33.3% vs 3% in mortality). These results also illustrate the interplay between a ColV plasmid and the chromosomally coded K1 antigen.

Further studies indicated that the plasmid ColV is present with a high frequency among E. coli K1 strains isolated from human extraintestinal infections, especially from bacteremia (22 of 42 vs 3 of 14). These results confirm that the ColV plasmids code for functions that increase the ability of the bacteria to invade the blood stream and tissues (4).

The virulence properties of Plasmid ColV

ColV plasmids belong to a heterogenous group of plasmids which are usually conjugative and belong to the F compatibility group. Plasmids of this group are able to express several properties that can increase the ability of E. coli to colonize and invade the human host such as serum and phagocytosis resistance and iron uptake (3,11,14).

Serum survival and phagocytosis experiments summarized in Table 2 indicate that the plasmid under study is able to increase the ability of E. coli to survive in concentrations of human serum that range from 20% to 50%, and that this increased survival is more evident when E. coli expresses the K1 antigen, underlining again the interaction between chromosomal and plasmid genes. The plasmid also produces a small but constant protection against phagocytosis which is clearer in the presence of the K1 antigen. Further experiments indicated that this plasmid also encodes for aerobactin and its receptor as evidenced by the susceptibility of the strain harboring it to Cloacin DF13 which detects the aerobactin receptor. Excretion of aerobactin by the plasmid containing strain is demonstrated by its ability to crossfeed an ent mutant. All these three properties can be transmitted to other laboratory and clinical isolates of E. coli after conjugation of the plasmid, confirming their location in the plasmid ColV genome, and their ability to be expressed in different bacterial hosts.

These results indicate that the ColV plasmid under study has accumulated several properties, namely serum resistance, phagocytosis resistance and iron uptake that may be responsible for its ability to endow E. coli with the ability to produce extraintestinal infections.

Genetics of the complement and phagocytosis resistance mediated by plasmid ColV

The genes involved in the expression of the complement and phagocytosis resistance phenotype mediated by plasmid ColV may be the traT, first described in R conjugative plasmids of the F compatibility group, and the increased survival in serum gene or iss described by Binns and Hardy in the plasmid ColV, I-K94 (5,10).

To detect the presence of these genes in the plasmid ColV DNA under study we performed DNA-DNA hybridization experiments with intragenic probes for these two genes. The traT probe was a 600bp BstEII DNA fragment and the iss probe, was a 446bp ThaI-BglI DNA fragment. Figure 1A and 1B shows that these probes are able to hybridize with undigested and digested

ColV plasmid DNA. Moreover, these two probes hybridize with different ColV fragments, confirming that the two genes do not share homologous sequences, (Data not shown). The ColV DNA, Fig. 1A, also hybridizes with aerobactin-synthesis and aerobactin-receptor DNA probes as expected.

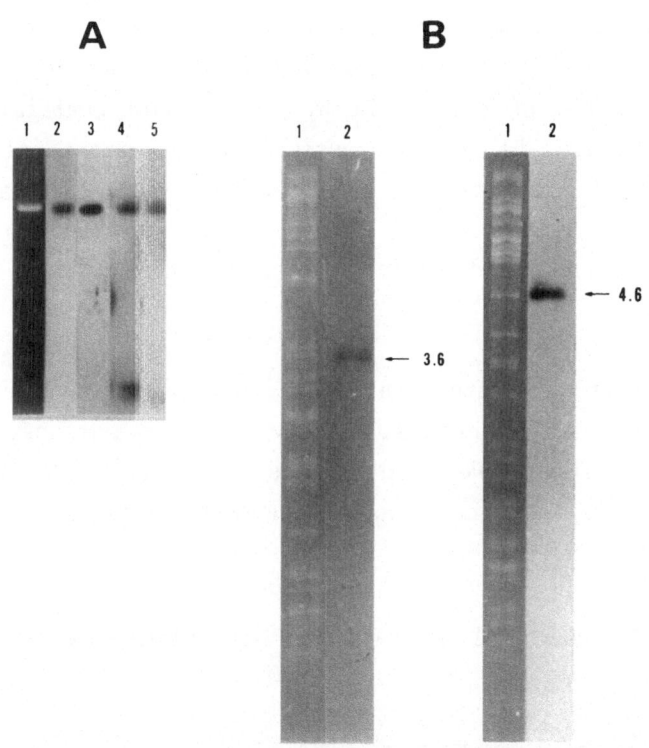

Fig. 1. Hybridization of ColV DNA with different DNA probes. A. 1) Undigested ColV DNA; 2) as 1, hybridized with the traT DNA probe; 3) hybridized with the iss DNA probe; 4) hybridized with the aerobactin-synthesis DNA probe; and 5) hybridized with the aerobactin-receptor DNA probe. B. Digested ColV DNA left; 1) with Pst1; 2) same as 1, hybridized with the iss DNA probe; right; ColV DNA digested with BstEII; 3) Same as 1, hybridized with the traT DNA probe.

We cloned the 4.6 Kb BstEII ColV DNA fragment, which gave a positive signal with the traT probe, into a cloning vector pCP14 developed by us. The 3.6 Kb PstI fragment, which gave a positive signal with the iss probe, was cloned into the pBR322 cloning vector. Table 3 indicates that the _iss_ and _traT_ genes of the plasmid ColV confer upon a wild type E. coli increased survival to concentrations of 10% to 50% of normal human serum.The same table shows that the phagocytic index of the E. coli strains containing the cloned genes decreases, indicating less phagocytosis of the strains by mouse peritoneal macrophages.

Table 3

Serum and phagocytosis resistance of the E. coli FC004 strains containing different plasmids with cloned ColV genes

Plasmids	Gene	phagocytic index[a]	% Serum survival[b]
None	-	2578 ± 330	5
pH2	_iss_[+] ColV	2108 ± 85	37
pCP15	_traT_[+] ColV	1739 ± 320	60

[a]The phagocytic index is the percentage of macrophages ingesting bacteria multiplied by the average number of E. coli cells ingested by macrophage. The value is the mean of at least three experiments ± the standard deviation.

[b]Viable counts after 90 minutes of exposure to 10% normal human serum.

Restriction enzyme analysis of the cloned DNA fragments containing these genes and their comparison with the probes indicate differences at the level of DNA sequences between them as evidenced by different restriction enzyme sites. Analysis of the outer membranes of bacteria containing these different clones indicate that, as expected, the clone which contains the ColV _traT_ gene contains an outer membrane protein of similar molecular weight to the one described in the plasmid R6-5. Analysis of the inner and outer membranes and maxicells of the clones containing the _iss_ gene has failed to demonstrate the presence of an outer membrane protein.

Mechanisms of traT mediated complement and phagocytosis resistance

Further experiments suggested that the decrease in phagocytosis mediated by the traT protein is due to a decreased and altered deposition of the complement opsonin C3 on the surface of traT⁺ bacterial cells (15).

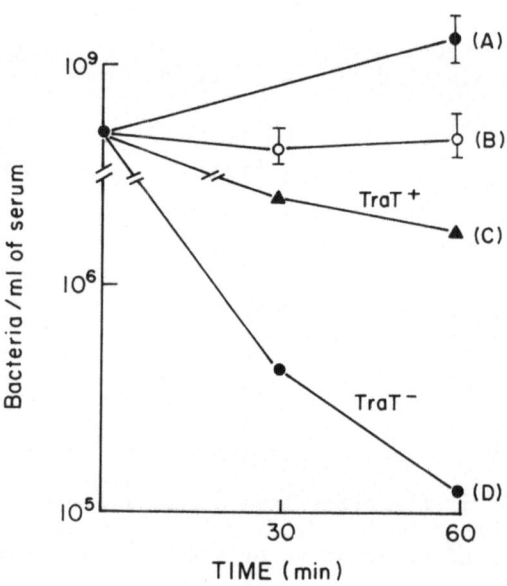

Fig. 2. Survival of isogenic E. coli 01 strains in the presence of different complement preparations A) E. coli 01 traT⁺ + C9 depleted serum; B) E. coli 01 traT⁺ + C9 depleted serum + C9 + EGTA Mg^{+2}; C) E. coli 01 traT⁺ + C9 depleted serum + C9; D) E. coli 01 + C9 depleted serum + C9.

To improve our understanding of the mechanisms behind the complement and phagocytosis resistance coded by the ColV plasmid we studied the interaction between the bacterial cells expressing these proteins and complement. Experiments with the wild type 01 E. coli expressing the traT gene product and 30% serum depleted in C9 and reconstituted with different supplements show that the protective effect of the traT gene product is only evident in the presence of C9 and the classical pathway of complement, Figure 2. The data suggest that the effect of the TraT protein may be due to an interference with the lytic effect of C9.

Experiments using a hemolytic assay to measure the consumption of C9 by the bacteria, indicate that the number of molecules of C9 consumed per bacterial cell is similar for the serum-resistant and the serum-sensitive strains. Preliminary experiments with radiolabeled C9 confirm these findings and suggest that the traT+ cells liberate faster the polymerized C9 into fluid phase. A similar deposition of C9 in serum-sensitive and serum-resistant strains has been previously found by others in experiments using radiolabeled C9 (16,17).

Frequency of the traT, iss and iu genes among clinical isolates

Colony hybridization of E. coli isolates from intestinal and extraintestinal infections using the probes for the iss and traT genes, indicates that there is no difference in the frequency of these genes in the two groups of strains (Table 4, Bacteremia, iss 45%, traT 44%; stools, iss 42%, traT 41%). Nonetheless, if the comparison is done among strains harboring ColV plasmids, significant differences appear between the intestinal strains and those isolated from extraintestinal infection.

Table 4

Frequency of iss, traT and aerobactin iron-uptake
genes in clinical isolates of ColV$^+$ E. coli[a]

Source of isolates	iss$^+$	traT$^+$	iu$^+$	iss$^+$, traT$^+$, iu$^+$
Blood	21/22 (95.4%)	19/22 (86.3%)	20/22 (90.9%)	19/22 (86.3%)
Stools	11/16 (68.8%)	6/16 (37.5%)	13/16 (81.2%)	6/16 (37.5%)

[a]Detected by colony hybridization with the corresponding probe.

For example, 95.4% of the extraintestinal ColV⁺ strains hybridize with the
iss probe, compared with 68.8% of the ColV⁺ strains isolated from the
intestine. This difference is even more pronounced if the strains
hybridizing with both probes are considered, 86.3% of extraintestinal
strains versus 37.5% of the intestinal. Moreover, positive hybridization
in the intestinal strains seems to be correlated more with the presence of
other plasmids such as Enterotoxin and R plasmids than with ColV per se,
suggesting that these genes are found on all types of virulence plasmids
and not just on ColV plasmids.

Similar analyses, using probes for the iron uptake genes, suggest that
the traT and iss gene are better markers than the iron uptake genes for
detecting strains with ability to invade the blood stream, because there
is no difference in the hybridization with an iron uptake probe between
blood and intestinal isolates. These results are consistent with the idea
that these plasmid encoded properties endow pathogenic E. coli with a
selective advantage in their relationship with the mammalian host. We have
also confirmed previous results that indicate that a high degree of
correlation exists between the presence of these properties and the
presence of the K1 antigen (3,18).

Conclusion

These studies, and previous studies by other investigators and by us,
suggest that plasmid coded functions are involved in the genesis of
extraintestinal infections by Enterobacteriaceae such as E. coli and
Salmonella (19,20,21). The linkage of several genes coding for different
pathogenic properties in one plasmid is an indication of the ability of
plasmids to recombine and generate evolutionary advantageous structures
that permit bacterial growth in different environments. In this manner, the
ColV plasmids appear to have accumulated a range of propertienvironments
as dissimilar as the intestinal tract, tissues and the blood stream. The
accumulation of properties, capable of increasing the survival of E. coli,
in one plasmid is probably not due to periodic selection or 'piggy back'
effect and underlines the evolutionary significance of these phenomena. We
also suggest that the ability of these plasmids to be transmitted by

conjugation among different strains may undermine the relevance of the concept of clonal distribution of virulent strains (22). The clonal distribution analysis is based on the study of, at most, 30 to 40 properties, whereas the acquisiton or loss of one of these large plasmids changes the cell in at least 80 to 100 properties, drastically altering the properties of a stable clone and its ability to adapt (23).

References

1. Smith. H.W. and M. Linggood, 1971. J. Med. Microbiol. 4:467-485.
2. Bark, A.L., G. Christiansen, C. Christiansen, A. Stenderup, I. Ørskov and F. Ørskov, 1972. J. Gen. Microbiol. 73:342.
3. Agüero, M.E., H. Harrison and F.C. Cabello, 1983. Increased frequency of ColV plasmids and mannose-resistance hemagglutinating activity in an Escherichia coli K1 population. J. Clin. Microbiol. 18:1413-1416.
4. Agüero, M.E., G. de la Fuente, E. Vivaldi and F.C. Cabello, 1987. The ColV plasmid increases the virulence of an E. coli K1 strain in two animal models. Submitted for publication.
5. Moll, A., P.A. Manning and K.N. Timmis, 1980. Plasmid-determined resistance to serum bactericidal activity: a major gene product, is responsible for plasmid-specified serum resistance in Escherichia coli. Infect. Immun. 28:259-367.
6. Lian, C., W.S. Hwang, and C.H. Pai, 1987. Plasmid-mediated resistance to phagocytosis in Yersimia enterocolitica. Infect. Immun. 55:1176-1183.
7. Hackett, J., P. Wyk, P. Reeves and V. Mathan, 1987. Mediation of serum resistance in Salmonella typhimurium by an 11-kilodalton polypeptide encoded by the cryptic plasmid. J. Infect. Diseases 155:540-549.
8. Vanderbosch, J.L. and G.W. Jones, 1987. Abstracts of the annual meeting of the American Society for Microbiology p. 53.
9. Mäkela, P.H., Houi, M., Saxen, H., Valtonen, M. and V. Valtonen. Ability to activate the alternative complement pathway as a virulence determinant in Salmonella (this volume).
10. Binns, M.M., D.L. Davis and K.G. Hardy, 1979. Cloned fragments of the plasmid ColV, I-K94 specifying virulence and serum resistance. Nature (London) 279:778-781.
11. Williams, P.H. and H.K. George, 1979. ColV plasmid-mediated iron uptake and the enhanced virulence of invasive strains of Escherichia coli in plasmids of medical environmental and commercial importance, K.N. Timmis and A. Puhler (eds.) Elsevier/North Holland, Biomedical Press, Amsterdam p. 161-172.
12. Colonna, B., M., Ricoletti, P., Visca, M. Casalino, P. Valenti and F. Maimone, 1985. Complete IS, elements encoding hydroxamate-mediated iron uptake in FIme plasmids from epidemic Salmonella spp. J. Bacteriol. 162:307-316.
13. Pollack, C., S.C. Straley and M.S. Klempner, 1986. Probing the phogolysosomal environment of human macrophages with a Ca^{+2}-response operon fusion in Yersinia pestes. Nature 322:834-836.

14. Aguero, M.E. and F.C. Cabello, 1983. Relative contribution of ColV plasmid and K1 antigen the pathogenicity of Escherichia coli. Infect. Immun. 40:359-368.
15. Aguero, M.E., L. Aron, A.G. DeLuca, K. Timmis and F.C. Cabello, 1984. A plasmid-encoded outer membrane protein, TraT, enhance resistance of Escherichia coli to phagocytosiş. Infect. Immun. 46:740-746.
16. Joiner, K.A., C.H. Hammer, E.J. Brown, R.J. Cole and M.M. Frank. 1982. Studies on the mechanism of bacterial resistance to complement-mediated killing. I. Terminal complement components deposited and released from Salmonella minnesota S218 without causing bacterial death. J. Exp. Med. 155:797-808.
17. Joiner, K.A., C. Hammer, E.J. Brown and M.M. Frank, 1982. Studies on the mechanism of bacterial resistance to complement-mediated killing. II. C8 and C9 release C5b67 from the surface of Salmonella minnesota S218 because the terminal complex does not insert into bacterial outer membrane. J. Exp. Med. 155:809-919.
18. Timmis, K.N., G.J., Boulnois, D., Bitter-Suermann and F.C. Cabello. 1985. Surface components of Escherichia coli that mediate resistance to the bactercidal activities of serum phagocytes. Curr. Top. in Microb. Immun. 118:197-218.
19. Montenegro, M.A., D. Bitter-Suermann, J.K. Timmis, M.E. Aguero, F.C. Cabello, S.C. Sanyol and K.N. Timmis. 1985. traT gene sequences, serum resistance and pathogenicity-related factors in clinical isolated of Escherichia coli and other gram-negative bacteria. J. Gen. Microbiol. 131:1511-1521.
20. Jones, G.W., P.K. Rabert, D.M. Svinavich and H.J. Whitfield. 1982. Association of adhesive, invasive and virulent phenotypes of Salmonella typhimurium with autonomous 60-megadalton plasmids. Infect. Immun. 38:476-486.
21. Elwell, L.P. 1980. Plasmid-mediated factors associated with virulence of bacteria to animals. Ann. Rev. Microbiol. 34:65-96.
22. Achtman, M. 1986. Clonal analysis of descent and virulence among selected Escherichia coli. Ann. Rev. Microbiol. 40: 185-210.
23. Harth, D.L. and D.E. Dykhuizen. 1984. The population genetics of Escherichia coli. Ann. Rev. Genet. 18: 31-68.

THE RELATIVE ROLE OF LIPOPOLYSACCHARIDE AND CAPSULE IN THE VIRULENCE OF E. COLI

A. Cross[1], J. Sadoff[1], P. Gemski[2], Kwang Sik Kim[3]

[1] Department of Bacterial Diseases
Walter Reed Army Institute of Research
Washington, DC 20307-5100
USA

[2] Department of Biological Chemistry
Walter Reed Army Institute of Research
Washington, DC 20307-5100
USA

[3] Department of Pediatrics
Childrens Hospital of Los Angeles
Los Angeles, California 90027
USA

E. coli is the leading cause of Gram-negative bacteremia and of neonatal meningitis. In order to cause these extra-intestinally invasive infections, E. coli must traverse the gut, its normal habitat, (or from a common site of infection, the genitourinary tract), into the blood and multiply. These events require a diverse array of bacterial factors, including adhesins, iron-seeking organelles and perhaps enzymes. Extra-intestinal invasion by E. coli also requires that the organism be able to evade normal host defenses, which include serum bacteriolytic and phagocytic mechanisms. For these latter events two moieties of the bacterial outer envelope, the capsule and lipopolysaccharide (LPS) have long been considered to be important virulence factors.*

In this discussion, the contribution of capsule and LPS to the virulence of E. coli, as measured by in vitro tests of serum bacteriolysis and phagocytosis, as well as by animal virulence studies, will be reviewed.

---

*Virulence factor in this study refers to bacterial substances that permit the evasion of host defense mechanisms.

NATO ASI Series, Vol. H24
Bacteria, Complement and the Phagocytic Cell
Edited by F.C. Cabello und C. Pruzzo
© Springer-Verlag Berlin Heidelberg 1988

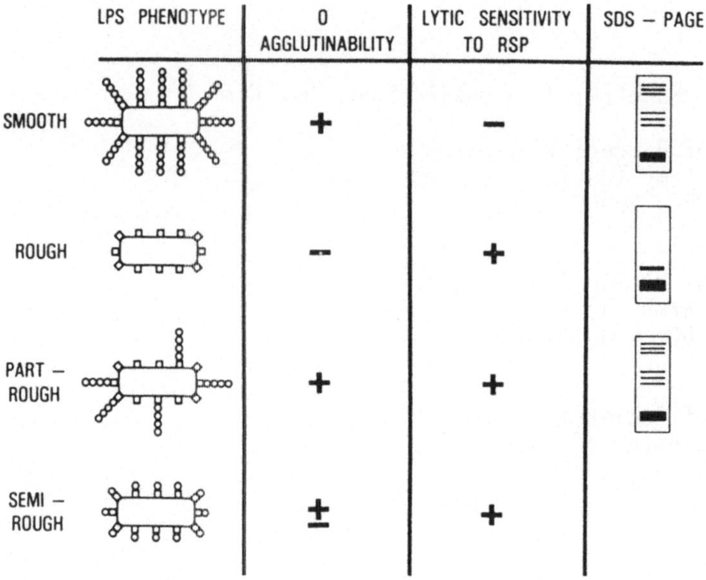

LPS PHENOTYPE	O AGGLUTINABILITY	LYTIC SENSITIVITY TO RSP	SDS — PAGE
SMOOTH	+	−	
ROUGH	−−	+	
PART − ROUGH	+	+	
SEMI − ROUGH	±	+	

Fig. 1. Lipopolysaccharide phenotypes may be characterized by susceptibility to rough LPS-specific bacteriophages and agglutinability
by O-specific antisera. Differences between smooth and part-rough
LPS phenotype are not apparent on SDS-PAGE pattern.

We hope to demonstrate that the highly ordered association of O and K
antigens seen in extra-intestinal infections with E. coli may have evolved
in some instances from necessary pathogenetic relationships between
specific O and K serotypes.

As part of a study of the core portion of LPS, we decided to screen for
LPS phenotype a collection of over 500 clinical isolates of E. coli
obtained from different anatomic sites. Since chemical or SDS-PAGE
analysis of so many strains was not feasible, we adapted for use in E.

coli a set of rough-LPS specific bacteriophages originally described by Subbaiah and Stocker in the analysis of Salmonella LPS (1). E. coli isolates not susceptible to lysis by these phages and which were agglutinated by O-specific antisera were considered to have a smooth LPS phenotype (Figure 1). In contrast, isolates that were lysed by at least one of these rough LPS-specific phages (RSP) and that were not agglutinated by O-specific antisera were considered to have a rough LPS phenotype. A part-rough phenotype was characterized by lysis with RSP, but O-agglutinability. The SDS-PAGE patterns of these part-rough E. coli were indistinguishable from patterns of smooth E. coli, while the E. coli with a rough LPS phenotype lacked the typical step ladder appearance by this analysis.

Since the K1 antigen had already been shown to be an important surface component of E. coli, we also screened the strains for this capsular polysaccharide with K1-specific phages originally described by Cross et al. and Gross, et al. (2,3). The clinical isolates tested were obtained prospectively from consecutive E. coli isolates cultured from the blood, cerebrospinal fluid, urine and wounds of patients seen at the Walter Reed Army Medical center over a 5 year period. Fifty consecutive E. coli isolates were obtained from the stool of healthy patients who were undergoing routine proctoscopic examination. Thus, these strains of E. coli were collected in an epidemiologically significant way. Analysis of these strains revealed that over a quarter of bacteremic isolates of E. coli had a rough or part-rough LPS phenotype (Table 1). This was not significantly different from what was observed for isolates cultured from stool or wounds. Moreover, the frequency of K1-capsular phenotype was similar from those three sites. In contrast, urinary isolates of E. coli had a higher likelihood of lysis by RSP and lower incidence of K1 encapsulation. When one analyzed the association of K1-encapsulation with rough or part-rough LPS phenotype, however, there was a highly significant association of two phenotypes with E. coli retrieved from the blood, a trend toward significance in wound isolates, and no significant association between the two phenotypes in isolates obtained from stool or urine (Table 1). This suggests that K1 encapsulation is required by E. coli with a rough or part-rough LPS phenotype to successfully invade the bloodstream.

Table 1. Association of K1-positive phenotype with rough or part-rough
(RSP⁺) phenotype

Site	Total no. isolates	%RSP⁺	%K1⁺	K1⁺ associated with:
Blood	248	28	22	33/70 (47%) rough strains[a]
				22/172 (12%) smooth
Urine	193	50	11	12/97 (12%) rough strains[b]
				9/96 (9%) smooth
Wounds	57	30	21	6/17 (35%) rough strains[c]
				6/40 (15%) smooth
Stool	50	20	18	1/10 (10%) rough strains[b]
				8/40 (20%) smooth

[a] $p = 0.0001$
[b] p = not significant
[c] $p = 0.10$

We also compared the serogroups of these bacteremic isolates of E. coli
with K1 capsular and LPS phenotype (Table 2). This analysis revealed that
(1) only 10 O-serogroups accounted for nearly 60% of these extra-
intestinally invasive strains, (2) the K1 capsule was usually associated
with the O1, O16 and O18 serogroups, those serogroups that had a higher
proportion of isolates susceptible to RSP, and (3) that certain O
serogroups such as the O6 and O75 were rarely associated with K1 capsules
and these strains were rarely associated with lysis by RSP. Since E. coli
with rough LPS phenotype were previously shown to be susceptible to lysis

Table 2. Correlation of O group, K antigen and LPS phenotype in clinical
isolates from blood and CSF

| | RSP[+a] | | | RSP[-] | | | |
| | Capsule antigen | | | Capsule antigen | | | |
O group	K1	Non-K1	Nontypable	K1	Non-K1	Nontypable	Total
O1	4	0	1	10	3	0	18
O2	0	1	0	3	2	2	7
O4	0	4	0	0	13	1	18
O6	0	4	3	0	18	5	30
O8	0	0	1	0	3	6	10
O12	0	0	0	4	0	0	4
O15	0	0	0	0	5	1	6
O16	5	0	1	6	1	0	13
O18ac	5	1	1	3	1	3	14
O75	0	2	3	0	4	1	10
Other Os	2	4	3	2	10	40	61
Rough	5	0	5	2	3	2	17
O non-typable	0	1	2	0	2	6	11
Total	21	17	20	30	65	66	219

[a]Susceptibility to rough LPS-specific phages (RSP)

by mammalian serum, we speculated that such isolates required K1
encapsulation for optimal virulence, and in contrast, that other O
serogroups, which usually had a smooth LPS phenotype, did not require this
capsule for full virulence.

To test this hypothesis we selected K1-negative mutants from strains of
E. coli that were susceptible or not susceptible to lysis by RSP, and
tested these parent-mutant pairs in both serum bactericidal and animal

Table 3. Comparison of LD_{50} in neonatal rats between K1- or K5-
encapsulated $\underline{E}$. $\underline{coli}$ and unencapsulated mutants

Strain	Serotype	RSP[a]	LD_{50} (CFU)
EC5	018:K1+	-	3.1×10^1
	018:K1-	-	2.8×10^6
A90	01:K1+	+	4.9×10^4
	01:K1-	+	5.0×10^6
RS167	016:K1+	+	9.2×10^1
	016:K1-	+	2.6×10^7
E457	06:K5+	-	4.5×10^4
	06:K5-	-	1.0×10^5
A63	018:K5+	-	1.5×10^6
	018:K5-	-	1.4×10^6

[a]Susceptibility of rough-specific phages (RSP).

challenge assays (4-6). Whereas the parental forms of both types of
strains were resistant to bacteriolysis in up to 80% serum, the
unencapsulated mutants of strains susceptible to RSP were easily killed by
as little as 1% serum (data not shown). The RSP-insensitive K1+ isolates
remained insensitive to kill by up to 50% serum (but were lysed by 90%
human serum). Inoculation of these parent-mutant pairs into neonatal rats
revealed a 2-5 log difference in LD_{50} between the K1-positive and negative
pairs regardless of susceptibility to RSP (Table 3). Thus the K1 capsule
did appear to provide increased virulence in both these assays to $\underline{E}$. $\underline{coli}$
of a particular 0 serogroup or LPS phenotype.

With the availability of a K5-specific bacteriophage, we similarly
tested K5+ parent - K5 negative mutants of the 06 serogroup, a serogroup
rarely lysed by RSP and rarely associated with K1 capsule. These strains

revealed no difference in either susceptibility to serum killing or in LD_{50} between parent and unencapsuled mutant (Table 3). We concluded that the 06 serogroup might be a prototype of an LPS phenotype that does not require a capsule for optimal virulence. For some other 0 serogroups, however, the K5 capsule may provide some protection from serum killing (02,04,015) (5).

Table 4. Bactericidal effect of serum against rough, K1[+]
E. coli (E412) and its hybrid derivatives expressing
both K1 and K27 capsules

Strain	Phenotype			% kill[b]
	K1	RSP[a]	K27	
F639 (Hfr donor)	-	ND	++	100
E412 (recipient)	+	+	-	2
E412 X K27i	+	+	+	28
E412 X K27c	+	+	++	60
E412mu X K27c	-	+	++	100

[a]RSP = susceptibility to rough LPS-specific bacteriophages;
 ND = not done;
 K27i = received one chromosomal locus;
 K27c = received two chromosomal loci;
 E412mu = K1-negative mutant of E412.
[b]Kill in 10% normal human serum at 60 minutes.

During the course of these studies, Taylor found that the K27 capsule failed to inhibit serum killing of E. coli and therefore questioned the important role of encapsulation in resistance to serum killing postulated by Glynn and Howard (7). Opal therefore constructed transformants that expressed either one or both of the K27 genetic loci in a rough, K1 encapsulated E. coli recipient (Table 4) (8). While the donor strain that carried the Hfr plasmid for the K27 loci was, indeed, serum sensitive and the K1-bearing recipient was serum resistant, a transformant that received one chromosomal locus of K27 was killed 28% at 60 minutes and a transformant that received two of the K27 chromosomal loci was killed 60% compared to the original inoculum. A K1-negative mutant of the recipient strain that had received both K27 loci was completely killed at 30 minutes. These data suggest that the K27 capsule differs from the K1 capsule in not being able to provide protection to the organism from serum lysis. This K27 capsule, found in only 2 of 219 bacteremic isolated, is an example of an unusual capsule that does not provide any apparent benefit to the bacteria. It also demonstrates that resistance to the serum bactericidal reaction does not correlate completely with the ability to isolate E. coli from the blood. Perhaps other unidentified determinants may be important for extra-intestinal invasion in these select strains.

In sum, there appears to be a close correlation between certain types of capsules and certain O serogroups. Further, those O serogroups associated with K1 encapsulation are characterized by an increased frequency of rough or part-rough LPS phenotypes and require the K1 capsule for optimal virulence, whereas other serogroups, such as the O6, are predominantly smooth and do not appear to require encapsulation for resistance to serum killing or enhanced animal virulence. Finally, the data with the K1, K5 and K27 capsules suggest a functional heterogeneity among capsules of E. coli.

Opsonophagocytosis may be considered to be another level of host defenses. Fifty-one of 61 (84%) E. coli with K1 encapsulation were resistant to opsonophagocytosis by PMN compared to 90 of 173 (52%) of E. coli with other capsules, including the common K5, K2 and K95 capsular polysaccharides (2). Thus, the K1 capsule is unique in its ability to impart resistance to opsonophagocytosis. The rough and part-rough LPS phenotypes also increase the likelihood of opsonophagocytosis among both the K1-positive and K1-negative groups of bacteria. For example, 93% (26/28) of smooth K1-positive E. coli strains were completely resistant to phagocytosis vs. 74% (17/23) of rough K1-positive strains (p <0.05). Thus, it is also necessary to consider simultaneously both the LPS and capsular phenotypes in the opsonophagocytosis of E. coli. We also noted that the K12 and K52 capsular polysaccharides reduced the opsonophagocytic kill in

Table 5. Correlation of amount of K1 capsule bound to E. coli[a] with smooth (018:K1) or rough (E412) LPS phenotype with kill in serum bactericidal and opsonophagocytic (OP) assays

	E412 (R:K1$^+$)		018:K1$^+$		018:K1$^-$ [b]	
	log	stn[c]	log	stn	log	stn
% serum kill	0	98±1	0	0	0	0
% OP kill	52±22	99±0	0	93±5	96±2	99±1
mcg K1/10^{10} E. coli	164±47	15±5	382±52	24±8	0	0

[a]K1 antigen levels measured by rocket immunoelectrophoresis
[b]018:K1$^-$ strain is isogenic unencapsulated mutant from 018:K1$^+$
[c]Bacteria studies at mid-log (log) and stationary (stn) phases

5 of 9 of the former and 4 of 6 of the latter isolates. This suggests that other select capsules may also be important to the evasion of this host defense mechanism by E. coli.

Since many researchers have suggested that the amount of capsular polysaccharide may explain the variation in bacterial killing observed within a given serogroup, Vermeulen correlated the amount of K1 encapsulation with likelihood of killing in both a rough and smooth strain of K1-encapsulated E. coli (9). He measured the amount of K1 polysaccharide bound to the organism directly with a rocket immunoelectrophoresis. By varying the phase of growth, nutrient quality or pH of growth medium, it was possible to generate E. coli that differed in the amount of K1 capsule bound to the organism, while not changing the LPS phenotype. The E. coli strain having a rough LPS phenotype was resistant to lysis by 20% serum down to a level of 50 mcg K1/10^{10} CFU, whereas the smooth E. coli strain remained serum resistant even in the absence of K1 capsule. This smooth K1 encapsulated strain, however, was killed in an opsonophagocytic assay when there was approximately ≤ 80 μg K1/10^{10} CFU.

At a similar level of K1 capsular polysaccharide $\geq$ 80 μg/10^{10} CFU, the rough strain was more easily killed. Under normal laboratory growth conditions, however, both the rough and smooth strains of E. coli produce K1 capsule so far in excess of that amount required to impart resistance to killing such that it seems unlikely that differences in killing among K1-positive strains observed earlier would be due solely to differences in the amount of capsule (Table 5).

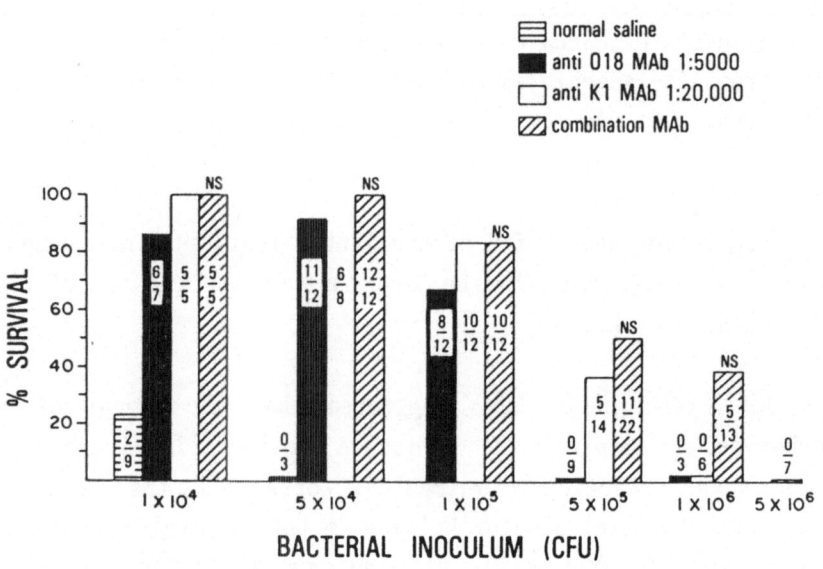

Fig. 2. Ability of optimal amounts of anti-0 and anti-K monoclonal antibody, either alone or in combination, to prevent lethal infection from increasing inocula of E. coli 018:K1:H7.

The relative role of LPS and capsule in the virulence of E. coli might also be delineated with the use of monoclonal antibodies. Earlier studies using polyclonal antisera suggested that antibodies to capsule were more efficacious than anti-O antibodies in both in vitro and in vivo assays of E. coli virulence (10,11). We have previously shown that an anti-K1 monoclonal antibody (MAb) of the IgM class was able to mediate the opsonophagocytosis of K1 encapsulated bacteria and to protect mice from lethal infection (12). Further, it has also been shown that MAb directed toward the polysaccharide portion of LPS was also protective against lethal infection (13,15). We, therefore, studied whether a combination of the two MAbs, the anti K1-capsular IgM MAb and anti-O18 IgG3 MAb, each directed against a different epitope on the same bacteria was more effective in protecting mice from lethal infection than either MAb alone (16). When outbred mice were pre-treated with combinations of sub-optimal amounts of MAb (data not shown) or with a combination of MAb at concentrations of MAb that each provided 100% protection at a challenge of 5 LD_{50}'s (Figure 2) there was no enhanced protection against increasing inocula of E. coli. Thus, under these two conditions, it did not appear that the epitope to which the MAb was directed was as important as the fact that there was sufficient protective antibody against any epitope. It may be more important that the host "see" sufficient Fc portions of any immunoglobulin rather than the Fc portion of a MAb of particular antigenic specificity for the killing of E. coli. Perhaps MAbs of different isotypes may exhibit a synergy not found here.

When the anti-K1 MAb was combined with preparation of MAb directed against a determinant not found on the O18:K1 strain tested, pre-treatment with this combination resulted in a significantly decreased survival than that observed with the anti-capsular MAb alone. If the potential for this adverse interaction between MAbs is generally true, then great care may be required in the use of polyvalent preparations of MAb for passive therapy.

Further study with the anti-O18 MAb was pursued with an E. coli O18-K5. No difference in virulence for neonatal rats was observed between the K5 encapsulated, O18 parent and a K5-negative mutant (1.5 x 10^5 CFU LD_{50} for each). Pre-treatment with the same anti-O18 MAb that was previously shown to protect against lethal challenge with E. coli O18:K1 resulted in no protection of these mice when challenged with the K5+ parent, but was fully protective against challenge by a similar inoculum of the K5-negative mutant. The K5-positive bacteria retrieved from the blood of moribund mice were still capable of reacting with the anti-018 MAb on a dot-blot immunoassay, suggesting that this anti-018 MAb was directed towards a common epitope on the 018 serogroup, and not toward one of the 4 subtypes of 018 LPS previously described (17). These data suggest that there may be important morphologic differences in the relation between LPS and capsule in the 018:K5 strain versus 018:K1 strain of E. coli.

An important relationship between LPS and capsule in pathogenetic mechanisms may also be elicited with the use of C3H mice (18). The C3H/HeN sub-line is normally responsive to the biological effects of LPS, whereas the C3H/HeJ sub-line, differing only at one locus on chromosome 4, is relatively hypo-responsive. C3H/HeJ mice had an LD_{50} of 10 CFU when challenged with an 018:K1 E. coli compared to 10^4 CFU for the C3H/HeN mice (Table 6). Removal of the K1 capsule, however, resulted in a similar LD_{50} for the 2 different mice. This implies that the K1 capsule is important in the virulence of this strain of E. coli for both mice. Pre-treatment of C3H/HeJ mice with MAb to either K1 capsule or LPS provided 90% protection against lethal infection in C3H/HeJ mice.

Table 6. Susceptibility of LPS-responsive (C3H/HeN) and LPS-hyporesponsive (C3H/HeJ) mice to live E. coli or its LPS[a]

| Challenge with: | LD_{50} | |
	C3H/HeH	C3H/HeJ
018 LPS	400 mcg	3750 mcg
E. coli 018:K1 +	10^4 CFU[b]	10^1 CFU
E. coli 018:K1 -	10^7 CFU	10^7 CFU

[a]Mice challenged intraperitoneally with E. coli or homologous LPS
[b]CFU = colony-forming units

In order to see whether these observations were confined to one unusual strain, we performed LD_{50} determinants with other wild type strains of $\underline{E}$. $\underline{coli}$ retrieved from the blood of patients. $\underline{E}$. $\underline{coli}$ of serogroups 01 and 018 had LD_{50}'s below 5×10^4 if there was also a smooth LPS phenotype and K1 encapsulation (data not shown). If these 0 serogroups were associated with non-K1 capsules, or with part-rough LPS phenotype, then the LD_{50}'s were significantly increased. Moreover, a deep rough, K1$^+$ $\underline{E}$. $\underline{coli}$ and a nitrosoguanidine mutagenized isolate of $\underline{E}$. $\underline{coli}$ 018:K1 that retained its K1 capsule but was now susceptible to RSP both had an $LD_{50} > 10^8$ CFU. Thus, both K1 encapsulation and smooth LPS phenotype were important determinants of this increased virulence of $\underline{E}$. $\underline{coli}$ for the C3H/HeJ.

Kinetic studies of bacterial clearance following intravenous injection of 10^5 CFU $\underline{E}$. $\underline{coli}$ demonstrated that during the first 4 hours there was little difference in clearance between the 2 sub-lines of mice; however, during the next 24 hours the $\underline{E}$. $\underline{coli}$ entered the blood of C3H/HeJ mice, probably from Kupffer cells, until lethal levels were attained, while in C3H/HeN mice a significantly lower level of bacteremia occurred. Over the next 7 days, the LPS normo-responsive mice continued to have a low-grade bacteremia which was well tolerated until complete eradication occurred. There were no inflammatory cells in light microscopic views of the liver of moribund C3H/HeJ mice; however, Kupffer cells did contain visible bacteria. Electron microscopic examination of the Kupffer cells revealed intact bacteria inside lysosomes, with some bacteria dividing, and an electron lucent area suggestive of retained capsule.

We believe the hypersusceptibility of the LPS-hyporesponsive mice is best explained by an inability of this host to generate important LPS-mediated host defense mechanisms. For example, in the absence of polyclonal activation or macrophage activation it may be difficult for the host to clear the invading bacteria. The efficacy of the passive administration of antibody in preventing lethal infection may be due to by-passing a block in the former mechanism. The presence of $\underline{E}$. $\underline{coli}$ in Kupffer cells for 48 hr following infection may be due to a lack of activation of these phagocytes, as has been described for $\underline{Rickettsia}$ and $\underline{Leishmania}$. Thus, LPS induced mechanisms may be an important part of normal host defense mechanisms.

In addition, it appears that invading bacteria must also evade initial host defenses in the blood before the replication to lethal levels of bacteria can occur. We believe that these defenses are functioning in C3H/HeJ mice as they are in C3H/HeN mice since bacteria lacking a smooth LPS phenotype or K1 capsule are less lethal in both of these mice, implying a greater facility in clearing these organisms. Thus, the LPS may work at two levels: the O chains affecting the fate of the organism early after invasion of the blood, and the lipid A portion affecting the ability of the host to later mount important defenses. The unique requirement for K1 encapsulation in this model is not entirely clear; however, the presence of E. coli inside lysosomes of phagocytes suggests that the K1 capsule simply may not be an anti-phagocytic determinant.

We conclude:

(1)   there are functionally significant differences in LPS and capsular phenotypes among E. coli isolated from different anatomic sites; therefore, care must be taken in selecting strains for study of pathogenesis appropriate to that site.

(2)   there is a heterogeneity in functional properties conferred to E. coli by specific O and K antigens such that associations between specific O and K antigens may have necessarily evolved. For example, the K1 capsule is uniquely associated with strains having a rough or part-rough phenotype.

(3)   susceptibility to rough-specific phages may correlate with the ability of terminal complement components to insert into and kill strains of E. coli. The behavior of E. coli with a part-rough phenotype may explain the previous paradox of apparently smooth strains of E. coli being serum sensitive.

(4)   both LPS and capsule are important in determining virulence. Further, the LPS O side chain and capsule may be important in the initial phase of infection where bacteriolytic and phagocytic mechanisms are most important. A second phase of infection that may determine the ultimate outcome is the host's ability to mount lipid A-inducible defense mechanisms.

References

1. Subbaiah, T.V. and Stocker, B.A.D, Rough mutants of Salmonella typhimurium. 1. Genetics. Nature London 201:1298-1299 (1964).
2. Cross, A.S., Gemshi, P., Sadoff, J.C., Orskov, F., and Orskov, I., The importance of the K1 capsule in invasive infections caused by Escherichia coli. J Infect Dis 149:184-193 (1984).
3. Gross, R.J., Cheasty, T., and Rowe, B., Isolation of bacteriophages specific for the K1 polysaccharide antigen of Escherichia coli. J Clin Microbiol 6:548-550 (1977).
4. Gemski, P., Cross, A.S., and Sadoff, J.C., K1 antigen-associated resistance to the bactericidal activity of serum. FEMS Microbiol Letts 9:193-197 (1980).
5. Cross, A.S., Kim, K.S., Wright, D.W., Sadoff, J.C., Gemski, P., Role of lipopolysaccharide and capsule in the serum resistance of bacteremic strains E. coli. J Infect Dis 154:497-503 (1986).
6. Kim, K.S., Kang, J.H., Cross, A.S., The role of capsular antigens in serum resistance and in vivo virulence of Escherichia coli. FEMS Microbiol Letts 35:275-278 (1986).
7. Taylor, P.W.and Robinson, M.D., Determinants that increase the serum resistance of Escherichia coli. Infect Immun 29:278-280 (1980).
8. Opal, S., Cross, A., and Gemski, P., K antigen and serum sensitivity of rough Escherichia coli. Infect Immun 37:956-960 (1982).
9. Vermeulen, C., Cross, A., Byrne, W.R., and Zollinger, W., Quantitative relationship between capsular content and killing of K1-encapsulated E. coli. Abstr B247, American Society for Microbiology, National Meeting, Atlanta (1987).
10. Welch, W.D., Martin, W.J., Stevens, P., and Young, L.S., Relative opsonic and protective activities of antibodies against K1, O and lipid A antigens of Escherichia coli. Scand J Infect Dis 11:291-301 (1979).
11. VanDijk, W.C., Verbrugh, H.A., van Erne-van der Tol, M.E., Peters, R., and Verhoef, J., Escherichia coli antibodies in opsonisation and protection against infection. J Med Microbiol 14:381-389 (1981).
12. Cross, A.S., Zollinger, W., Mandrell, R., Gemski, P., and Sadoff, J., Evaluation of immunotherapeutic approaches for the potential treatment of infections caused by K1-positive Escherichia coli. J Infect Dis 147:68-76 (1983).
13. Kirkland, T.N. and Ziegler, E.J., An immunoprotective monoclonal antibody to lipopolysaccharide. J Immunol 132:2590-2592 (1984).
14. Pluschke, G. and Achtman, M., Antibodies to O antigen of lipopoly saccharide are protective against neonatal infection with Escherichia coli K1. Infect Immun 49:365-370 (1985).
15. Kaufman, B.M., Cross, A.S., Futrovsky, S.L., Sidberry, H.F., and Sadoff, J.C., Monoclonal antibodies reactive with K1-encapsulated E. coli lipopolysaccharide are opsonic and protect mice against lethal challenge. Infect Immun 52:617-619 (1986).
16. Cross, A.S., Sadoff, J.C., Kaufman, B.M., and Zollinger, W., The use of monoclonal antibody combinations in the treatment of E. coli bacteremia. Abstr 124, 4th International Symposium on Infections in the Compromised Host, Ronneby Brunn, Sweden (1986).

17. Pluschke, G., Moll, A., Kusecek, B., Achtman, M., Sodium dodecyl sulfate-polyacrylamide gel electrophoresis and monoclonal antibodies as tools for the subgrouping of _Escherichia coli_ lipopolysaccharide 018 and 023 antigens. Infect Immun 51:286-293 (1986).

18. Cross, A.S., Gemski, P., Sadoff, J., Byrne, W., and Zollinger, W., Lipopolysaccharide as a virulence factor in LPS-hyporesponsive (C3H/HeJ) mice infected with K1-encapsulated _E. coli_. Abstr. 2151, FASEB National Meeting, Anaheim, California (1985).

LECTINOPHAGOCYTOSIS OF BACTERIA MEDIATED BY CARBOHYDRATE-LECTIN INTERACTIONS

Itzhak Ofek and Alex Perry

Department of Human Microbiology
Tel-Aviv University
Tel-Aviv   69978
Israel

During the last decade considerable evidence has accumulated showing that specific recognition between phagocytic cells and their targets may be accomplished by the interaction of carbohydrate binding proteins, i.e., lectins on the surface of one type of cell with complementary sugars on the surface of the other, in a lock and key manner. This type of recognition, which also leads to phagocytosis, has been termed by us *lectinophagocytosis*. In this review we summarize the evidence for such a mechanism and discuss its possible significance in the defense against bacterial infection.

Lectinophagocytosis of bacteria occurs in two major modes (figure 1). In the first mode, bacteria which carry surface lectins bind to complementary carbohydrates on the surface of the phagocytic cells. In the second mode, lectins that are integral components of the phagocytic cell membrane bind to carbohydrates on the bacterial surface.

Although other modes of lectinophagocytosis are possible, e.g., when a lectin forms a bridge between bacteria and phagocytic cells by binding to sugar residues exposed on both types of cell (1), these will not be discussed here.

Lectinophagocytosis mediated by bacterial surface lectins

Binding of many bacterial species to animal cells is mediated by lectin-like molecules that are present on the bacterial surfaces and are specific for sugar residues present on the animal cell (2). Most of these lectins, often called "adhesins" or agglutinins" are in the form of

NATO ASI Series, Vol. H24
Bacteria, Complement and the Phagocytic Cell
Edited by F. C. Cabello und C. Pruzzo
© Springer-Verlag Berlin Heidelberg 1988

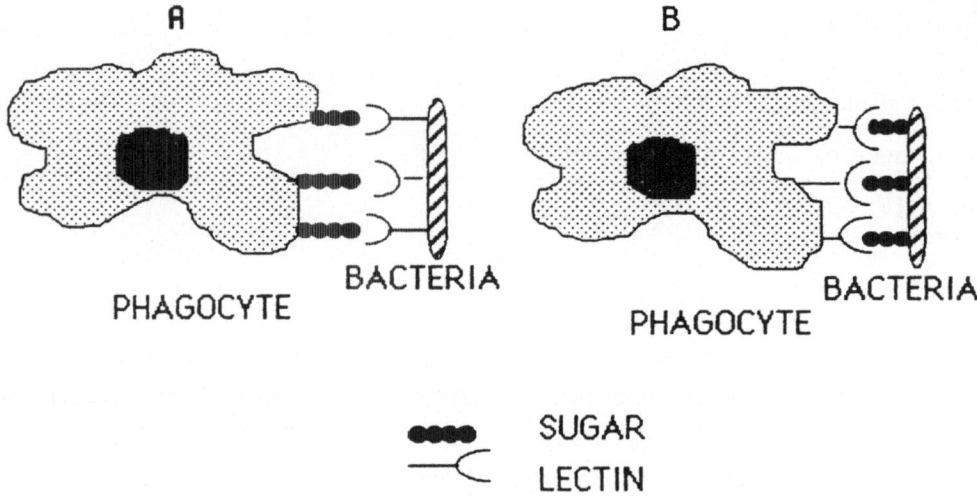

Fig. 1.   Two modes of recognition in lectinophagocytosis of bacteria. In mode A bacterial surface lectins bind corresponding sugars on the surface of the phagocyte. In mode B lectins on the surface of the phagocyte bind corresponding sugar residues on the bacterial surface.

fimbriae (or pili) and in some cases they can mediate binding of the bacteria to different types of phagocytic cells bearing sugar residues specific for the bacterial lectins. The fimbria is a heteropolymer composed of a major fimbrial subunit and minor distinct subunit(s) that either possess or that are required for sugar binding activity (3-7) with a possible location at the fimbrial tip (7).

The most thoroughly investigated lectinophagocytosis of E. coli is that mediated by the mannose specific (MS) lectin, associated with type 1 fimbriae. The evidence that the recognition of type 1 fimbriated E. coli by phagocytes is mediated by interaction of the fimbrial lectin with mannose containing glycoproteins on the surface of the phagocyte is based on several lines of evidence: (a) the specificity pattern of inhibition of bacteria-host cell interaction observed is the same for phagocytic and non-phagocytic target cells (8,9); (b) whenever sugars other than mannose were examined, they inhibited poorly, if at all, the interaction of type 1 fimbriated E. coli with mouse peritoneal macrophages or human polymorphonuclear leukocytes (PMNL) (10,11); (c) a very good correlation was found between the mannose binding activity of the bacteria and the extent of their attachment to mouse peritoneal macrophages (12); (d) the finding that pretreatment of type 1 fimbriated bacteria with yeast mannan inhibited their attachment to mouse and human phagocytes, whereas pretreatment of the phagocytes did not have such an effect (10), shows that the receptor for the bacterial lectin is on the surface of the phagocytes, and (e) latex particles coated with purified type 1 fimbriae stimulated human PMNL and this activity was inhibited by D-mannose (13).

Since the only class of mannose-containing compounds in animal membranes are glycoproteins (14), the receptor for mannose specific bacteria must belong to this class.

The MS lectin-mediated binding to phagocytic cells leads to ingestion, stimulation of antimicrobial systems (e.g., oxygen burst, degranulation) and killing of the bacteria (Table 1).

All of the activities listed in Table 1 are initiated by specific interaction of the MS fimbrial lectin with mannose-containing receptors on the phagocytic cell. This is confirmed from experiments showing that (a) methyl-α mannoside (Me-Man) inhibits activation of phagocytes by MS fimbriated E. coli; (b) the sugar does not inhibit activation of phagocytes induced by opsonized bacteria; (c) non-fimbriated bacteria do not induce activation of phagocytes; (d) latex particles coated with MS fimbriae stimulate high phagocytic activities that are inhibited by Me-Man (13).

Maximum stimulation of antimicrobial systems in phagocytes appeared to require cross-linking of the MS fimbriae on the bacterial surfaces. Such cross linking may cause aggregation of the receptors on the phagocytic cells (20, 22). Aggregation of receptors is important for many other membrane initiated events.

The phagocytic receptors mediating MS lectinophagocytosis were recently isolated from membranes of human PMNL on a column of immobilized type 1 fimbriae (24). Three surface glycoproteins were isolated. One of these glycoproteins is highly glycosylated, reacts strongly with Concanavalin A, and has a molecular weight of 150,000 daltons. It may be related to the Mac.1 glycoprotein which is the receptor for the C3bi fragment of complement, also known as CR3 (24).

Lectinophagocytosis mediated by macrophage surface lectins

The second mechanism of lectinophagocytosis involves recognition of sugar residues on the bacterial surfaces by lectins that are integral components of the phagocytic cell membrane.

Evidence for the presence of lectins on phagocytic cells began in the late 1960's with the work of Ashwell and Morell who studied blood clearance of asialoglycoproteins, mediated by receptors on the surfaces of liver cells (hepatocyte and Kupffer cells). At least three types of lectins expressed on the surfaces of liver cells and on tissue macrophages were isolated and characterized (25,26). All three are glycoproteins. The GAL type lectin (or asialoglycoprotein receptor) is found on Kupffer cells, hepatocytes and subpopulations of peritoneal macrophages. The human hepatic receptor appears to be a single polypeptide with a molecular weight of 41,000 (27) or 46,000 daltons (28) as determined by SDS-PAGE. The GAL type lectin is specific for N-acetylgalactosamine and galactose. The MAN type lectin (or GlcNAc/Man lectin) is found on tissue macrophages (e.g., Kupffer cells and alveolar macrophages), it has a molecular weight of 175,000 daltons (29) and is specific for mannose, N-acetylglucosamine, glucose and fucose. The FUC type lectin is found on Kupffer cells, it contains two subunits of 88,000 and 77,000 daltons (30) and is specific for L-fucose.

Table 1 Stages of phagocytosis of E. coli mediated by mannose specific
type 1 fimbriae

Phagocytic stages	Temp	Phagocytic cell	Assay system	Reference
Attachment	4°	Macrophage, mouse	CFU	Pocino (pers. communication)
Ingestion	37°	PMNL, human	EM	11, 15
			FITC-E. coli	16, 17
		Macrophage, mouse	FITC-anti E. coli	12
Stimulation	37°C	PMNL, human	O_2 consumption	13, 15
			Chemiluminescence	18, 19, 20
			lysozyme release protein iodination	18, 21 22
		Macrophage, rat	chemiluminescence	23
Killing	37°	PMNL, human	CFU	11, 16
		Macrophage, mouse	CFU	Pocino et al. (pers. communication)
		Macrophage, human	CFU	Boner et al. (pers. communication)

EM: Electron microscopy, FITC: Fluoresceine isothiocyanate, CFU: colony
forming units, PMNL: polymorphonuclear leukocytes.

These lectins were found to serve as receptors for serum asialoglycoproteins, various glycoproteins and perhaps asialoerythrocytes and thus, possess a physiologic role in clearance of these elements from the blood.

The idea that bacterial surface sugars are important for non-opsonic binding to macrophages was first suggested by Ogmundsdottir and Weir (31) and by Freimer et al. (32). However, the exact sugar specificities of these interactions were not defined and the authors did not refer to any of the macrophage lectins as being involved in phagocytosis of a particular strain of bacteria. Warr (33) was the first to implicate the involvement of the MAN type lectin of alveolar macrophages in the binding of mannan- containing yeast cells.

Recently, we studied the blood clearance, in mice, of gram negative (non-fimbriated E. coli) and gram positive (Group B Streptococci (GBS)) bacteria which resisted killing by whole blood of mice in vitro and were sequestered in the liver after intraveneous injection (34,35). The E. coli strain employed agglutinated with Concanavalin A but not with wheat germ, peanut and ricinus communis lectins, indicating that glucose or mannose residues, or both, were exposed on the bacterial surfaces. The blood clearance of the bacteria in mice was strongly inhibited by derivatives of D-mannose, D-glucose or L-fucose, but not of D-galactose or L-rhamnose (Tabel 2) suggesting that the blood clearance of the E. coli was mediated by the MAN type receptor of liver phagocytes (e.g., Kupffer cells). In contrast, blood clearance of antibody-coated E. coli was not inhibited by D-mannose (Table 2).

To investigate the involvement of the GAL type lectin in the blood clearance of bacteria we employed a strain of type Ib GBS. The structure of the type Ib capsular polysaccharide is composed of repeating units in which all the side chain β-D-galactose residues are masked by terminal sialic acid residues (36). Removal of the sialic acid with neuraminidase results in exposure of the galactose residues. Indeed, recinus communis agglutinin, a galactose-specific lectin, reacted only with neuraminidase treated bacteria. The blood clearance of desialylated GBS, which had terminal galactose residues exposed on their surface, was strongly inhibited by galactosylated, but not by mannosylated, or fucosylated BSA derivates (Table 2), suggesting the involvement of the GAL type lectin of liver cells in the blood clearance of these bacteria.

In another set of experiments the blood clearance of type II GBS, whose surface sugar structure contains both galactose and glucose residues (36), was investigated. Strong inhibition of blood clearance was obtained only with a combination of galactosyl-BSA and mannosyl-BSA but not with each neoglycoprotein alone (Table 2), indicating that either one of the liver macrophage lectins (i.e., the GAL type or the MAN type) can mediate blood clearance of the bacteria which have exposed residues of galactose and glucose.

To obtain more direct evidence that lectins on the surface of macrophages mediate binding of bacteria, we studied the attachment of type II GBS, expressing both galactose and mannose residues on their surface, to a monolayer of mouse thioglycolate-elicited peritoneal macrophages. As with blood clearance of these organisms, only the combination of Gal-BSA and Man-BSA or Gal-BSA and Glc-BSA strongly inhibited the attachment of the bacteria to the macrophage monolayer (Table 2). The specificity of the inhibition of attachment by the mixture of neoglycoproteins is further shown by the inability of these compounds to inhibit attachment to phagocytes of bacteria coated with antibodies (Table 2). The involvement of liver lectins in binding of bacterial surface sugars was further demonstrated by co-aggregation experiments in which mouse liver homogenates induced agglutination of E. coli and this agglutination was inhibited by the same sugars that inhibited the blood clearance of the bacteria, as well as by lipopolysaccharide extracted from the E. coli employed (34).

In summary, several lines of evidence implicate the interactions of macrophage and liver cell lectins with bacterial surface sugars as mechanisms for blood clearance and bacteria-macrophage attachment in mice. First, the pattern of sugar derivatives and neoglycoproteins inhibiting blood clearance and attachment of the bacteria to macrophages corresponded to the sugar residues exposed on the surface of the organisms and to the sugar specificity of the type of liver or macrophage lectin involved

Table 2   Inhibition of blood clearance and phagocytic attachment of E. coli
and GBS by sugars and neoglycoproteins

| Inhibitor | Percent inhibition of blood clearance in mice of: | | | Percent inhibition of attachment of type II GBS to elicited mouse peritoneal macrophages[b] |
	E. coli[a]	type Ib GBS[b]	type II GBS[b] (desialylated)	
Gal	<20	>70	<20	23
Man	>70 (<20)[c]	<20	<20	44
Fuc	>70	<20	<20	ND
Glc	>70	ND	ND	40
Gal+Glc	ND	ND	ND	71 (0.5)[c]
Gal+Man	ND	ND	>70	67

Gal: methyl α-galactoside or galactosyl-bovine serum albumin (BSA) Man:
methyl α-mannoside or mannosyl-BSA. Fuc: methyl α-fucoside or fucosyl-BSA.
Glc: methyl α-glucoside or glucosyl-BSA. GBS: group B streptococci. ND:
not done.
Inhibition of blood clearance of E. coli was measured with the
corresponding methyl glycoside and inhibition of blood clearance and
attachment to macrophages of GBS was measured with the BSA derivatives. a
and b: data adopted from Perry and Ofek (34) and Perry et al. (35),
respectively. c: values in brackets were obtained with bacteria opsonized
with specific antibody. The degree of attachment of opsonized GBS (type
II) was 7 times higher than that of non-opsonized GBS.

(34,35). Second, the inhibitory sugars or glycoproteins did not inhibit
the blood clearance or attachment to macrophages of bacteria precoated
with antibodies, which modify the bacterial surface, and hence the
mechanisms of attachment (34,35), and third, mouse liver preparations
induced agglutination of E. coli and this agglutination was inhibited by
the same sugars that inhibited the blood clearance of the bacteria, as
well as by lipopolysaccharide extracted from the bacterial strain employed
(34).

The galactose- and mannose-mediated attachment of the type II GBS to the thioglycolate-elicited peritoneal macrophages was 7-fold lower than that mediated by the Fc portion of antibody coating the bacteria (35). This could be attributed to either a low proportion of cells expressing the GAL and MAN type lectins (10-20 percent) as compared to that expressing Fc receptors, or to the fact that the efficiency of attachment to elicited macrophages via the MAN and GAL type lectins is much lower than that via the Fc receptor. Furthermore, in the elicited macrophage population there may be macrophages, each expressing a different type of surface lectin. The additive effect of inhibition of attachment of the GBS obtained by neoglycoproteins tends to favour the latter possibility, but further experiments are needed to clarify these points. It is now clear that resident macrophages are heterogeneous with respect to phagocytic ability (37) and the expression of the macrophage lectins, especially the MAN type, is dependent on the degree of differentiation of the mononuclear phagocytic cell (38,39).

As far as the _in vivo_ situation is concerned, it appears that the population of macrophages in the liver is sufficient to eliminate most of the non-opsonized type II GBS from the blood via one type of lectin since only the combination of two types of neoglycoproteins inhibited blood clearance (Table 2).

These results, taken together with those of other investigators (31-33) suggest that macrophage surface lectins may bind sugars on the surface of microorganisms and, thus, mediate phagocytosis.

Although there is no direct proof yet that binding of bacteria to macrophage lectins is followed by ingestion and killing of the organisms, recent studies show clearly that both the GAL type (40,41) and the MAN type (38,42) lectins are capable of mediating internalization of particles coated with sugars or glycoproteins. The promastigotes of _Leishmania_ sp. bind and penetrate peritoneal macrophages _in vivo_ by interaction of sugars on the surface of the parasite with the MAN type lectin on the surface of the phagocyte (43,44). The data suggest that the macrophage lectins are capable of functioning as receptors for lectinophagocytosis both _in vitro_ and _in vivo_.

Conclusions

The major conclusion arising from the evidence presented here is that one of the non-opsonic mechanisms of phagocytosis is that mediated by lectin-carbohydrate interaction, and that this mechanism may function in vivo. Thus, there appear to be three principal receptor-mediated mechanisms that are responsible for recognition, binding and in vivo clearance of microorganisms (and foreign particles) by phagocytes: (i) lectin-carbohydrate interactions between integral constituents of the cell surfaces; (ii) bridging via IgG molecules, and (iii) bridging via C3b molecules.

The phagocytic process initiated by the first mechanism was termed "lectinophagocytosis" to distinguish it from the two other mechanisms that require serum opsonins as bridging molecules, and have been termed "opsonophagocytosis". It is possible, however, that other receptor-mediated forces also take place, as many bacterial species can undergo phagocytosis in the absence of opsonins. The term "interactions of unknown mechanism", rather than "non-specific", may be applied, therefore, to such non-opsonic phagocytosis until the surface molecules responsible for cell-cell binding are identified.

Since amebae recognize and ingest bacteria via lectin-carbohydrate interactions (45), the lectinophagocytosis of micororganisms by mammalian phagocytes may represent a conservation of a primitive system of host defense which acts against many saprophytic or opportunistic microorganisms encountered by humans and animals. It is especially important in defense against bacteria which evade opsonophagocytosis. For example, bacteria that do not activate the alternate pathway of complement, or bacteria invading serum-poor sites or a complement-deficient host. However, bacteria may escape lectinophagocytosis as well. For example, the phenomenon of phase variation in the expression of fimbrial lectin during the infectious process was described in a number of bacterial species expressing MS fimbrial lectin (46-48). Also, lectinophagocytosis may be inhibited by host factors. For example, lectinophagocytosis of MS E. coli is inhibited by Tamm-Horsfall, a mannose containing glycoprotein found in the urinary tract (49), and blood clearance of bacteria by the MAN type lectin of liver macrophages may be blocked by high concentrations of glucose in the blood, such as those present in diabetes. Thus, high local concentrations of soluble glycoconjugates bearing sugars specific for the cell-associated lectin may render a host susceptible to infection by bacteria whose phagocytic clearance is dependent on this lectin.

Opsonophagocytosis may have developed in mammals as a potent defense mechanism to protect against infections caused by bacteria which survived throughout evolution as ones that evaded lectinophagocytosis. Consequently, opsono- and lectinophagocytosis seem to be functioning in mammals in a redundancy to provide efficient protection against invading microorganisms.

Further studies on the lectinophagocytosis process _in vitro_ and _in vivo_ should enable us to define the conditions which render a host susceptible to certain bacterial infections on one hand, and to have a better understanding of the mechanisms through which organisms evade lectinophagocytosis on the other. Such studies may provide a more knowledgable approach to prevent or treat bacterial and perhaps other microbial infections.

References

1. R. Gallily, B. Vray, I. Stain, and N. Sharon, Wheat germ agglutinin potentiates uptake of bacteria by murine peritoneal macrophages. Immunol. 52:679-686 (1984).
2. D. Mirelman and I. Ofek, Microbial lectins and agglutinins (D. Mirelman, ed.) J Wiley and Sons, New York, pg. 1-19 (1986).
3. B. E. Uhlin, M. Norgen, M. Baga, and S. Normak, Adhesion to human cells by Escherichia coli lacking the major subunit of a oligolactoside-specific pilus adhesion. Proc. Natl. Acad. Sci. USA 82:1800-1804 (1985).
4. M. Norgren, S. Normark, D. Lark, P. O'Hanley, G. Schoolnik, S. Falkow, C. Svanborg-Eden, M. Baga, and B. E. Uhlin, Mutations in E. coli cistrons affecting adhesion to human cells do not abolish Pap pili fiber formation. EMBO J. 3:1159-1165 (1984).
5. L. Maurer and P. E. Orndorff, A new locus, pil E required for the binding of type 1 piliated Escherichia coli to erythrocytes. FEMS Microbiol. Lett. 30:59-66 (1985).

6. F. C. Minion, S. N. Abraham, E. H. Beachey, and J. D. Goguen, The genetic determinant of adhesive function in type 1 fimbriae of Escherichia coli is distinct from the gene encoding the fimbrail subunit. J. Bacteriol. 165:1033-1036 (1986).

7. T. Moch, H. Hoschirtzky, J. Hacker, K. D. Kroncke, and K. Jann, Isolation and characterization of the α-sialyl-β-2, 3 galactosyl (s-) specific adhesin from fimbriated Escherichia coli. Proc. Natl. Acad. Sci. (in press, 1987).

8. N. Firon, I. Ofek, and N. Sharon, Carbohydrate specificity of the surface lectins of Escherichia coli, Klebsiella pneumoniae and Salmonella typhimurium. Carbohydr. Res. 120:235-249 (1983).

9. N. Firon, D. Duksin, and N. Sharon, Mannose specific adherence of Escherichia coli to BHK cells that differ in their glycosylation patterns. FEMS Microbiol. Lett. 27:161-165 (1985).

10. Z. Bar-Shavit, I. Ofek, R. Goldman, D. Mirelman, and N. Sharon, Mannose residues on phagocytes as receptors for the attachment of Escherichia coli and Salmonella typhi. Biochem. Biophys. Res. Commun. 78:455-460 (1977).

11. F. J. Silverblatt, J. S. Dreyer, and S. Schauer, Effect of pili on susceptibility of Escherichia coli to phagocytosis. Infect. Immun. 24:218-223 (1979).

12. Z. Bar-Shavit, R. Goldman, I. Ofek, N. Sharon, and D. Mirelman, Mannose-binding activity of Escherichia coli: a determinant of attachment and ingestion of the bacteria by macrophages. Infect. Immun. 29:417-424 (1980).

13. M. B. Goetz and F. J. Silverblatt, Simulation of human polymorphonuclear leukocyte oxidative metabolism by type 1 pili from Escherichia coli. Infect. Immun. 55:534-540 (1987).

14. H. Ranvala and J. Finne, Structural similarity of the terminal carbohydrate sequence of glycoproteins and glycolipids. FEBS Lett 97:1-8 (1979).

15. G. Rottini, F. Cian, H. R. Soranzy, R. Albriga, and P. Patriarca, Evidence for the involvement of human polymorphonuclear leukocyte mannose-like receptors in the phagocytosis. Infect. Immun. 24:218-223 (1979).

16. L. Ohman, J. Hed, and O. Stendahl, Interaction between human polymorphonuclear leukocytes and two different strains of type 1 fimbriae-bearing Escherichia coli. J. Infect. Dis. 146:751-757 (1982).

17. L. Ohman, K. E. Magnusson, and O. Stendahl, Mannose-specific and hydrophobic interaction between Escherichia coli and polymorphonuclear leukocytes-influence of bacterial culture period. Acta Path. Microbiol. Immunol. Scand. Sect. B 93:125-131 (1985).

18. D. F. Mangan and J. S. Synder, Mannose-sensitive interactions of Escherichia coli with human peripheral leukocytes in vitro. Infect. Immun. 26:520-527 (1979).

19. B. Bjorkten and T. Wadstrom, Interaction of Escherichia coli with different fimbriae and polymorphonuclear leukocytes. Infect. Immun. 38:298-305 (1982).

20. T. Soderstrom and L. Ohman, The effect of monoclonal antibodies against Escherichia coli type 1 pili and capsular polysaccharides on the interaction between bacteria and human granulocytes. Scand. J. Immunol. 20:299-305 (1984).

21. C. F. Svanborg-Eden, L M. Bjursten, R. Hull, S. Hull, K. E. Magnusson, Z. Meldoveno, and H. Leffler, Influence of adhesins on the interaction of Escherichia coli with human phagocytes. Infect. Immun. 44:672-680 (1984).

22. A. Perry, I. Ofek, and F. J. Silverblatt, Enhancement of mannose-mediated stimulation of human granulocytes by type 1 fimbriae aggregated with antibodies on Escherichia coli surfaces. Infect. Immun. 39:1334-1345 (1983).

23. E. Blumenstock and K. Jann, Adhesion of pilated Escherichia coli strains to phagocytes. Differences between bacteria with mannose-sensitive pili and those with mannose-resistant pili. Infect. Immun. 35:264-269 (1982).

24. M. Rodriguez-Ortega, I. Ofek, and N. Sharon, Membrane glycoproteins of human polymorphonuclear leukocytes that act as receptors for mannose-specific Escherichia coli. Infect. Immun. 55:968-973 (1987).

25. G. Ashwell and A. G. Morell, The role of surface carbohydrates in the hepatic recognition and transport of circulating glycoproteins. In: A. Meiser (ed) Adv. Enzymology 99-128 (1974).

26. G. Ashwell and H. Harford, Carbohydrate-specific receptors of the liver. Annu. Rev. Biochem. 51:531-554 (1982).

27. J. N. Baenziger and Y. Maynard, Human hepatic lectin. J. Biol. Chem. 255:4607-4613 (1980).

28. A. L. Schwartz and D. Rup, Biosynthesis of the human asiaologlycoprotein receptor. J. Biol. Chem. 258:11249-11252 (1983).

29. T. E. Wileman, M. R. Lennazitz, and P. D. Stahl, Identification of the macrophage mannose receptor as a 175-KDa membrane protein. Proc. Natl. Acad. Sci. USA 83:2501-2505 (1986).

30. A. M. Lehrman and R. L. Hill, The binding of fucose-containing glycoproteins by hepatic lectins. Purification of a fucose-binding lectin from rat liver. J. Biol. Chem. 16:7419-7425 (1986).

31. H. M. Ogmundsdottir and D. M. Weir, The characteristics of binding of Corynebacterium parvum to glass-adherent mouse peritoneal exudate cells. Clin. Exp. Immunol. 26:334-339 (1976).

32. I. W. Sutherland, L. Graham, and D. M. Weir, The role of cell wall carbohydrates in binding of microorganisms to mouse peritoneal exudate macrophages. Acta Path. Microbiol. Scand. Sect. B 86:53-57 (1978).

33. G. A. Warr, A macrophage receptor (mannose/glucosamine) - glycoproteins of potential importance in phagocytic activity. Biochem. Biophys. Res. Commun. 93:737-745 (1980).

34. A. Perry and I. Ofek, Inhibition of blood clearance and hepatic tissue binding of Escherichia coli by liver lectin-specific sugars and glycoproteins. Infect. Immun. 43:257-262 (1984).

35. A. Perry, Y. Keisari, and I. Ofek, Liver cell and macrophage surface lectins as determinants of recognition in blood clearance and cellular attachment of bacteria. FEMS Microbiol. Lett. 27:345-350 (1985).

36. D. L. Kasper, C. J. Baker, B. Galdes, and B. E. Katzenellenbogen, Immunochemical analysis and immunogenicity of the type II group B streptococcal capsular polysaccharide. J. Clin. Invest. 72:260-269 (1983).

37. D. B. Chandler, J. J. Kennedy, and J.D. Fulmer, Studies of membrane receptors, phagocytosis, and morphology of subpopulations of rat lung interstitial macrophages. Am. Rev. Respir. Dis. 134:542-547 (1986).

38. M. Kataoka and M. Tavassoli, Development of specific surface receptors recognizing mannose-terminal glycoconjugates in cultured monocytes: a possible early marker for differentiation of monocytes into macrophage. Exp. Hematol. 13:44-50 (1985).

39. V. L. Shepherd, E. J. Campbell, R. M. Senior, and P. D. Stahl, Characterization of the mannose/fucose receptor on human mononuclear phagocytes. J. Ret. Soc. 32:423-431 (1982).

40. V. Kolb-Bachofen, J. Schlepper-Schaffer, P. Roos, D. Hulsmann, and H. Kolb, GalNAc/Gal-specific rat liver lectins: their role in cellular recognition. Biol. Cell. 51:219-226 (1984).

41. J. Schlepper-Schafer, D. Hulsmann, A. Djovkar, H. E. Meyer, L. Herbertz, H. Kolb, and V. Kolb-Bachofen, Endocytosis via galactose receptors in vivo. Ligand size directs uptake by hepatocytes and/or liver macrophages. Exp. Cell. Res. 165:494-506 (1986).

42. S. J. Sung, R. S. Nelson, and S. C. Silverstein, Yeast mannans inhibit binding and phagocytosis of zymosan by mouse peritoneal macrophages. J. Cell. Biol. 96:160-166 (1983).

43. K. P. Chang, Leishmania donovani-macrophage binding mediated by surface glycoproteins/antigens: characterization in vitro by a radioisotopic assay. Molecular and Biochem. Parasitol. 4:67-76 (1981).

44. E. Handman, J. W. Goding, The Leishmania receptor for macrophages is a lipid-containing glycoconjugate. EMBO J. 4:329-336 (1985)

45. D. Mirelman, Ameba-bacteria relationship in amoebiasis. Microbiol. Rev. (June 1987 issue, in press).

46. I. Ofek and F. J. Silverblatt, Bacterial surface structures involved in adhesion to phagocytic and epithelial cells. In: Microbiology (D Schlesinger ed.) ASM Publications, Washington D.C. pg. 296-300 (1982).

47. A. J. Schaeffer, W. R. Schwan, S. J. Hultgren, and J. L. Duncan, Relationship of type 1 pilus expression in Escherichia coli to ascending urinary tract infection. Infect. Immun. 55:373-380 (1987).

48. N. G. Guerina, T. W. Kessler, V. J. Guerina, M. R. Neutra, H. W. Clegg, S. Langermann, F. A. Scannapicco, and D. A. Goldman, The role of pili and capsule in the pathogenesis of neonatal infection with Escherichia coli K1. J. Infect. Dis. 148:395-405 (1983).

49. S. M. Kuriyama and F. J. Silverblatt, Effect of Tamm-Horsfall urinary glycoprotein on phagocytosis and killing of type 1-fimbriated Escherichia coli. Infect. Immun. 51:193-198 (1986).

RESPONSES OF PHAGOCYTOSIS TO BACTERIA

Shiro Kanegasaki

Institute of Medical Science
University of Tokyo
Shirokanedai 4-6-1
Minatoku
Tokyo 108
Japan

Requirement of physical impact for induction of efficient responses of phagocytes (1-3)

One of the earliest responses of host upon bacterial infection may be mobilization of phagocytes such as neutrophils and macrophages to the site of infection. These phagocytes are armed well in various ways and defend the body against infectious agents. If this forefront is broken through, an SOS message passes to the immune systems. The reinforcements, which make the work of the phagocytes more efficient, are sent back from the systems to the front. Whether the phagocytes respond to the infectious agents well or not is, therefore, firstly important for host defense against infections.

Many physical and chemical properties of potentially ingested particles as well as various opsonins are known to influence response of phagocytes. Beside opsonins and adhesive properties of particles, there is another important factor which influences the interaction between particles and phagocytes: i.e., we found that a certain level of physical impact-force is required to induce efficient responses of phagocytes. This is like when one closes up an envelope. One spreads glue on the flap, folds it and presses it on to the envelope to seal it. The glue corresponds to opsonins. Like the flap of an adhesive envelope, certain bacteria have surface nature ready to bind to the cells. In any case, one must press the glued flap of the envelope to close it. Similarly, to induce efficient responses of phagocytes, bacteria or particles must be pressed to the cells with a certain kind of forces.

NATO ASI Series, Vol. H24
Bacteria, Complement and the Phagocytic Cell
Edited by F.C. Cabello und C. Pruzzo
© Springer-Verlag Berlin Heidelberg 1988

To demonstrate this fact, we have employed centrifugation. If monolayered macrophages (obtained from the mouse peritoneal cavity) were centrifuged together with bacterial suspension, a dramatic increase was observed in the number of bacteria ingested per single macrophage. The increase did not take place in the presence of cytochalasin B or at low temperature. The number of serum-opsonized or non-opsonized bacteria ingested increased linearly with increments of centrifugal forces and the increase became gradual. The second gradual increase above the threshold velocity is considered to be caused by increased collision frequency between macrophages and bacteria which increases proportionally as the sedimentation velocity increases.

Under the conditions where collision frequency between bacteria and macrophages was constant for different centrifugal forces tested (namely, the number of bacteria added was reduced in inverse proportion with increased centrifugal forces), the number of bacteria ingested per single cell increased linearly up to 5 xg and then became constant. The second gradual increase was no longer observed. The results indicate that the threshold force which induced the maximum response of phagocytes is 5 xg whether the particles were opsonized or not.

Sometimes, bacterial motility satisfies the impact force since motile bacteria can collide with phagocytes at least 10 times faster than at the rate required, but phagocytes themselves move much more slowly. How fast bacteria must collide with phagocytes can be calculated by using the equation shown below.

$$V = 2R^2 (\rho - \rho_0) F / 9\eta$$

The equation is based on Stokes' law where v is the sedimentation velocity, R is the radius of the bacterium, rho is the density of the bacterium, rho-zero is the density of medium at 37^0C, ita is the viscosity of medium at 37^0C and F is the centrifugal force. Supposing the radius of bacteria is 1 micrometer, the 2.5 sedimentation velocity of centrifuged bacteria 5 xg is calculated to be 2.5 micrometer per second if we employ a known dimension for rho (1.15 xg/cm^3), rho-zero (0.994 g/cm^3) ita (0.00694 poise). Therefore, the bacteria which collide against a phagocyte surface with velocity above 2.5 micrometer per second, should induce full responses of phagocytes. The velocity of movement of motile <u>Salmonella</u> <u>typhimurium</u>, <u>Escherichia</u> <u>coli</u> and <u>Pseudomonas</u> <u>aeruginosa</u> was within the range of 20 to 50 micrometers per second, which was far greater than the required value.

In our experiments, we compared susceptibility of parent, motility mutant and revertant strains of <u>Salmonella</u> <u>typhimurium</u> to phagocytosis by mouse peritoneal macrophages. The bacterial strains used were isogenic and the mutant bacteria did not move although they had flagella. The motility mutant was phagocytized far less efficiently than its parent and revertant strains. We observed the similar different response of phagocytes to the motility mutant and the parent strains of <u>Escherichia</u> <u>coli</u> and <u>Pseudomonas</u> <u>aeruginosa</u>.

Furthermore, a higher phagocytic response of macrophages to motile bacteria was also observed in mouse peritoneal cavities. In this experiment, a certain number of bacteria were injected into the cavity and after 15 min, macrophages were washed out from the cavity and the number of bacteria ingested were counted. The number of bacteria phagocytized was much higher when motile bacteria instead of non-motile bacteria were used.

Physical impact of bacteria to phagocytes enhances not only phagocytosis but also some other metabolic functions of the cells, such as chemiluminescent response. Chemiluminescence is a strong indication of active oxygen generation by the cells during phagocytosis. Much lower responses were induced by non-motile and UV-killed strains as compared with those induced by the motile parental. Moreover, great difference between living and killed bacteria for induction of chemiluminescence was only observed with motile bacteria such as <u>Salmonella</u>, <u>Escherichia</u>, <u>Pseudomonas</u> and <u>Enterobacter</u> but not with non-motile bacteria including <u>Shigella</u>, <u>Klebsiella</u> and <u>Propionibacterium</u>.

Only a slight increase in chemiluminescent response and practically no increase in phagocytic response could be observed upon exposure to quite large numbers of non-motile Salmonella. The level of the responses never reached those induced by 10 motile bacteria per cell, even at a 200 times or greater number of bacteria. All these results indicate the requirement of physical impact for induction of efficient responses of phagocytes into bacteria and other particles.

Quantitative analysis of bacterial binding sites of macrophage and their turnover rate (4)

Under the conditions where bacteria collide with phagocytes stronger than the threshold velocity required, we studied a number of binding sites of non-opsonized <u>Salmonella</u> and those opsonized with antiserum or normal serum on mouse peritoneal macrophages at 4^0C. The number increased with an increased number of any bacteria added. We plotted ratios of free and attached bacterial numbers on the vertical axis and bacterial numbers attached to a single macrophage on the horizontal axis. The plotting gave straight lines, indicating that affinities between bacteria and macrophage were constant whether or not the bacteria were opsonized by antiserum or normal serum. The inclines of the lines show how tightly bacteria attach to macrophages and the data indicated that opsonized bacteria (by antiserum and normal serum as well) were more tightly bound to the cells than non-opsonized bacteria. The intercepts on vertical axis exhibit maximum numbers of bacteria which could attach to a single macrophage. The results are summarized in Table 1.

Table 1: Number of bacterial binding sites and their turnover rate

Opsonized with	Number of bacteria attached or ingested at		Ratio (37^0C/4^0C)
	4^0C	37^0C	
None	3.0	15	5.0
Normal serum	9.5	44	4.6
Antiserum	8.5	70	8.2

Under similar conditions, we studied the number of opsonized and non-opsonized <u>Salmonella</u> ingested by macrophages at 37°C. Scatchard plotting may be applicable to this case also and we plotted ratios of the number of free and ingested bacteria against the number of bacteria ingested by a single macrophage. The plotting gave straight lines, suggesting that phagocytic activity of macrophages was constant whether or not the bacteria were opsonized by antiserum or normal serum. The inclines of the lines show how efficiently macrophage can ingest bacteria and the data indicated that opsonization by antiserum enhances phagocytosis most efficiently. The intercepts on the vertical axis shows maximum numbers of bacteria ingested by a single macrophage and the results are shown in Table 1. Both normal and antiserum enhanced binding but only antiserum enhanced efficiency of phagocytosis. The results are consistent with the view that the complement receptors mediate only bacterial binding to the macrophages while IgG receptors mediate both attachment and ingestion. These results were obtained by using resident macrophages.

Oxygen metabolism of phagocytes during phagocytosis (5)

Oxygen metabolism, so-called the "respiratory burst", is activated during phagocytosis. One can recognize this change by increased oxygen consumption, production of active oxygen such as superoxide anion and hydrogen peroxide as well as enhanced glucose metabolism through the glucose-monophosphate-shunt. Since the electron used for the reduction of molecular oxygen is supplied from NADPH, the enzyme system is designated as NADPH oxidase. A defect in this special function such as in chronic granulomatous disease results in severe recurrent infections of bacteria and fungi, indicating that active oxygen plays an essential role in microbicidal activity of the cells.

We recently established that the initial biochemical event of the respiratory burst of phagocytes is a conversion of molecular oxygen to superoxide anion. Hydrogen peroxide is formed within the phagosomes. The difficulties in the determination of the initial product of oxygen metabolism came from the fact that superoxide generates molecular oxygen and hydrogen peroxide through either enzymatic or non enzymatic reactions and further reactions among these species yield other active oxygen species including hydroxy radical. Although superoxide anion has been supposed to be a primary product, a considerable amount of hydrogen peroxide is always observed and there was no definite evidence that hydrogen peroxide was not formed directly from molecular oxygen by NADPH oxidase.

The most popular method for the determination of superoxide anion is to measure reduction of cytochrome c given into the reaction mixture. The reaction gives back molecular oxygen to the system and even when oxygen consumption can be measured separately by using an oxygen electrode, the values must be corrected for the amounts of the molecular oxygen evolved from dismutation of superoxide and from decomposition of hydrogen peroxide. Therefore, the stoichiometry between oxygen consumption and formation of the reactive oxygen species has never been established. We cannot distinguish whether hydrogen peroxide is produced directly by the cells or derived from superoxide by spontaneous dismutation.

To cirvumvent these difficulties, we employed horseradish peroxidase as a trapping agent for both superoxide and hydrogen peroxide. The peroxidase reacts stoichiometrically with superoxide to form compound III and with hydrogen peroxide to form compound II, both compounds can be distinguished from each other by their spectra. Thus, either of the oxygen metabolites can be trapped in a form of respective compound without giving back oxygen to the system. Hence, their conversion to other reactive oxygen species is prohibited. The main problem was the stability of the compounds. This was overcome by substitution of the protoheme of the enzyme by diacetyldeuteroheme. The heme-substituted enzyme forms a large quantity of stable compound II and III by reacting with hydrogen peroxide and superoxide anion, respectively, and, once formed, the compounds hardly dissociate to regenerate oxygen derivatives.

When neutrophils (obtained from peripheral blood) were stimulated by phorbol myristate acetate, the spectrum of the ferric form of the heme-substituted peroxidase present in the reaction mixture changed into another spectral species, giving a set of isosbestic points at 480, 530 and 605 nm. The new spectra is characteristic of compound III, and compounds I and II were not detected. The rates of oxygen uptake and compound III formation were one to one. The results indicate that superoxide anion but not hydrogen peroxide is produced by neutrophils stimulated by the phorbol ester.

When an excess amount of superoxide dismutase, which converts superoxide into hydrogen peroxide, was present in the reaction mixture, the formation of compound III was completely inhibited and compound II, instead, was produced, giving clear isosbestic points, different from compound III. The rate of oxygen consumption decreased compared to that in the absence of the enzyme and the rates of oxygen consumption and compound II formation again agreed well with each other. The results are consistent in that superoxide released from neutrophils is converted by superoxide dismutase to a half mole each of hydrogen peroxide and molecular oxygen.

When superoxide dismutase was added during the course, the resultant spectra were those of a mixture of compounds II and III which can be judged from the shifts of the peak positions as well as from the shift of isosbestic point from 605 to near 626 nm. This means if hydrogen peroxide was generated simultaneously with superoxide, such spectrum should be detectable by the present method. On the basis of these results, we concluded that the phorbol ester-stimulated neutrophils released exclusively superoxide anion into the medium.

Similar results were also obtained when neutrophils were stimulated by formyl-methionyl-leucyl-phenylalanine, a soluble chemotactic peptide. It is now well established that only superoxide anion is formed during the respiratory burst.

However, the situation was very much different when the cells were exposed to particulate stimuli such as opsonized zymosan or bacteria. The rate of formation of compound III did not correspond to that of oxygen consumption. Furthermore, compound III formation occurred only within the initial stage of oxygen consumption and thereafter ceased. Under these conditions, only 10 to 20% of oxygen consumed was found to be released into extracellular medium as superoxide anion. Similar results were obtained when cytochrome c was employed as an indicator of superoxide. Compound II was not detected throughout the experiment.

When azide was present in the reaction mixture, however, formation of compound II was observed with an increased rate of oxygen consumption. Compound II formation occurred after the completion of compound III formation. The results indicate that the remaining portion of the total oxygen consumed is recovered at least in part as hydrogen peroxide.

These results may be explained as follows. Superoxide which is released into the phagosome is not accessible to the trapping agent, the heme-substituted peroxidase, probably because superoxide cannot pass through the membrane or is destroyed before it diffuses out of the cells. Thus, hydrogen peroxide derived from superoxide by dismutation (probably by a non-enzymatic reaction), is secreted into the medium only when a sufficient amount of azide, an inhibitor of peroxidases and catalase, is present in the system. An oxidase reaction in which superoxide is the sole reaction product seems to be unique to the respiratory burst system. Since almost all of the superoxide produced was recovered in the extracellular medium, the NADPH oxidase system seems to be located in the cytoplasmic membrane where the reactive site with oxygen faces outside of the cells. The mechanism of superoxide production is now under investigation.

References

1. Tomita, T., Blumenstock, E., and Kanegasaki, S. Infect. Immun., 32:1242-1248 (1981).
2. Tomita, T., and Kanegasaki, S. Infect. Immun., 38:865-870 (1982).
3. Kanegasaki, S., Tomita, T., and Yasuda, T. Recent Advances in Res., 22:180-186 (1983).
4. Tomita, T., Kobayashi, S., and Kanegasaki, S. In preparation.
5. Makino, R., Tanaka, T., Iizuka, T., Ishimura, Y., and Kanegasaki, S. J. Biol. Chem. 261:11444-11447 (1986).

NON-SECRETION OF ABO BLOOD GROUP ANTIGENS AND SUSCEPTIBILITY TO INFECTIOUS DISEASES

C.C. Blackwell[1], D.M. Weir[1], V.S. James[1] and K.A.V. Cartwright[2], J. Stuart[3] and D. Jones[4]

[1]Department of Bacteriology
University of Edinburgh Medical School
Teviot Place
Edinburgh EH8 9AG
Scotland

[2]Public Health Laboratory, Gloucester

[3]Department of Community Medicine, Gloucester

[4]Public Health Laboratory, Manchester

Introduction

"Much more attention should be given to the combined effects of blood group and secretor state on susceptibility to bacterial infections."(1).

The innate host defenses that resulted in one non-immune individual developing protection when exposed to a particular pathogen while another develops disease have not been clearly defined. We have examined the ABO blood group and secretor state of patients as the first step in the investigation of a number of infectious diseases. From our studies and those of other groups there is evidence that, regardless of ABO blood group, the ability to secrete these antigens appears to play a role in protection against infections. A higher proportion of non-secretors, those individuals who are unable to secrete the water-soluble glycoprotein form of their ABO antigens into body fluids, is found among patients with cholera (2), rheumatic fever (summary of studies by 3), carries (4,5), recurrent urinary tract infections in women (6,7,8), development of kidney scarring following urinary tract infections (9) bacterial meningitis (10,11,12) superficial candida (13,14) and ankylosing spondylitis, a rheumatic condition for which an infectious aetiology has been postulated (15).

NATO ASI Series, Vol. H24
Bacteria, Complement and the Phagocytic Cell
Edited by F.C. Cabello und C. Pruzzo
© Springer-Verlag Berlin Heidelberg 1988

The ability to secrete the glycoprotein form of the ABO antigens is controlled by a single gene inherited in a Mendelian dominant pattern (16). It is also associated with expression of the Lewis blood group system. The glycoproteins found in the body fluids of secretors and non-secretors are shown in Table 1.

Table 1: Blood Group Glycoproteins Present in Secretor and Non-Secretor Body Fluids

	A[1]	B	H[3]	Lewis[a]	Lewis[b]
Secretor	+/-	+/-	+	$\pm$[2]	+
Non-secretor	-	-	-	+	-

[1] Depending on presence of A or B blood group genes
[2] Present in small quantities
[3] H antigen is the antigen of blood group O

In order to investigate the host-parasite interactions underlying these epidemiological observations, we have proposed and begun to test several hypotheses.

1. The carbohydrate moieties of the ABH and Lewis glycoproteins in body fluids of secretors can bind to lectin adhesins on the surface of microoganisms and reduce their binding to target cells.

Although there is no direct evidence that the ABH or Lewis antigens can act as receptors for microbial adhesins, the P, M and N blood groups have been found to act as receptors for fimbrial adhesins of enteric organisms (see Jann, this volume), and the Duffy blood group as the receptor for the malaria parasite Plasmodium knowlesii (17).

Inhibition of binding of Streptococcus salivarius to human epithelial cells by salivary glycoproteins with blood group activity has been demonstrated (18). We have been able to inhibit binding of Candida blastospores to epithelial cells by pretreating the yeast with boiled saliva from secretors. The saliva of non-secretors did not inhibit the binding but often enhanced it (13). The latter set of experiments and the glycoproteins found in the body fluids of secretors and non-secretors led to the next hypothesis.

2. The Lewis[a] antigen, found predominantly in non-secretors, is one of the receptors for microbial adhesins.

Unlike the ABH antigens, the Lewis antigens do not form part of the structure of the host cell. They are absorbed onto the cell surface from either plasma or body fluids. If these antigens could act as receptors, the organism might bind directly to the antigen on the surface of the epithelial cell. Alternatively, it might bind to the carbohydrate moiety of the antigen in a body fluid and the microorganism-Lewis[a] complex might then be absorbed onto the surface of this host cell. Pretreatment of non-secretor epithelial cells with anti-Lewis[a] antisera inhibited binding of blastospores, but pretreatment of secretor cells with anti-Lewis[b] antisera did not result in inhibition (19).

3. The product of the secretor gene, a fucosyl transferase alters the receptor for the adhesin or a site near the receptor so that attachment is inhibited (20).

This hypothesis by Lomberg et al is based on their observation that uroepithelial cells from non-secretors bind significantly higher number of uropathogenic bacteria than do those from secretors.

4. The lower levels of secretory (21) and serum IgA (22) reported for non-secretors might result in a compromised state at their mucosal surfaces.

In an earlier study we tested the hypothesis that the increased proportion of non-secretors among women with recurrent urinary tract infections (7,8) might be due to a generally impaired IgA response among non-secretors. We found the opposite. The non-secretors who had been followed by the local pyelonephritis clinic for 20 years had significantly higher levels of IgA than the secretors in the same group of women. Reduction in the number of infectious episodes was also associated with increased levels of IgA. From this study we suggested that non-secretors were more dependent on their specific immune responses than secretors for dealing with colonization of infection of the mucosal surfaces (8).

5. As the gene for the C3 complement is on the same chromosome (19) as the secretor gene, the expression of C3 might be altered in non-secretors.

We have recently had the opportunity to examine the last two hypothesis which fall into the interests of this workshop in the context of bacterial meningitis.

Protection against bacterial meningitis is associated with antibodies specific for the invading strain and an intact complement system. Regardless of age, individuals who lack these antibodies or in whom the complement system is impaired are compromised. The innate defenses that influence susceptibility or resistance of the non-immune host to these infections have not been fully defined.

We have found a significant increase in the proportion of non-secretors among patients with invasive diseases due to the three species responsible for 75% of these infections - N. meningitidis, Haemophilus influenzae type b, Streptococcus pneumoniae (Table 2).

Table 2. Proportions of Non-secretors Among Patients with Bacterial Meningitis

Organism	Source	total no.	non-secretors no.	(%)	P
N. meningitidis	Scotland				
	controls	334	89	(26.6)	
	patients	26	18	(69)	<0.005
	Iceland				
	controls	228	94	(41.2)	
	patients	98	53	(54)	<0.05
	Nigeria				
	controls	186	92	(49.5)	
	patients	42	31	(73.8)	<0.01
S. pneumoniae	Scotland				
	controls	337	89	(26.6)	
	patients	47	22	(47)	<0.01
H. influenzae type b	Iceland				
	controls	228	94	(41.2)	
	patients	43	29	(67)	<0.005

Between October 1981 and December 1986 there were 89 cases of meningococcal disease in the Gloucester Health district. These infections were due mainly to a serogroup B, serotype 15 p1.16, sulphonamide resistant strain (B15R). The attack rate in this area was at least 5 times the national average throughout this 5 year period. Between January 1983 and December 1986 there were 14 cases in the town of Stonehouse. Most of these 14 were localized to one predominantly council-owned housing estate.

Subjects and Materials

Secretor state was determined by the haemagglutination inhibition method described by Mollison (25). Levels of total serum IgA, IgG, IgM and C3c were determined from plasma specimens and total salivary IgA from saliva collected at the time of blood samples. Behring NOR-Partigen plates were used for determination of IgA, IgG, IgM and C3c levels and their LC-Partigen plates for determination of salivary IgA. Results obtained for meningoccoccal carriers were compared with those obtained for non-carrier controls. The controls were chosen at random from the age range in which the majority of these infections occurred, participants under the age of 20 from whom no meningococci were isolated. They ranged in age from less than 1 year to 19 years with a median age of 15. The carriers ranged in age from less than 1 year to 78 years with a median age of 15 years. Results for the specimens and the Behring standard serum control were expressed as international units (IU) determined from the table accompanying the plates. The logs of these values were compared with the Student's test.

Results

C3c levels

The geometric mean of the levels of C3c in the plasma of 59 non-secretors (122 IU/ml) was not statistically different from that found in 41 secretors (130 IU/ml); however, among the 8 subjects in whom the C3c levels were below the lower limits of the normal range (63 IU/ml), 1 was a secretor and 7 non-secretors. The individuals in whom these values were below the normal range were not the very young children; the secretor was 15 and the 7 non-secretors were between 6 and 17 years.

Salivary IgA levels

The results shown in Table 3 indicate that there was no difference in the salivary IgA levels of secretors and non-secretors in either the control group or in the carriers. There was, however, a significant increase in the levels of salivary IgA among carriers compared with the controls ($p < 0.005$) within both the secretor and non-secretor groups (Table 3). There were 17 specimens in which there was no detectable IgA or the levels were too low to be calculated accurately. Of these, 5 were from secretors (3 of whom were under the age of 5) and 12 from non-secretors (of whom 2 were under the age of 5).

Table 3. Geometric Means of Salivary IgA Levels of Controls and Carriers

Secretors		Non-secretors	
Controls	Carriers	Controls	Carriers
(n = 80)	(n = 146)	(n = 87)	(n = 101)
Iu/ml	Iu/ml	Iu/ml	Iu/ml
1.8	3.4	1.7	3.4

Serum immunoglobulin levels

There were no significant differences between levels of the three isotypes of secretors and non-secretors of the carriers and controls. Within the non-secretor group, the IgA levels of the carriers were significantly higher than those of the controls and their IgG levels significantly lower (Table 4).

Table 4. Geometric Means of Total Serum Immunoglobulin of Controls and Carriers of N. meningitidis

	Secretors		Non-secretors	
	Controls (n = 64)	Carriers (n = 133)	Controls (n = 72)	Carriers (n = 84)
	Iu/ml	Iu/ml	Iu/ml	Iu/ml
IgA	101	113	98*	116*
IgG	137	135	150**	128**
IgM	191	182	186	175

*p < 0.0125
**p < 0.0025

Because the non-serogroupable strains of meningococci are usually non-pathogenic, the levels of the three isotypes for carriers of the non-serogroupable strains were compared with those of carriers of the serogroupable, pathogenic strains. The only significant difference was the increased levels of IgA among the carriers of the serogroupable strains (Table 5).

Table 5. Geometric Means of Total Serum Immunoglobulin Levels of Carriers of Serogroupable and Non-serogroupable Strains of N. meningitidis

	Secretors		Non-secretors	
	serogroupable (n = 86)	non-serogroupable (n = 47)	serogroupable (n = 50)	non-serogroupable (n = 34)
	Iu/ml	Iu/ml	Iu/ml	Iu/ml
IgA	112	113	127*	101*
IgG	143	123	131	124
IgM	184	178	183	179

*p < .025

As the significant differences were found among the carriers of the serogroupable strains, the immunoglobulin levels in the sera of secretor and non-secretor carriers of the two major serogroups, B and C, were compared. The IgA levels were increased for non-secretors carrying both serogroups, but these differences were significant only for serogroup B. The levels of the complement-fixing isotypes, IgG and IgM, were significantly lower for non-secretors carrying serogroup B strains. The IgG and IgM levels were increased for the non-secretor carriers of serogroup C, but the increase was significant only for IgM (Table 6).

Table 6: Geometric Means of Total Serum Immunoglobulin Levels of Carriers of Serogroups B or C N. meningitidis

	Secretors (n = 52) Iu/ml	Non-secretors (n = 37) Iu/ml	P
Serogroup B			
IgA	118	130	NS
IgG	149	128	< 0.0125
IgM	188	163	< 0.05
Serogroup C	(n = 21)	(n = 9)	
IgA	114	164	< 0.025
IgG	143	160	NS
IgM	194	245	< 0.05

The ratios of IgA to IgG and IgA to IgM were calculated to determine if non-secretors had increased levels of IgA compared with those of the complement fixing isotypes. In Table 7 the numbers of individuals with IgA levels greater than IgG or IgM are presented in the numerator and the total number of individuals in each group in the denominator. Compared with the controls the proportion of those with higher levels of IgA was increased among the carriers. The increase was not significant for carriers of non-serogroupable strains. The increase was significant for both secretor and non-secretor carriers of serogroup B. This was not found for secretor carriers of serogroup C. Although the proportion of these individuals is increased among the non-secretor carriers of serogroup C, the numbers were too small for analysis by the X^2 method used for the other categories.

Table 7. Proportion of Carriers of Serogroup B or C <u>N. meningitidis</u> in Whom Total Serum IgA Levels were Greater than their Levels of IgG or IgM or Both IgG and IgM

	Secetors		Non-secretors	
	No.	(%)	No.	(%)
Controls	12/64	(18.8)	16/71	(22.5)
Carriers				
Non-groupable/				
non-typable	9/25	(36)	8/22	(36.4)[*]
Serogroup B	22/51	(43.1)[*]	18/37	(48.6)[*]
Serogroup C	4/21	(19)	6/9	(66.7)

[*] p < 0.02

The proportion of individuals in whom serum IgA was greater than IgM is shown in Table 8. The increase in the proportion of these individuals was significant only for the non-secretor carriers of serogroup B. Among the non-secretor carriers of the B15R strains 56% had increased levels of IgA compared with IgM, but the numbers were again too small for statistical analysis.

Table 8. Proportion of Carriers in Whom Total Serum IgA Levels were Greater than their Levels of IgM

	Secretors		Non-secretors	
	No.	(%)	No.	(%)
Controls	5/64	(7.8)	5/71	(7)
Carriers				
Non-serogroupable	5/20	(25)	2/22	(9.1)
Serogroup B	11/51	(21.6)	10/37	(27)*
B15 P1.16	4/19	(21)	4/7	(57)
Serogroup C	1/21	(4.8)	2/9	(22.2)

* $p < 0.05$

Discussion

In the context of this workshop we have looked at 3 questions related to secretor state and susceptibility to bacterial meningitis.

1. As the secretor gene is on the same chromosome as the gene for the C3 component of complement, are there differences in the C3 levels of secretors and non-secretors?

Although no statistically significant differences in the levels of C3c were observed between secretors and non-secretors, of the 8 specimens in which these levels were below the lower limits of the normal range, 1 was from a secretor and 7 from non-secretors. Individuals with C3 deficiencies are very rare, but non-secretors comprise about 20-25% of the general Caucasian populations in Western Europe and North America. The work reported in this workshop regarding the role of C3 in dealing with infections due to type b $\underline{H}$. $\underline{influenzae}$ infections suggest these studies should be extended.

2. Are the levels of salivary IgA in non-secretors lower than those of secretors?

Because IgA is thought to play a role in prevention of colonization, the lower levels of salivary IgA reported for non-secretors (21) suggested their specific immune responses at the mucosal surfaces might be compromised compared with those of secretors. We were unable to confirm the earlier report of lower levels of IgA in the saliva of non-secretors in either the control group or the carriers. The significant increase in the levels of IgA in the saliva of carriers compared with non-carriers suggest that meningococci do elicit a strong response from the mucosal lymphoid system (Table 3). Although there was no statistically significant difference in the salivary IgA levels of the secretor and non-secretor controls, the greater number of non-secretors with very low or undetectable levels of IgA merits further investigation.

Because meningococci can produce IgA1 protease and the specimens were collected under conditions of stress, the serum IgA levels are probably a more accurate reflection of any differences between IgA levels of secretor and non-secretor carriers.

3. Are there differences in levels of serum immunoglobulin classes of secretors and non-secretors?

There were no significant differences in the levels of IgA, IgG or IgM between secretors and non-secretors in either the control or carrier groups. Within the non-secretor group, the carriers had significantly higher levels of IgA and significantly lower levels of IgG than the controls (Table 4). The increase in the non-complement fixing IgA immunoglobulins was found only among non-secretor carriers of the capsulate serogroupable strains, not among the carriers of the non-pathogenic, non-serogroupable strains.

Much of the bactericidal activity against serogroup B strains in normal serum is against the B polysaccharide (26,27) and almost all of these antibodies to the B polysaccharide are IgM (28). The results in Table 8 show that among the non-secretors carrying serogroup B strains, there is a significant proportion with higher levels of total serum IgA compared with their levels of IgM. Although the numbers are small, 57% of the non-secretors carrying the B15R strain fall into this category. If non-secretors respond to meningococcal colonization by producing high levels of IgA antibodies to the capsular material of the invading strain they might act as "blocking antibodies" competing with the complement fixing classes as suggested by Griffiss (23).

The major subclass found among adults who have "natural" antibodies to type b Haemophilus capsule is IgG2 (29). It also appears to be the predominant form produced in response to streptococcal polysaccharide (30). The increase in the proportion of carriers of serogroupable meningococci with increased levels of IgA compared with IgG and IgM suggest that the non-complement fixing isotypes and subclasses might form a significant part of the host's responses to these polysaccharide antigens. The mechanisms underlying selective induction or suppression of particular isotypes or subclasses of immunoglobulins are not yet understood. No information is yet available on the abilities of secretors and non-secretors to preferentially synthesize particular subclasses or isotypes in response to polysaccharides or other antigens. This area merits further investigation in view of the differing bactericidal and opsonic abilities of the various isotypes and subclasses and these preliminary findings of altered ratios of IgA to IgM.

Our early observations on the general immune response of secretor and non-secretor carriers of N. meningitidis indicate that further investigation might provide insights not only into the pathogenesis of bacterial meningitis but also into the development of natural immunity to these capsulate pathogens. They also suggest some cautionary thoughts regarding development of vaccines to these pathogens.

References

1. Mourant, A.E. Blood Relations--Blood Groups and Anthropology. Oxford University Press. Oxford (1983).
2. Chaudhuri, A. and Das Adhikary, C.R. Possible role of blood-group secretory substances in the aetiology of cholera. Trans R Soc Trop Med Hyg 72:664-665 (1978).
3. Haverkorn, M.J. and Goslings, W.R.O. Streptococci, ABO blood groups and secretor status. Amer J Hum Gent 21:360-75 (1969).
4. Arneberg, P., Kornstad, L., Nordbo, H., Gjermo, P. Less dental carries among secretors than among non-secretors of blood group substance. Scand J Dent Res 84:362-366 (1976).

5. Holbrook, W.P. and Blackwell, C.C. Secretor state and dental caries in Iceland. Dis. Markers (submitted for publication).

6. Kinane, D.F., Blackwell, C.C., Brettle, R.P., Weir, D.M., Winstanley, F.P., Elton, R.A. ABO blood group, secretor state and susceptibility to recurrent urinary tract infection in women. Br Med J 285:7-9 (1982).

7. Blackwell, C.C., May, S.J., Brettle, R.P., MacCallum, C.J., Weir, D.M. Host-parasite interactions underlying non-secretion of blood group antigens and susceptibility to recurrent urinary tract infection. Protein Carbohydrate Interactions in Biological Systems ed. D. Lark, Academic Press, London (1986).

8. Blackwell, C.C., May, S.J., Brettle, R.P., MacCallum, C.J., Weir, D.M. Secretor state and immunoglobulin levels among women with recurrent urinary tract infections. J Clin Lab Immunol (in press) (1987).

9. May, S.J., Blackwell, C.C., Brettle, R.P., MacCallum, C.J., Weir, D.M. Non-secretion of ABO blood group antigens: a host factor predisposing to urinary tract infections and renal scarring. Dis Markers (submitted for publication).

10. Blackwell, C.C., Jonsdottir, K., Hanson, M.S.P., Todd, W.T.A., Chaudhuri, A.K.R., Mathew, B., Brettle, R.P., Weir, D.M. Non-secretion of ABO antigens predisposing to infection by Neisseria meningitidis and Streptococcus pneumoniae. Lancet 2:284-285 (1986).

11. Blackwell, C.C., Jonsdottir, K., Hanson, M.F., Weir, D.M. Non-secretion of ABO antigen predisposing to infection by Haemophilus influenzae. Lancet 2:687 (1986).

12. Blackwell, C.C., Jonsdottir, K., Mohammed, I., Weir, D.M. Non-secretion of blood group antigens: a genetic factor predisposing to infection by Neisseria meningitidis. V Int Symp on the Pathogenic Neisseria species (in press).

13. Blackwell, C.C., Thom, S.M., Weir, D.M., Kinane, D.F. and Johnstone, F.D. Host-parasite interactions underlying non-secretion of blood group antigens and susceptibility to infections by Candida albicans. Protein-Carbohydrate Interactions in Biological Systems ed. D. Lark, Academic Press, London (1986).

14. Blackwell, C.C., Thom, S.M., Lawrie, O.R., Weir, D.M., Wray, D., Kinane, D.F. Non-secretion of blood group antigens and susceptibility to oral infections by Candida albicans. J Dent Res 64:502 (1986).

15. Shinebaum, R., Blackwell, C.C., Forster, P.J.G., Hurst, M.P., Weir, D.M. and Nuki, G. Non-secretion of ABO blood group antigens as a host susceptibility factor in the spondyloarthropathies. Br Med J 294:208-210 (1987).

16. Race, R.R., Sanger, R. Blood Groups in Man, 6th edition. Blackwell Scientific Publications, Oxford (1975).

17. Miller, L.H., Mason, S.J., Dvorak, J.A., McGinniss, M.H. and Rothman, I.K. Erythrocyte receptors for (Plasmodium knowelsii) malaria: Duffy blood group determinants. Science 189:561-563 (1975).

18. Williams, R.C. and Gibbons, R.J. Inhibition of streptococcal attachment to receptors on human buccal epithelial cells by antigenically similar salivary glycoproteins. Infect Immun 11:711-718 (1975).

19. May, S.J., Blackwell, C.C., Weir, D.M. Non-secretion of blood group antigens and susceptibility to <u>Candida</u> <u>albicans</u>: the role of Lewis blood group antigens. J Den Res 64:503 (1986).
20. Lomberg, H., Cedergren, B., Leffler, H., Nilsson, B., Carlstrom, A-S, Svanborg-Eden, C. Influence of blood group on the availability of receptors for attachment of uropathogenic <u>Escherichia</u> <u>coli</u>. Infect Immun 51:919-926 (1986).
21. Waissbluth, J.G., Langman, J.S. ABO blood groups, secretor status, salivary protein, and serum and salivary immunoglobulin concentrations. Gut 12:646-649 (1971).
22. Grundbacher, F.J. Immunoglobulins, secretor status and the incidence of rheumatic fever and rheumatic heart disease. Hum Hered 25:399-404 (1972).
23. Griffiss, J.M. Bactericidal activity of meningococcal antisera: blocking by IgA of lytic antibody in human convalescent sera. J Immunol 114:1779-1784 (1975).
24. Griffiss, J.M., Bertram, A.M. Immunoepidemiology of meningococcal disease in military recruits. II Blocking of serum bactericidal activity by circulating IgA early in the course of invasive disease. J Infect Dis 136:733-739 (1977).
25. Mollison, P.L. Blood transfusion in clinical medicine. 6th edition. Blackwell Scientific Publications, Oxford (1979).
26. Kasper, D.L., Winkelhake, J.C., Brandt, B.L., Artenstein, M.S. Antigenic specificity of bactericidal antibodies in antisera to <u>Neisseria</u> <u>meningitidis</u>. J Infect Dis 127:378-387 (1973).
27. Zollinger, W.D., Mandrell, R.E., Altieri, P., Berman, S., Lowenthal, J., Artsenstein, M.S. Safety and immunogenicity of a <u>Neisseria</u> <u>meningitidis</u> type 2 protein vaccine in animals and humans. J Infect Dis 137:728-739 (1978).
28. Skevakis, L., Frasch, C.E., Zahradnik, J.M., Dolin, R. Class-specific human bactericidal antibodies to capsular surface antigens of <u>Neisseria</u> <u>meningitidis</u>. J Inf Dis 149:387-396 (1984).
29. Shackelkford, P.G., Granoff, D.M., Nelson, S.J., Scott,, M.G., Smith, D.S., Nahm, M.H. Subclass distribution of human antibodies to <u>Haemophilus</u> <u>influenzae</u> type b capsular polysaccharide. J Immunol 138:587-592 (1987).
30. Riesen, W.F., Skvaril, F., Braun, D.G. Natural infection of man with group A streptococci. Levels, restruction in class, subclass and type; and clonal appearance of polysaccharide group-specific antibodies. Scand J Immunol 5:383-390 (1976).

INDEX

NATO ASI Series H

NATO ASI Series H